MINERALS &
GEMSTONES
of Southern Africa

Botswana, Eswatini, Lesotho, Namibia, South
Africa, southern Mozambique and Zimbabwe

Bruce Cairncross

Published by Struik Nature
(an imprint of Penguin Random House South Africa (Pty) Ltd)
Reg. No. 1953/000441/07
The Estuaries No. 4, Oxbow Crescent, Century City, 7441 South Africa
PO Box 1144, Cape Town, 8000 South Africa

Visit www.penguinrandomhouse.co.za and join the Struik Nature Club
for updates, news, events and special offers.

First published 2022
1 3 5 7 9 10 8 6 4 2

Publisher: Pippa Parker
Managing editor: Roelien Theron
Editor: Colette Alves
Designer: Gillian Black
Proofreader and indexer: Emsie du Plessis

Reproduction by Studio Repro
Printed and bound in China by C&C Offset Printing Co., Ltd

ISBN 978 1 77584 753 3 (Print)
ISBN 978 1 77584 754 0 (ePub)

Front cover: Malachite crystals in a cavity of blue shattuckite, 5.2 cm – Kaokoveld Plateau, Namibia – Bruce Cairncross specimen and photo; **Back cover:** (top) Cluster of quartz crystals, some included with blue ajoite, 6.2 cm – Messina No. 5 shaft, South Africa – Bruce Cairncross specimen and photo; (bottom) Bright yellow baryte crystals, field of view 1.5 cm – Tsumeb mine, Namibia – Bruce Cairncross specimen and photo; **Title page:** Clusters of bright red rhodochrosite, 9.5 cm – N'Chwaning I mine, South Africa – Desmond Sacco specimen, Bruce Cairncross photo; **Page 3:** Ettringite, calcite and black metallic manganite, 7.2 cm – N'Chwaning II mine, South Africa – Desmond Sacco specimen, Bruce Cairncross photo

Penguin Random House independently submitted this book for critical peer review prior to its publication.
This independent peer review was co-ordinated by Pippa Parker: Publisher, Struik Nature.
The reviewers were Professor Carl Anhaeusser, School of Geosciences, University of the Witwatersrand,
1 Jan Smuts Avenue, Braamfontein, Johannesburg 2001, and
Dr Lauren Blignaut, 168 Ross Lane, Oakville, Ontario L6H5K3, Canada

CONTENTS

FOREWORD

Minerals & Gemstones of Southern Africa is a fitting successor to previous publications on the topic by Bruce Cairncross, starting with his first book with Roger Dixon entitled *Minerals of South Africa,* which was prepared for the Centennial Congress of the Geological Society of South Africa in 1995. In more recent books and articles, Professor Cairncross has expanded further on the topic.

In this latest publication, the scope has been broadened to include not only South African examples, but comprehensive descriptions of the more important mineral and gem localities from the six contiguous southern African countries. It focuses on the beauty and variety of the more important minerals and gemstones to be found in the region, and their localities, and is illustrated with photographs, maps and diagrams.

The book is well laid out and very readable, the important minerals and gems being described alphabetically with brief overviews of the major mining fields of southern Africa, as well as other mines, small workings and localities where some of the finest specimens are to be found. A series of superb photographs of crystals and/or faceted or polished specimens concludes the description of every mineral species.

The age of each mineral and/or the host rock in which it occurs is also given and placed into the timescale of geological formations of southern Africa. It is of interest that the minerals and gems discussed were formed by varied processes throughout geological time, and preserved in rocks varying in age from some 3.6 billion years ago to the present. Attention is drawn to the evolutionary trend of the various mineral species over this huge time span.

A hallmark of the book is undoubtedly the hundreds of magnificent colour photographs of minerals and gems, some appearing in published form for the first time. Most of the photographs have been taken by the author himself, who is a world-renowned photographer of minerals and gems. He has won a number of prizes and awards at various mineral and gem shows around the world, both as a photographer and author. His enthusiasm and passion for the collection, photography and documentation of minerals and gems is evident throughout this book and it is a pleasure to have the subject described and illustrated in such a convenient and readable fashion.

The book will be of great interest and value to a wide variety of readers, not the least of whom are members of gem and mineral clubs, the wider mineral collectors fraternity, interested laypersons and learners of all ages, for whom books such as this have often ignited a lifetime interest in the subject. Minerals and gemstones are an important part of our outstanding geological and mining heritage, and a better understanding of them will help in engendering a more controlled approach to future exploitation and conservation.

Dr Richard Viljoen
(PhD (Wits), FGSSA, FSAIMM, FSEG, FGSI, Pr. Sci. Nat.)
Technical Advisor, VM Investments, Bushveld Minerals and Afritin Mining

PREFACE

The minerals and gemstones of southern Africa have been featured in many publications and on numerous online resources. In almost all instances, these publications have concentrated on specific mineral occurrences, mineral regions, geological formations associated with specific deposits, individual countries or even regions within a single country, rather than the southern African region as a whole. An early exception is McIver (1966), *Gems, Minerals and Rocks of Southern Africa*, a book now long out of print and somewhat outdated, and the more recent Cairncross (2004a), *Field Guide to Rocks & Minerals of Southern Africa*. The latter publication was in field guide format and is now superseded by this larger format edition, which features fully updated text and more than 900 specimen and locality photographs, most of which are new.

This book is not only the most up-to-date publication on the interesting, collectable and economically relevant minerals of Botswana, Eswatini, Lesotho, Namibia, South Africa, southern Mozambique and Zimbabwe, but also the first to feature the region's gemstones. These gemstones do not only include diamonds, but also fancy faceted stones of unusual and often rare minerals, most of which were in the 'Rainbow of Africa' collection amassed over many years by Warren Taylor. Most of these have not been seen before.

Each year, new mineral species are discovered, even though over 5,700 are already known to exist. Southern Africa, in particular Namibia and South Africa, has provided numerous type-locality species. Most times, these discoveries are microscopic and esoteric, with very few examples. Occasionally there are exceptions, and in early 2000, the new species olmiite was found in the N'Chwaning mine in South Africa. Thousands of specimens have since been collected, yet the species remains, to date, unique to the locality, a somewhat unusual situation for such a plentiful mineral.

During the past two decades, some southern African localities have impacted on the international mineral market by producing world-class specimens. Examples include the Erongo Mountains in Namibia, from which outstanding aquamarine, schorl tourmaline and jeremejevite have been collected, and the remote Kaokoveld in northern Namibia, which has been a source of excellent gem-quality spessartine garnet, dioptase and shattuckite. All of these are featured in this book, together with recent discoveries from other well-established mineral localities.

A book of this nature cannot present or document every mineral species in the region. Some minerals are mundane, while others exist only as microscopic minerals imbedded in rock or economic ore samples. Discretion (and possibly an element of personal bias) was used to select which minerals to include in this publication. A complete list of minerals for each country can be found on Mindat.org, the most comprehensive online source of mineral occurrences.

Bruce Cairncross
(PhD (Wits), FGSSA, Pr. Sci. Nat.)

INTRODUCTION

Figure 1 The enormous opencast Rössing uranium mine in Namibia, started in 1976. BRUCE CAIRNCROSS PHOTO, 2014.

Southern Africa – defined here as Botswana, Eswatini, Lesotho, Namibia, South Africa, southern Mozambique and Zimbabwe – is famous for its wealth of minerals and gemstones (Cairncross, 2016d). In fact, the subcontinent has some of the largest known resources and reserves of gold, manganese, diamonds, platinum, chrome and iron ore in the world.

Namibia and South Africa have the greatest diversity and abundance of notable collectable minerals and gemstones, and the emphasis in this book is therefore skewed in their favour. Zimbabwe is perhaps second most important, and Eswatini, Botswana, Lesotho and southern Mozambique are less well endowed, but do have some notable exceptions. The mineral wealth of each country is determined by economic and geological factors.

Exploitation of these resources varies from some of the largest and deepest mines in the world, to one-man artisanal workings of a small surface outcrop. Both of these scales of operations, and all others, sporadically produce mineral specimens. Some of the smaller-scale workings are specifically aimed at producing mineral specimens and do not explore ore for processing.

All of the minerals and gemstones featured in this book originate from two general sources: those found with economic deposits during mining operations and those that occur in host rocks with no large economically important minerals and, hence, no major mining activity. The latter are normally exploited by the informal sector. The term mining is

used here *sensu stricto*, referring to large, commercially viable operations and not artisanal diggers.

Certain minerals are restricted to specific geological settings, and some of the southern African countries have greater geological potential than others to host these commodities. For example, sand-bedecked Botswana has a notable paucity of collectable minerals and gemstones (apart from diamonds), while Namibia, with its well-exposed and accessible rocky outcrop, has far more occurrences.

In the past, several mineral provinces were described by Söhnge (1986). He used the age of the geological formations to subdivide the important mineral deposits in southern Africa, starting with oldest Archaean Eon (greater than 2,500 million years), followed by the Proterozoic Eon (2,500–541 million years), and ending with the younger Phanerozoic Eon (less than 541 million years). Depending on the geological fabric of the various countries, the concentration of economic deposits shifts geologically, and geographically, with time. For example, the most ancient Archaean greenstone belt rocks occur in South Africa and Zimbabwe but are absent from Namibia (Anhaeusser, 1976a). Therefore, some of the mineralization found in these rocks – e.g. gold, antimony, asbestos – is common during this ancient time period. As progressively younger rocks evolved, so did different varieties of mineral deposits. During the early Proterozoic, the huge Northern Cape iron and manganese deposits formed; they never existed before or

Southern African Mineral Provinces (*pro parte* after Söhnge, 1986)			
Time Period	Select Geology / Stratigraphy	Examples of Important Economic Commodities	Some Examples of Associated Collectable Minerals/Gemstones
Phanerozoic (541 million years to present day)	Kimberlites	Diamonds	Calcite, celestine, garnet, ilmenite, diopside
	Karoo and Etendeka lavas	Gemstones	Quartz – amethyst, agate, prehnite
	Erongo/Spitzkoppe granite	Gemstones	Topaz, aquamarine, fluorite, schorl
	Other: Musina	Copper	Quartz, ajoite, papagoite
	Other: Okoruso	Fluorine	Fluorite
Neoproterozoic (1,000–541 million years)	Damara Orogen	Copper, lead, zinc, vanadium, tin, tungsten	Cuprite, galena, sphalerite, descloizite, cassiterite, ferberite
Mesoproterozoic (1,600–1,000 million years)	Natal-Namaqua Metamorphic Province	Copper, tungsten	Chalcopyrite, chalcocite, quartz, various pegmatite mineral and gemstones
	Pilanesberg	Rare-earth elements, fluorine	Fluorite
	Other	Gemstones	Sodalite
Palaeoproterozoic (2,500–1,600 million years)	Phalaborwa Complex	Copper, phosphate, gold, uranium, zirconium	Baddeleyite, magnetite, diopside, apatite, calcite, mesolite
	Great Dyke	Chromium, platinum, nickel, asbestos	Chromite, brucite, mtorolite
	Bushveld Complex	Platinum, chrome, vanadium, tin, fluorine	Sperrylite, chromite, cassiterite, fluorite, andalusite
	Transvaal Supergroup	Manganese, iron, lead-zinc	Rhodochrosite, hematite, galena, sphalerite, andalusite
Archaean (greater than 2,500 million years)	Limpopo Belt	Industrial minerals	Corundum
	Witwatersrand goldfield	Gold, uranium	Quartz, calcite, sphalerite, baryte
	Greenstone belts and granite/gneiss	Gold, nickel, chromium, antimony, tungsten, asbestos	Stichtite, stibnite, magnesite, ferberite, corundum, emerald, garnet, alexandrite, various pegmatite minerals

after this period, while in the late Proterozoic, the copper-lead-zinc deposits of northern Namibia were formed.

Söhnge's subdivision of the subcontinent's important economic deposits remains largely applicable today still, and has been partly summarized in the table above.

As can been seen, there is an evolutionary trend over time in the genesis of mineral deposits, and certain mineral species, in southern Africa. For example, stibnite, the main mineral containing antimony, is known only from ancient Archaean formations, while the giant gold deposit of the Witwatersrand conglomerates is unique. Similarly, the massive layered complexes such the Bushveld Complex in South Africa and Great Dyke in Zimbabwe appeared during the early Proterozoic Eon, not before and not since. Therefore, certain associated minerals, and some gemstones, are time specific, such as the alexandrite from Zimbabwe greenstone belts, or corundum in some southern

African Archaean granite/gneiss. In contrast, some minerals that are well known to collectors, span the entire geological time scale and are not confined to specific time periods or specific host rocks. Examples of these are quartz and calcite. But some, particularly quartz, can have certain associated minerals and characteristics that link them exclusively to certain deposits. For example, the quartz that contains spectacular blue ajoite and papagoite inclusions is unique to the Musina copper deposits in South Africa, while multicoloured calcite, although not unique, is perhaps most famous from the Tsumeb mine in Namibia. Comparing the various geological time periods in the table (above) with the corresponding geological formations on the geological map (page 11) can illustrate why some important mineral deposits and/or associated minerals and gemstones are confined to certain geographic regions in southern Africa: time and the geological formations determine this distribution.

Figure 2 The southern African region, showing places closest to sites of mineralogical and/or gemological interest.

Geological overview

The geology of southern Africa is highly varied (see map), with a variety of rock formations spanning many geological time periods. The region has large tracts of relatively young sedimentary strata, specifically in South Africa and southern Mozambique, as well as spectacular rock formations that formed over 2,500 million years ago.

Starting with the oldest formations, South Africa, Zimbabwe and Eswatini have some of the most ancient rocks on Earth. These are found in various **Archaean** greenstone belts and contain numerous gold deposits. One of these belts, the Murchison greenstone belt in South Africa, also boasts the oldest known antimony deposits. Later in geological time, but still in the Archaean Eon, the Witwatersrand gold deposits contain the largest accumulation of this metal on Earth. By a strange quirk in geological history, these rocks also contain the largest baryte and sphalerite crystals to be found in southern Africa.

In South Africa, the dolomites and associated strata of the **Palaeoproterozoic** Transvaal Supergroup contain enormous resources of iron and manganese, the latter being the largest known land-based occurrence. Intruding into these sedimentary rocks is the slightly younger Bushveld Complex, the largest known mafic layered igneous body, which contains huge deposits of platinum and chromium. The Great Dyke in neighbouring Zimbabwe is equally important. Allied to the Bushveld Complex mafic rocks are intrusive granites containing tin, fluorite and other metallic deposits and non-metallic deposits that accompanied the intrusion.

Namibia has important **Neoproterozoic** copper-lead-zinc deposits in the Otavi mountainland, and also in the south close to the border with South Africa, at Rosh Pinah and the Skorpion mine.

Younger **Phanerozoic** granites such as the Klein Spitzkoppe and Brandberg massif host important gemstones such as topaz and aquamarine. Much of the southern African Phanerozoic strata do not contain notable collectable minerals, but there are exceptions, such as the basalt-hosted minerals of the Jurassic lavas. These are basalts and rhyolites that in Lesotho contain beautiful crystals of stilbite, quartz and apophyllite. The same rocks form the Drakensberg escarpment in South Africa, as well as the country's eastern border with Mozambique and extend into Zimbabwe, and host agate, chalcedony, carnelian, quartz crystals and amethyst, particularly in Zimbabwe. Internationally famous specimens of quartz, amethyst and prehnite are found in the Etendeka volcanic rocks in north-western Namibia, notably the Goboboseb Mountains, which have produced world-class examples of these minerals.

Southern Africa's coal is found in the Phanerozoic Karoo Supergroup rocks. Of the region's younger rock formations, the Cretaceous kimberlites are extremely relevant, as some of these contain diamonds, for which Botswana, Lesotho, South Africa and Zimbabwe are renowned.

In summary, southern Africa possesses major economic deposits in the form of the Witwatersrand goldfield, the Kalahari manganese field, kimberlitic and alluvial diamonds, massive sulphide Cu-Pb-Zn deposits and W-Sn-gem pegmatites, particularly in Namibia and Zimbabwe, the mafic layered complexes such as the Bushveld Complex in South Africa and Great Dyke in Zimbabwe, carbonatites, and greenstone belt deposits.

Important literature reviews and compilation sources of information of the geology and economic deposits of southern Africa can be found in Hunter (1961, 1962), McIver (1966), Baldock (1977), Anderson S.M. (1980), Anhaeusser and Maske (1986), *Mineral Resources of Namibia* (1992), Carney *et al.* (1994), Wilson and Anhaeusser (1998), and Johnson *et al.* (2006). Zimbabwe, Botswana, Eswatini, and Lesotho have less published information available. The most comprehensive references for Namibian minerals can be found in Von Bezing *et al.* (2008, 2014, 2016) and for South Africa in Cairncross and Dixon (1995) and Cairncross (2004a).

Figure 3 Rugged outcrops in the Northern Cape near the Orange River. Complex geology can be seen in the foreground; original horizontal rocks have now been tilted up and dip steeply to the right. BRUCE CAIRNCROSS PHOTO, 2010.

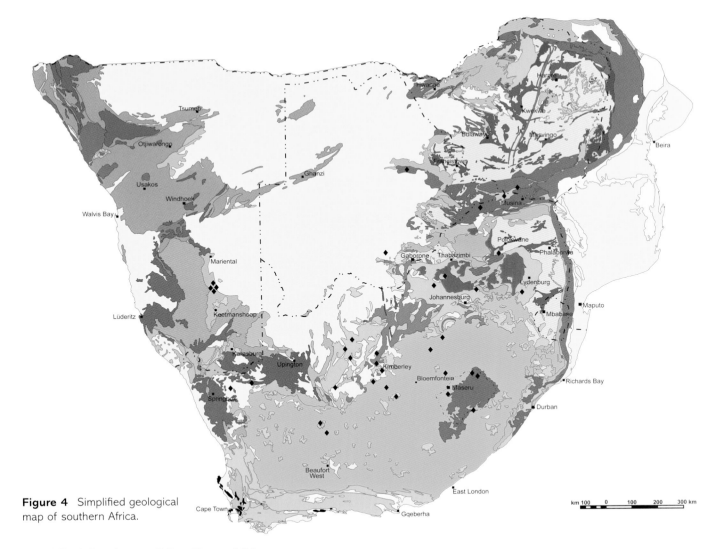

Figure 4 Simplified geological map of southern Africa.

Simplified Geology of Southern Africa

Main Rock Types

		Selected Examples of Geological Formations
◆	Kimberlite	Orapa, Venetia, Bultfontein
	Unconsolidated sand, soil	Recent Kalahari and coastal sands
	Sandstone, conglomerate, limestone	Cretaceous to Early Tertiary deposits
	Dolerite	Karoo dolerite
	Basalt, rhyolite	Karoo and Etendeka lavas
	Sandstone, shale, siltstone, conglomerate, (coal)	Karoo Supergroup
	Quartzite, sandstone, shale	Cape Supergroup
	Granite	Cape granite
	Quartzites sandstone, shale, limestone	Nama Group
	Sandstone, siltstone, schist, conglomerate, limestone	Gariep and Malmesbury Groups
	Conglomerate, quartzite, dolomite, schist, marble, amphibolite and granite	Damara Group

Main Rock Types

	Selected Examples of Geological Formations
Alkaline igneous rocks	Pilanesberg
Sandstone, shale, conglomerate	Waterberg, Sijarira
Limestone, sandstone, basalt	Umkondo
Quartzite, conglomerate, mafic and felsic lavas, schist, amphibolite	Rehoboth
Gabbro, norite, pyroxenite, peridotite	Bushveld Complex, Tete Complex, Great Dyke
Quartzite, shale, limestone, dolomite, iron-formation	Transvaal, Deweras, Lomagundi, Piriwiri
Andesitic lava, porphyritic lava, agglomerate, shale, quartzite, conglomerate	Ventersdorp
Quartzite, shale, iron-formation, conglomerate, lava	Witwatersrand, Pongola
Metamorphic rocks of various ages – gneiss, amphibolite, granulite	Limpopo belt, Mozambique, Zambezi and Natal-Namaqua belts, Namibian metamorphic complexes
Granite, granite-gneiss	Ancient crustal rocks – Kaapval + Zimbabwe craton
Ultramafic + mafic lavas, iron-formation, conglomerate, chert, sandstone	Greenstone belts in South Africa, Botswana, Zimbabwe & Eswatini

Botswana

Large tracts of Botswana are covered by surficial unconsolidated sand. The hard-rock geological framework of Botswana comprises Archaean basement in the east and south-east of the country, and elsewhere mainly Phanerozoic sedimentary strata of the Karoo Supergroup that hosts coal and gas. In the north-west and west, rocks of the Proterozoic Damara Belt (see also Namibia) exist, although most are covered by younger Kalahari sediment.

Botswana's mining sector contributes almost 10% to the country's export earnings. The most important economic commodity in Botswana is diamonds (Janse, 1995; Richardson *et al.*, 1999), which are found in kimberlites that are mainly Cretaceous in age. There are nonetheless more diversified economic deposits, including gold and base metals (Baldock *et al.*, 1976). The country is the world's second-largest diamond producer, with Jwaneng mine the richest kimberlite known (Weldon and Shor, 2014). Other commodities are gold (Chatupa, 1999), copper and nickel, with lesser quantities of cobalt, soda ash and coal.

Although Botswana has diamonds, collectable minerals are generally scarce from the country. However, Botswana does have other economically viable mineral resources (Massey, 1973;

Figure 5 Typical scene from north-east Botswana showing the flat, semi-arid landscape. BRUCE CAIRNCROSS PHOTO, 1985.

Baldock, 1977), such as the famous agates from the Bobonong area in eastern Botswana, close to the border with South Africa. These are primarily alluvial deposits produced by weathering of the agates from the basalt host rocks (Zenz, 2005).

Mindat.org (accessed December 2021) lists 113 valid species in Botswana and four type-locality species. However, there are only 50 photos of specimens on the site, of which virtually all feature either diamonds or agates.

Figure 6 General scenic view from Kgale Hill, Gaborone, Botswana. The rock in the foreground is granite, which is extracted at the quarry for use as aggregate. ANTHONY TUMO SEBOLAI PHOTO, 2020.

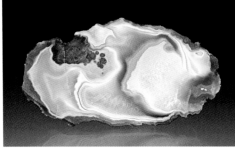

Figure 7 A cut and polished 9.7-cm agate from the Bobonong region in eastern Botswana. BRUCE CAIRNCROSS SPECIMEN AND PHOTO.

Figure 8 The Tsodilo Hills in north-west Botswana consist of metamorphic rocks, including quartzite, mica schist and meta-conglomerates. This is a World Heritage site (De Wit and Main, 2016) due to the plethora of rock art and the record of human settlement in the region. JEREMIE LEHMANN PHOTO, 2021.

Figure 9 Although a large portion of Botswana is semi-arid, the north-west region contains the permanent swamps of the Okavango Delta, seen here with luxurious papyrus vegetation. BRUCE CAIRNCROSS PHOTO, 1988.

Eswatini

As can be seen on the geological map (page 11), Eswatini can be simply divided into three geological terrains: Younger Phanerozoic (Karoo Supergroup) in the east, Archaean granite-gneiss in the central regions, and the Archaean greenstone belt in the north-west. Mining plays a relatively modest role in the economy of Eswatini, with gold, coal and dimension stone forming only a small percentage of income (Scott, 1950; Hunter, 1962; Barry, 2021). Mineral specimens and gemstones are equally scarce and are rarely seen in collections (Mountain, 1942). One exception is the hematite- and goethite-included quartz specimens that were extracted from the defunct Devil's Reef gold deposit in the north-west of the country.

Mindat.org (accessed December 2021) lists 59 valid species for Eswatini, but only 11 photographs of minerals are illustrated for the entire country.

Figure 10 The northern and western regions of Eswatini are the most important mineral-producing areas in the country, with gold and asbestos mines located here (Hunter, 1962). The Devil's Reef was a small but very rich gold occurrence that produced interesting quartz specimens such as this 4-cm crystal that contains hematite inclusions. BRUCE CAIRNCROSS SPECIMEN AND PHOTO.

▲ **Figure 11** Mountainous scenery in southern Eswatini. BRUCE CAIRNCROSS PHOTO, 1991.

Figure 12 Typical scenery in pastoral Eswatini with quartzite outcropping in the foreground. On the distant horizon are the Lebombo Mountains, which host quartz, amethyst and agate. BRUCE CAIRNCROSS PHOTO, 1992.

Lesotho

Landlocked Lesotho is characterized by little geological diversity: the country consists predominantly of Jurassic basaltic lavas with some Permo-Triassic sedimentary strata in the east and south. However, in terms of economic deposits, particularly diamonds, the country is notable for its kimberlites (Janse, 1995). The Kao and Letšeng-la-Terae kimberlites were discovered in 1954 and 1957, public diggings were proclaimed, and in 1967 a huge 601-carat diamond was found, the first of other large diamonds from the Letšeng pipe.

Apart from diamonds, the basalts of Lesotho (and neighbouring South Africa) yield beautiful mineral specimens from cavities, vugs and geodes contained in the rock. These include minerals such as quartz, and stilbite that is arguably the finest from southern Africa (Cairncross and Du Plessis, 2018).

Mindat.org (accessed December 2021) lists 30 valid species and two type-locality species for Lesotho. The site has 24 mineral photos, most of which are species found in the basalt – quartz, calcite, stilbite and apophyllite.

Figure 13 Lesotho has spectacular natural features, such as the Maletsunyane waterfall that was formed by headward erosion of the river into the layered basalt. HERMAN DU PLESSIS PHOTO, DECEMBER 2016.

Figure 14 Bladed stilbite crystals on columnar quartz, 4.9 cm. Butha-Buthe, Lesotho. BRUCE CAIRNCROSS SPECIMEN AND PHOTO.

Figure 15 Drusy, elongate finger-like stalagmites of quartz, 6 cm. Butha-Buthe, Lesotho. BRUCE CAIRNCROSS SPECIMEN AND PHOTO.

Figure 16 The verdant hills and valleys in Lesotho consist of dissected basalt landscapes. The basalts contain amygdales with stilbite, quartz and apophyllite. HERMAN DU PLESSIS PHOTO, 2017.

Namibia

Namibia is one of the more geologically complex countries of the subregion. It has no Archaean-age rocks, but rather a swathe of Proterozoic-hosted mineral deposits, and younger sedimentary cover in the form of the Nama Group, Karoo Supergroup and unconsolidated sand (Pirajno, 1994). Together with South Africa, Namibia is the premier source of minerals and gemstones in southern Africa, with world-famous localities having produced world-class mineral specimens. These localities, and their minerals and gemstones, have been featured in contemporary books such as Von Bezing *et al.* (2008, 2014, 2016) and Jahn *et al.* (2006), and details can be found therein. Some of the noteworthy mineral and gemstone localities are:

- The Otavi mountainland (Cairncross, 1997), notably the world-famous Tsumeb mine (Wilson, 1977; Gebhard 1999) that has produced some of the finest known specimens of azurite, cerussite, smithsonite and mimetite, to name a few. The Tsumeb mine currently (December 2021), has the third-largest number of approved mineral species of any locality in the world, 325, surpassed only by the Poudrette quarry, Mont Saint-Hilaire, Québec, Canada (432) and the Clara mine in Germany (440). Berg Aukas mine is also located in the Otavi mountainland and is the source of the finest descloizite specimens in the world (Cairncross, 2021c).
- Goboboseb Mountains, from which outstanding amethyst and prehnite have been collected (Cairncross and Bahmann, 2006b).

- Okorusu (Cairncross, 2018e) and the Erongo Mountains are world-class multicoloured fluorite localities, the latter is also the source of exceptional aquamarine, jeremejevite and schorl (Cairncross and Bahmann, 2006a; Falster *et al.*, 2018).
- The Karibib-Usakos districts' pegmatites from which outstanding tourmaline crystals have been produced (Ashworth, 2014).
- The Kaokoveld in northern Namibia has various copper prospects, which have produced dioptase, malachite and shattuckite (Bowell *et al.*, 2013).
- The pegmatites and quartz veins in southern Namibia, the source of highly colourful quartz and fluorite (Von Bezing *et al.*, 2014).
- Klein Spitzkoppe, the source of thousands of topaz crystals (Cairncross, 2005a).
- The southerly located Skorpion mine, with its outstanding tarbuttite, and nearby Rosh Pinah mine, the source of some of the finest southern African baryte (Cairncross and Fraser, 2012).
- Swartbooisdrif, a source of sodalite (Menge, 1986).
- Offshore marine diamond mining sites (Schneider and Miller, 1992).

In total, **Mindat.org** (accessed December 2021) lists 851 valid minerals from Namibia, including 108 type-locality species. There are 16,286 mineral photos, more than double the number of all the other countries featured in this book.

Figure 17 Aerial view of southern Namibia. The horizontally layered sedimentary rocks are clearly evident in the arid landscape. BRUCE CAIRNCROSS PHOTO, 2010.

Figure 18 Spectacular geology can be seen in many regions of Namibia. In this outcrop, a near-horizontal black sill intrusion is cut by a younger, steeply dipping dyke. BRUCE CAIRNCROSS PHOTO, 2017.

Figure 19 Dikker Willem, a subvolcanic carbonatite (on the right), seen looking north from the Aus–Lüderitz road, Namibia. Other ancient volcanic mountains, such as Gross Brukkaros north of Keetmanshoop, have produced interesting minerals. BRUCE CAIRNCROSS PHOTO, 1977.

◄ **Figure 20** The red dunes of the Namib Desert are found along most of the west coast of Namibia. BRUCE CAIRNCROSS PHOTO, 2017.

▼ **Figure 21** Typical Namibia Desert scenery west of Uis, with granite in the foreground and the Brandberg in the distance. The Brandberg massif contains interesting minerals, but is a protected heritage environment. BRUCE CAIRNCROSS PHOTO, 2016.

Figure 22 Tafelkop can be seen on the horizon on the right, with other distant peaks forming the Goboboseb Mountains west of the Brandberg. These ancient volcanic rocks contain high-quality quartz, smoky quartz, amethyst and prehnite. BRUCE CAIRNCROSS PHOTO, 2017.

Figure 23 Quartz with red hematite inclusions and small analcime crystals attached, 4.5 cm. Goboboseb Mountains, Namibia. BRUCE CAIRNCROSS SPECIMEN AND PHOTO.

Figure 24 Interlocking quartz and amethyst, 10.2 cm. Both crystals have numerous vapour and tiny red hematite inclusions. Goboboseb Mountains, Namibia. BRUCE CAIRNCROSS SPECIMEN AND PHOTO.

Figure 25 The Gross Spitzkoppe in the distance, viewed looking east from the granite foothills of the Klein Spitzkoppe. The Klein Spitzkoppe are well known for producing topaz crystals. BRUCE CAIRNCROSS PHOTO, 2017.

TSUMEB MINE RARE GEMSTONES:

The Tsumeb mine in Namibia is world famous for the variety and quality of minerals that were discovered and preserved during its 80-year history. Some of the minerals are of high enough quality to be faceted into 'fancy' gemstones. These are rare and are not often seen in collections. A selection is shown here.

Figure 26 A typical topaz crystal from Klein Spitzkoppe, Namibia, 5.3 cm. WARREN TAYLOR RAINBOW OF AFRICA COLLECTION, MARK MAUTHNER PHOTO.

Figure 27 (above left) 'Fancy' gemstones cut from Tsumeb minerals. From top to bottom: anglesite 4.65 carats (1.3 cm); azurite 8.21 carats (1.3 cm); wulfenite 3.69 carats (1 cm) and 9.67 carats (1.5 cm); and calcite 66.91 carats (2.9 cm). WARREN TAYLOR RAINBOW OF AFRICA COLLECTION, MARK MAUTHNER PHOTO.

Figure 28 (above right) 'Fancy' gemstones cut from Tsumeb minerals. From top right, down: wulfenite 3.84 carats (8 mm); baryte 1.74 carats (7 mm); mimetite 1.57 carats (6 mm); wulfenite 2.87 carats (8 mm); azurite 9.55 carats (1.2 cm); anglesite 18.41 carats (1.4 cm); and cobalt-rich smithsonite 25.36 carats (1.7 cm). WARREN TAYLOR RAINBOW OF AFRICA COLLECTION, MARK MAUTHNER PHOTO.

South Africa

South Africa contains some of the oldest rocks on Earth, including several greenstone belts, as shown on the geological map (Johnson *et al.*, 2006) on page 11. The country also contains Archaean sedimentary deposits, most now metamorphosed, and younger clastic and carbonate/chemical sedimentary strata that formed during the Proterozoic Eon. The bulk of these rocks form the ancient Kaapvaal Craton, a proto-continent precursor to the current landmass. Highly deformed and metamorphosed belts, such as the Limpopo Belt, border parts of the ancient craton. During the Phanerozoic Eon, younger sediments and lavas covered much of South Africa, and most of the kimberlite intrusions and volcanoes erupted through this younger cover.

South Africa contains the world's largest resources of gold, platinum and land-based manganese, plus important deposits of iron ore, chromite, diamonds and coal (Pretorius, 1976; Anhaeusser and Maske, 1986). All of these commodities have been exploited, and in the process have yielded mineral specimens and gemstones (Snyman, 1998).

The geological age of South Africa's rocks runs the gamut from the early Archaean to the Recent, and this extraordinary 3,500-million-year span, coupled with the state of preservation of these rocks, is the reason why South Africa is so well endowed with mineral wealth. The relevant geological formations that have produced important mineral specimens include:

- Archaean greenstone belts, such as Barberton and Murchison, from which gold has been one of the main commodities produced (Anhaeusser and Maske, 1986; Brandl and De Wit, 1997).
- The Limpopo Belt's copper mineralization in the Musina district (Söhnge, 1945; Bahnemann, 1986; Cairncross, 1991; Mayer and Moore, 2016).
- The Transvaal Supergroup strata that host huge manganese and iron ore deposits, and base metal Pb-Zn (Southwood, 1986; Cairncross *et al.*, 1997; Cairncross and Beukes, 2013; Cairncross *et al.*, 2017).
- The Phalaborwa Complex containing world-class baddeleyite crystals and other collectable minerals (Gliddon and Braithwaite, 1991; Southwood and Cairncross, 2017).
- The Bushveld Complex with its globally important platinum and chrome deposits, fluorite and cassiterite (Crocker, 1979; Anhaeusser and Maske, 1986; Vermaak and Von Gruenewaldt, 1986; Cawthorn *et al.*, 2002; Atanasova *et al.*, 2016).
- The Natal-Namaqua Belt containing the Okiep copper district, base metal mines at Aggeneys, fluorite at Riemvasmaak, and quartz from the Orange River region pegmatites (Hugo, 1970; Thomas *et al.*, 1990; Cairncross, 2004a; Minnaar and Theart, 2006).
- Kimberlites that have produced extraordinary diamonds, including the Cullinan diamond, the largest known to date (Wagner, 1914; Williams, 1932; Gurney, 1990; Gurney *et al.*, 1991; Lynn *et al.*, 1998).

Mindat.org (accessed December 2021) lists 938 valid species and 81 type-species for South Africa. It features 5,973 mineral photos.

Figure 29 The Amphitheatre in the Drakensberg mountain range, which consists mainly of basalt. The basalt boulders in the foreground contain white amygdales of minerals. BRUCE CAIRNCROSS PHOTO, 2014.

Figure 30 The arid Northern Cape scenery at Riemvasmaak, close to the Orange River. Excellent green fluorite has been collected here. BRUCE CAIRNCROSS PHOTO, 2010.

Figure 31
Chapman's Peak Drive in the Western Cape. The horizontal sedimentary rocks of the Cape Supergroup were deposited on the grey granite, and the road runs along the point of contact. Although the Western Cape has an abundance of spectacular geological scenery and numerous economic mineral resources (Cole *et al.*, 2014), the province is not known to produce a great variety of high-quality minerals.
BRUCE CAIRNCROSS PHOTO, 2005.

Figure 32 Fluorite and quartz are two of the most common minerals from Riemvasmaak, South Africa. This 13.4-cm specimen is colour-zoned octahedral fluorite with quartz. BRUCE CAIRNCROSS SPECIMEN AND PHOTO.

Figure 33 Clusters of pink ephesite crystals from Lohathla mine, Postmasburg, South Africa. Field of view 2.2 cm. BRUCE CAIRNCROSS SPECIMEN AND PHOTO.

Figure 34 Weathered dolomite pinnacles exposed in an old mine in the Postmasburg district, Northern Cape. Important deposits of iron and manganese are associated with rocks and structures such as these. BRUCE CAIRNCROSS PHOTO, 1992.

Figure 35 Highly folded and metamorphosed rocks are visible in a dry river bed close to South Africa's border with Zimbabwe. These form part of the high-grade Limpopo Belt that separates two ancient continents, the Zimbabwe Craton and the Kaapvaal Craton. The rocks host important copper deposits. STEVE MCCOURT PHOTO, 1990.

Figure 36 Quartz crystals with green epidote, 11.4 cm. Messina mine No. 5 Shaft, Limpopo, South Africa. BRUCE CAIRNCROSS SPECIMEN AND PHOTO.

Figure 37 The rugged Barberton mountainland is one of the oldest geological terrains on Earth and hosts numerous important economic deposits, notably gold. BRUCE CAIRNCROSS PHOTO, 1996.

Figure 38 The Pietersburg greenstone belt in Limpopo hosts mineral deposits, and the historic Eersteling gold mine, started in 1874, is located in this area. BRUCE CAIRNCROSS PHOTO, 1995.

Figure 39 Crystals of gold are rare in nature. These gold crystals, together with white calcite, come from the Barberton greenstone belt, South Africa. Field of view 2 cm. BRUCE CAIRNCROSS PHOTO.

Southern Mozambique

For the purposes of this book, southern Mozambique encompasses the part of the country adjacent to Zimbawe. Much of this terrain is covered by geologically young sediment and sand (Lächelt, 2004). However, the rocks that outcrop in a 200-km-wide zone bordering eastern Zimbabwe have more complex geology. This incorporates remnant Archaean greenstones and Proterozoic granite and gneisses, and some younger Phanerozoic strata.

Most of Mozamique's famous mineral and gemstone deposits occur in the northern part of the country, notably Nampula Province, which produces tourmalines and other economic minerals (Yager, 2019). Similarly, high-quality gem ruby deposits are mined in the northern and northeastern Niassa and Cabo Delgado provinces (Vertriest and Pardieu, 2016; Vertriest and Saeseaw, 2019). There is little in the way of major collectable mineral specimens or gemstones in southern Mozambique, but the region is not without interest. Quartz and agate are found in geodes in the Jurassic-aged volcanics in Maputo and Sofala provinces. Alluvial deposits containing agates that have weathered from lavas occur in the Gaza Province, and attractive blue dumortierite comes from the Tete Province. There are some alluvial gold deposits in the Manica Province near the border with Zimbabwe (Yager, 2019).

Figure 40 General view of rocky outcrops on the Songo River in the Tete Province. TERESA COTRIM/PIXABAY PHOTO.

Figure 41 Much of southern Mozambique is covered with sand. TERESA COTRIM/PIXABAY PHOTO.

Zimbabwe

The geology of Zimbabwe consists of ancient gold-bearing Archaean greenstone belts and relatively younger strata, ranging from the Proterozoic through the Phanerozoic to the present day (Stagman, 1978). Zimbabwe is well endowed with mineral resources and was once the world's largest producer of asbestos. It has also produced antimony, copper, diamonds, gold, iron ore, lithium, nickel, platinum, tin, tungsten, vanadium and vermiculite. Mining operations to extract these commodities have resulted in the discovery of interesting mineral specimens. The geological formations that have produced important mineral specimens and gemstones include:

- Greenstone belts; the Midlands greenstone belt in particular has produced excellent gold specimens (Blenkinsop *et al.*, 1997).
- The Great Dyke, primarily a source of chromite, which does not occur as aesthetic mineral specimens, although fine specimens of other secondary minerals, such as brucite, have been unearthed (Worst, 1960; Prendergast and Wilson, 1989).
- Numerous pegmatites in the north-western Mwami district (Wiles, 1961) contain gemstones; others elsewhere like in the Bikita area, Masvingo, produce lithium (Ackerman *et al.*, 1966).
- Novello claim, Masvingo district, where outstanding alexandrite crystals occur in ultramafic serpentinites (Schmetzer *et al.*, 2011).

Figure 42 View of Ngomakurira from Domboshava, approximately 40 km north-east of Harare. KEITH BEGG/IOA PHOTO.

- Sandawana (Zeus) mine, Mweza greenstone belt, has world-class emeralds (Zwaan *et al.*, 1997; Giuliani and Groat, 2020).
- Jurassic-aged basalts in the south-east produce amethyst and sceptered amethyst.
- Marange area in eastern Zimbabwe, where diamonds have been mined (Stocklmayer, 1981; Smit *et al.*, 2018).

Mindat.org (accessed December 2021) lists 328 valid species from Zimbabwe, including three type-locality species. There are 392 mineral specimen photos.

Figure 43 The geological formations in the region of Mount Belingwe, south-west of Masvingo in Zimbabwe. The ridge is composed of banded iron-formations. Pegmatites are known to produce important economic minerals in this region. BRUCE CAIRNCROSS PHOTO, 1996.

Figure 44 Chimanimani Mountains along the border between Zimbabwe and Mozambique, in the southern region of the Eastern Highlands. ROGER DE LA HARPE/IOA PHOTO.

Figure 45 A quartz crystal with an amethyst sceptre, 5.4 cm. Masvingo district, Zimbabwe. PAUL BOTHA SPECIMEN, BRUCE CAIRNCROSS PHOTO.

Figure 47 Granitic balancing rock near Epworth, approximately 10 km south-east of Harare. KEITH BEGG/IOA PHOTO.

Figure 46 Blue topaz crystal, 4.1 cm. St Ann's mine, Zimbabwe. JIM AND GAIL SPANN SPECIMEN, TOM SPANN PHOTO.

Aegirine ◆ $NaFe^{3+}Si_2O_6$

Aegirine crystallizes in the monoclinic system, has a hardness of 6, specific gravity of 3.55 to 3.6, pale yellowish-grey streak, and vitreous to resinous lustre. It is a sodium-rich clinopyroxene and is typically dark green to black in colour. Aegirine is a common rock-forming mineral of alkali rocks such as carbonatites and syenites, and is prismatic to fibrous. Associated minerals can include alkali feldspars, nepheline, riebeckite and arfvedsonite. Large aegirine crystals famously come from Mount Malosa in Zomba, Malawi.

Aegirine is common in many of the southern African alkaline complexes, such as the Pilanesberg in **South Africa** and Chishanya in **Zimbabwe**. Some rare crystals were found in the mines of the Kalahari manganese field, associated with fluorapophyllite-(K) and schizolite.

Tiny (less than 1 cm) but well-formed crystals of aegirine are found in the Aris phonolite in **Namibia**. Aegirine crystals up to 1.5 cm have been found in the Otjosondu manganese deposit in the Omatako district. Prismatic crystals are known from syenite in the Okorusu carbonatite complex in the Otjiwarongo district.

Figure 48 Black aegirine, pink schizolite and cream hydroxyapophyllite-(K). N'Chwaning II mine, South Africa. Field of view 2 cm. BRUCE CAIRNCROSS SPECIMEN AND PHOTO.

Figure 49 A polished sample of black aegirine crystals, intergrown with 'herringbone' fluorite and apatite, 10 cm. Pilanesberg, South Africa. BRUCE CAIRNCROSS SPECIMEN AND PHOTO.

Figure 50 Close-up of a 7.5-cm vug in phonolite. Small black aegirine crystals are associated with platy white microcline, brown tuperssuatsiaite and white needles of natrolite(?), possibly makatite. Aris quarry, Namibia. Field of view 2.2 cm. BRUCE CAIRNCROSS SPECIMEN AND PHOTO.

Ajoite ◆ (K,Na)Cu$_7$AlSi$_9$O$_{24}$(OH)$_6$·3H$_2$O

Ajoite (pronounced ah-hoa-ite) crystallizes in the triclinic system, has a hardness of 3.05, specific gravity of 2.96, a greenish-white streak and silky lustre. Ajoite is a rare secondary copper mineral that was originally described from the New Cornelia mine in Ajo, Arizona, in the USA. It typically forms radiating sprays of tiny prismatic crystals, never large crystals. Ajoite is pale blue and forms most commonly in the oxidized zones of copper deposits.

World-famous specimens come from the Musina district in **South Africa**, where copper was mined from the early part of the twentieth century until the mines closed in 1992 (Bahnemann, 1986; Cairncross, 1991; Mayer and Moore, 2016). The ajoite specimens, and others such as quartz, occur in breccia pipes that contain the copper mineralization (Jacobsen *et al.*, 1976).

Ajoite appears to have been a relatively rare mineral from the mines until July 1985, when a major discovery of quartz with ajoite inclusions was made, and again during the final few years of the mines' operation. These inclusions of blue-green ajoite, which commonly delineates phantom faces at or near the terminations of the quartz, can partially or wholly fill the quartz, turning it vibrant blue. The specimens range from small, loose crystals in thumbnail sizes to plates over a metre composed of large crystals. These specimens created a sensation amongst mineral collectors and it was thought that once the mines closed in 1992, no more would be forthcoming. However, between 2006 and 2011 some exceptional specimens were found near the surface at the old Artonvilla mine located east of the main Messina mine. The earlier discovered Messina mine No. 5 shaft specimens consisted primarily of cloudy, translucent quartz that also contained other inclusions, such as chloritoid, hematite, kaolinite and, more rarely, other dark blue copper silicate minerals shattuckite and papagoite. The Artonvilla quartz was more transparent and in some cases water-clear, creating spectacular specimens. The largest Artonvilla specimen extracted measured over a metre across and was ultimately trimmed into several smaller specimens.

The only notable occurrence of ajoite elsewhere in southern Africa is from the copper mine on the farm Mesopotamia 504 in the Khorixas district in **Namibia**. Here, tiny, free-standing pale blue-green crystals form dense sprays that cluster together and line the edges of tiny cavities in hard, massive quartz matrix (Cairncross, 2016a).

Figure 51 The Messina copper mine in 1991, when it was still operating. The No. 5 shaft headgear can be seen in the distance. This mine, and the others nearby, produced world-class ajoite specimens. BRUCE CAIRNCROSS PHOTO.

Figure 52 A cluster of quartz crystals, some included with blue ajoite. The orange-brown crystals are iron-rich albite feldspar and the green minerals are epidote and chlorite, 6.2 cm. Messina No. 5 shaft, South Africa. BRUCE CAIRNCROSS SPECIMEN AND PHOTO.

Figure 53 A doubly terminated quartz crystal included by blue ajoite and red hematite, 8.9 cm. Artonvilla mine, Musina district, South Africa. RONNIE MCKENZIE SPECIMEN, BRUCE CAIRNCROSS PHOTO.

Figure 54 A quartz crystal that was partially encrusted by calcite and clay, 11 cm. Some of the crust has been manually removed to reveal the blue ajoite inclusions. BRUCE CAIRNCROSS SPECIMEN AND PHOTO.

Figure 55 A cluster of quartz crystals, included with blue ajoite, white kaolinite and orange-red native copper. The shiny black crystals are specular hematite, 2.6 cm. Messina No. 5 shaft specimen, collected in 1986. BRUCE CAIRNCROSS SPECIMEN AND PHOTO.

Figure 56 A massive 400-kg quartz-ajoite specimen, measuring 1.35 m x 90 cm x 60 cm, from the Artonvilla mine area. The specimen was discovered in 2007 and was displayed at the 2012 Denver mineral show in Colorado. It has subsequently been trimmed into separate specimens. RONNIE MCKENZIE SPECIMEN AND PHOTO.

Figure 57 A large specimen of quartz enclosing ajoite, used as a garden feature, ±1.8 m. Messina mine, South Africa. BRUCE CAIRNCROSS PHOTO.

Albite ◆ NaAlSi$_3$O$_8$

Albite crystallizes in the triclinic system, has a hardness of 6 to 6.5, specific gravity of 2.6 to 2.63, white streak, and vitreous to pearly lustre. Albite is a sodium-bearing feldspar in the plagioclase group of feldspars. It is characteristically white to colourless and forms tabular crystals, but can also be found as massive lumps and granular aggregates. Cleavelandite (sometimes used as a synonym for albite) is the platy variety of albite often found in pegmatites. Albite is a common rock-forming mineral found in granites and other silica-rich igneous rocks. It is very common in granitic pegmatites.

Orange and white crystals of albite were found in the Messina copper mines, **South Africa**, the orange colour caused by traces of iron in the crystal lattice. In the Soutpansberg region of Limpopo, crystals have been found associated with lepidolite and quartz. In the Northern Cape and Namaqualand, albite is found in most pegmatites. Large crystals have been found in the Noumas pegmatite at the Blesberg mine north of Steinkopf, together with muscovite,

quartz, apatite and monazite-(Ce). Small, secondary white albite crystals occur in some alteration pockets in the Witwatersrand gold-bearing conglomerates.

Albite is a common species in **Namibian** pegmatites. It occurs as large crystals in many pegmatites in the Karibib-Uis-Usakos areas. Crystals of 2–3 cm were found in Klein Spitzkoppe, the Erongo Mountains and the Rubikon pegmatite. The well-known Gamsberg mineral localities at Natas and Kos also contain albite. Crystals are lustrous and white, up to 5 cm (Von Bezing *et al.*, 2016). Rosh Pinah mine in southern Namibia has produced clusters of highly lustrous white to colourless translucent albite crystals.

In **Zimbabwe**, albite is found at the Mistress mine, Salisbury, and Al Hayat claims in the Bikita area and other pegmatites. 'Moonstone' (a translucent green variety of calcium-rich albite) is found in the Karoi area.

Albite is found in pegmatites west of Francistown, **Botswana**, for example at the Prospect mine.

Figure 58 Orange iron-rich albite crystals on a doubly terminated quartz, 6.6 cm. Messina No. 5 shaft, South Africa. BRUCE CAIRNCROSS SPECIMEN AND PHOTO.

Figure 59 Transparent, colourless albite crystals with muscovite. Noumas 1 pegmatite, South Africa. Field of view 3.8 cm. BRUCE CAIRNCROSS SPECIMEN AND PHOTO.

Figure 60 Well-formed, lustrous albite crystals with quartz, 8.9 cm. Rosh Pinah mine, Namibia. BRUCE CAIRNCROSS SPECIMEN AND PHOTO.

Almandine ◆ $Fe^{2+}_3Al_2Si_3O_{12}$

Almandine crystallizes in the cubic system, has a hardness of 7 to 7.5, specific gravity of 4.1 to 4.3, white streak, and vitreous, resinous lustre. Almandine is a garnet and a member of the almandine-pyrope series of garnets. Like other garnets, almandine crystallizes as well-formed dodecahedral crystals. It is usually red but can occur as black crystals. It forms a chemical series with pyrope and spessartine, two other garnet species. It occurs in metamorphic rocks, schists and gneisses, and is common in pegmatites.

In **South Africa**, almandine is found in the Soutpansberg district of Limpopo, where crystals are sometimes clear and gemmy. Almandine occurs 30 km south of Musina on the farms Barend 523 MS and Piet 509 MS. Almandine associated with green epidote crystals occurs on the farms Rietvlei 375 JT and Vlakplaats 317 JT, close to Belfast in Mpumalanga. In the Northern Cape province, almandine is found on several farms in the Gordonia district.

Namibia has several almandine occurrences. Dark red almandine garnets are widespread in schists that outcrop in the Karibib-Usakos-Swakopmund districts and the Kuiseb schist belt. The Husab Mountains are a rich source of almandine garnet, as is the garnet-staurolite schist outcropping in the Gorob mine region, where sand is coloured red by alluvial garnet.

Almandine garnet has been mined primarily as the gemstone rhodolite in **Zimbabwe**. It was exploited mostly in the Beitbridge area, near Karoi and in the north-eastern parts of the country. At the Burgundy deposit (Beitbridge district), 275 kg of gem garnet was mined from 1973–1974. At the Lucky Fish prospect (Rushinga district), almandine garnet crystals averaged 4–5 cm in diameter and about 3–4% were deep red and of gem quality. At the nearby Treasure Casket deposit, 7,394 kg of gem-quality, dark ruby red almandine crystals about 1–2 cm in diameter were removed. Gem cordierite (iolite) was also mined at this prospect. Other noteworthy prospects are Sekuru (next to St Ann's pegmatite) and Manyuchi in the Mudzi district, which has produced almost 300 kg. Almandine garnet crystals up to 8 cm in diameter were found in the Miami mine in the Karoi district.

Almandine garnet is found in metamorphic rocks at many localities in the Tuli Block in north-eastern **Botswana**. It occurs on the farm Zanzibar, north-east of the Tuli Block, in schist in the Matsitama area and in amphibolite in the Moroka granite outcrop west of Bakaranga.

Figure 61
Almandine garnets partly enclosed in calc-silicate rock, 10 cm. Limpopo, South Africa.
BRUCE CAIRNCROSS SPECIMEN AND PHOTO.

Figure 62 A 3.8-cm dodecahedral almandine. Karoi district, Zimbabwe.
BRUCE CAIRNCROSS SPECIMEN AND PHOTO.

Figure 64 Almandine garnet in matrix, 5 cm. Limpopo, South Africa. BRUCE CAIRNCROSS SPECIMEN AND PHOTO.

Figure 63 A 7.5-cm rhombic dodecahedron almandine garnet from an undisclosed locality in Zimbabwe.
BRUCE CAIRNCROSS SPECIMEN AND PHOTO.

Anatase ♦ TiO$_2$

Anatase crystallizes in the tetragonal crystal system, commonly as dipyramids, has a hardness of 5.5 to 6, specific gravity of 3.79 to 3.97, a white to pale yellow streak, and adamantine to metallic lustre. It has perfect cleavage. Anatase can be transparent to opaque and colour is varied, from brown, yellow, pale green or blue to black. Rutile and brookite are also titanium oxide minerals, and hence chemically identical to anatase, but forming in different crystal systems. Anatase usually forms as a secondary mineral deriving titanium from other titanium-rich minerals, but does not need to be associated with such species. It occurs in alpine veins, granite and pegmatites.

Large collectable anatase crystals are not common in southern Africa but beautiful small ones are found in some localities. Tiny (1–2-mm) white crystals are found in the molybdenite deposit at Houtenbeck in the Groblersdal district, **South Africa**. These occur in the Bushveld granophyre, scattered on quartz crystals. Equally attractive dark blue to black crystals (microminerals less than 1 mm) are found at Vaalkop Dam in the Rustenburg district. These are scattered about together with cubic pyrite crystals and rhombohedral calcite, all as micro-crystals. They are hosted in tiny cavities and fractures in Bushveld Complex granite. What appears to be a once-off discovery of anatase on quartz was made at an undisclosed pegmatite/

quartz vein close to the Orange River in the Northern Cape. These crystals are also dark blue-black and are nested amongst tiny, well-formed quartz crystals.

To date, the best anatase crystals known from South Africa were collected in the Knysna district. These are associated with smoky quartz crystals and are hosted in Cape granite. The crystals are typically dipyramidal and up to 1.3 cm in size, making them the largest known from South Africa.

Fairly small, usually less than 1 mm, anatase crystals are reported from a few localities in **Namibia**, including the Erongo Mountains, the Gamsberg region and the Giftkuppe rutile occurrence in the Omaruru district, where it occurs together with rutile and albite.

Figure 65 Anatase crystal, 4 mm. Orange River region, Northern Cape, South Africa. BRUCE CAIRNCROSS SPECIMEN AND PHOTO.

Figure 66 Anatase crystal with tiny quartz crystals, 4 mm. Orange River region, Northern Cape, South Africa. Field of view 2.2 cm.
BRUCE CAIRNCROSS SPECIMEN AND PHOTO.

Figure 68 Minute white anatase crystals scattered on goethite-coated quartz. Houtenbeck molybdenum mine, Limpopo, South Africa. Field of view 1 cm. WOLF WINDISCH SPECIMEN AND PHOTO.

Figure 69 Scanning electron microscope backscatter image of elongate anatase crystals. Houtenbeck molybdenum mine, South Africa. Field of view 0.2 mm. WOLF WINDISCH SPECIMEN, MARIA ATANASOVA IMAGE.

Figure 67 A dipyramidal anatase crystal with smoky quartz on granite, 1.3 cm. Knysna district, South Africa. BRUCE CAIRNCROSS SPECIMEN AND PHOTO.

Andalusite ◆ Al$_2$SiO$_5$

Andalusite crystallizes in the orthorhombic system, has a hardness of 6.5 to 7.5, a specific gravity of 3.13 to 7.5, a white streak and vitreous lustre. Andalusite crystals are often elongate, like a pencil, and are usually red-brown to tan in colour. Andalusite is a common metamorphic mineral usually found in slates and schists. In areas where andalusite is common, the crystals easily weather out of the soft micaceous schist and can be found in stream sediment and soils. It forms under relatively low crustal pressure and low temperature conditions. Andalusite is referred to as a trimorphous mineral because it has a chemical composition identical to kyanite and sillimanite. Kyanite forms under high-pressure conditions, while sillimanite forms or crystallizes under high temperatures. Andalusite forms under relatively low crustal pressure and low-temperature conditions. A variety of andalusite, known as chiastolite, contains carbonaceous inclusions in the form of an 'X', which can be seen when the mineral is cut in cross section.

Andalusite is used primarily in the manufacture of high-alumina refractory products and by the ceramics industry.

Aluminium, which is extracted from andalusite, is lightweight and is used in the aircraft and aerospace industries. Aluminium is also used as an anti-corrosive agent.

The world's largest andalusite deposits are found in **South Africa**. Most of these economic deposits, which often have crystals over 10 cm long, formed when the Bushveld Complex intruded into surrounding country rock and the heat from the intrusion caused the contact metamorphic aureole that contains the andalusite (Hammerbeck, 1986). Alluvial andalusite deposits are common in the Marico district of the North West – the largest of these occur in the Thabazimbi/ Groot Marico/Zeerust area – and in Mpumalanga in the region around Chuniespoort/ Penge/Lydenburg. In the Parys-Vredefort districts in the Free State, andalusite occurs in metamorphic rocks in the aureole surrounding the Vredefort granite. Ancient sedimentary and mafic volcanic rocks outcrop in northern KwaZulu-Natal. In some areas, the upper portion of these lavas has been metamorphosed, producing large crystals of andalusite up to 3 cm in diameter, as on the

Figure 70 Andalusite crystals contained in a micaceous schist. The matrix is softer than the harder andalusite crystals and weathers faster, hence the positive relief of the crystals sticking out from the matrix. Penge mine, South Africa. Field of view 32 cm.

farm Nooitgedacht 620 in the Ngotshe district. Pegmatites in the Kakamas, Mottelsrivier and Bokvasmaak districts in the Northern Cape contain red-brown crystals, up to 15 cm long, which are often arranged in radiating, rosette-like clusters. Extraordinarily large crystals, up to a metre long, occur at Mottelsrivier.

Andalusite is relatively common in some schists in **Namibia**, notably in outcrops to the south-west of Okahandja (Schneider and Watson, 1992). It is also found in schists in the Kuiseb River valley close to the old Gorob mine.

In **Zimbabwe**, andalusite is found in metamorphic rocks such as schist and gneiss, particularly north of Mutoko and in the Mwami and Masvingo districts. Large chiastolite crystals, up to 5 cm in diameter, occur in the Karoi and Mwami districts.

Deposits of grey-white andalusite, together with diaspore, are found in **Eswatini** in the south near Sicunusa. They occur in pyrophyllite schist and were mined as a source of aluminium.

Andalusite is found in schists in the eastern part of **Botswana**.

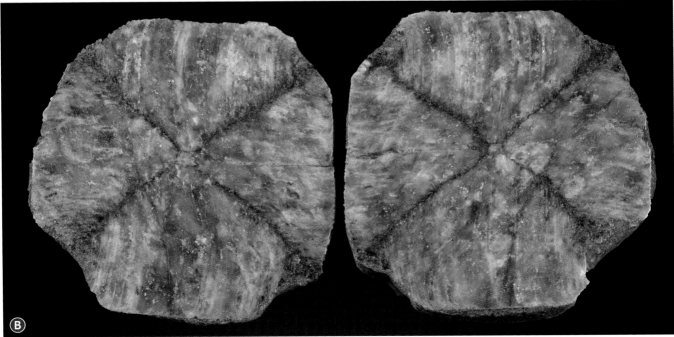

Figure 71 A specimen of andalusite, variety chiastolite: **A**, sawn into two halves **B** to reveal the internal carbon cross structure, 4.8 cm. Karoi district, Zimbabwe. BRUCE CAIRNCROSS SPECIMEN AND PHOTO.

Andradite ◆ $Ca_3Fe^{3+}_2(SiO_4)_3$

Andradite crystallizes in the cubic system, has a hardness of 6.5 to 7, specific gravity of 3.7 to 4.1, white streak and vitreous to resinous lustre. Andradite is a calcium-rich garnet and, like almandine, forms a chemical series with other garnet species, namely grossular, morimotoite and schorlomite. Andradite is usually found as red, yellow, orange or green dodecahedral crystals. The yellow-green to blue-green chrome-rich (Cr^{3+}) and iron-rich (Fe^{3+}) variety of andradite known as demantoid is prized as a gemstone. A black variety is sometimes called 'melanite' and a yellow variety 'topazolite'. These two latter names are used as commercial terms and are not recognized valid species. Andradite is most common in metamorphic marble and calc-silicate rocks, but can also occur as a secondary mineral in some altered economic ore deposits.

In **South Africa**, andradite occurs in the mines of the Kalahari manganese field in different hues of grey, brown, green, yellow, orange, red or black. Clusters of black crystals up to 10 cm in diameter have been found in the Soutpansberg, Limpopo. Green andradite occurs in marble at Kenkelbos 152 JQ near Rustenburg, North West. In the Northern Cape, specifically in Namaqualand, beautiful brown andradite crystals have been found at Doringkraal.

Namibia has notable andradite variety demantoid deposits west of the Erongo Mountains at the Green Dragon and Parrot mines, on the farms Tubussis 22, where attractive green crystals, some gem quality, occur in marble (Von Bezing *et al.*, 2016). Crystals are up to 3 cm and faceted stones weigh up to 10 carats. Andradite garnet, typically brown to black in colour, is found in the Otjosondu manganese deposits north-east of Okahandja. It occurs in marble 60 km east of Swakopmund, together with epidote and diopside. Andradite occurs in some pegmatites, such as Davib West 62 and Tsawisis 16. At Davib Ost 61, a tin-rich stannian andradite has been identified (McIver and Mihálik, 1975).

Demantoid is found in **Zimbabwe** on the Chimanda communal lands close to the Mazowe River and in the Bulawayo district on the farm Douglasdale.

Figure 72
Andradite garnet, 2.1 cm. N'Chwaning II mine, South Africa. BRUCE CAIRNCROSS SPECIMEN AND PHOTO.

⋏ Figure 73 Andradite variety demantoid, 1.8 cm. Tubussis, Erongo Region, Namibia.
BRUCE CAIRNCROSS SPECIMEN AND PHOTO.

Figure 74 ➤
Andradite crystals with metallic hausmannite. N'Chwaning II mine, South Africa. Field of view 2.9 cm.
BRUCE CAIRNCROSS SPECIMEN AND PHOTO.

Figure 75 Two matrix specimens of andradite variety demantoid, left crystal 5.5 cm. Tubussis, Erongo Region, Namibia.
WARREN TAYLOR RAINBOW OF AFRICA COLLECTION, MARK MAUTHNER PHOTO.

Figure 76 Faceted demantoid garnets, showing the diverse range of colours: **A** 9.64 carats (1.38 cm),
B 6.06 carats (1.13 cm), **C** 5.74 carats (1.1 cm), **D** 8.58 carats (1.3 cm). Tubussis, Erongo Region, Namibia.
WARREN TAYLOR RAINBOW OF AFRICA COLLECTION, MARK MAUTHNER PHOTOS.

Anglesite ◆ PbSO$_4$

Anglesite crystallizes in the orthorhombic system, has a hardness of 2.5 to 3, specific gravity of 6.38 and adamantine to vitreous lustre. Pure anglesite is colourless to white, but the mineral can become coloured by minor chemical impurities such as copper. Bright yellow anglesite can be formed by trace amounts of cadmium. Anglesite can form large, beautiful crystals that vary in form and habit – thick, tabular or prismatic crystals, or nodular or stalactitic masses. The crystals are typically heavy due to the high specific gravity. Anglesite forms as an oxidation product of lead ore, commonly galena, and therefore occurs as a secondary mineral associated with lead deposits. Apart from galena, associated minerals include cerussite, sphalerite, smithsonite and chalcopyrite.

Anglesite is relatively rare in **South Africa**. However, very aesthetic but small (under 1 cm) crystals are known from several lead deposits. Crystals have been collected from the defunct Argent mine, near Delmas, Gauteng, where the anglesite normally forms crusts around oxidized galena (lead sulphide) crystals and may itself be coated by another secondary lead mineral, cerussite. Similar tiny anglesite crystals are found at Houtenbeck and Stavoren

in Limpopo. Many small, scattered lead-zinc deposits that formed in dolomite in the North West province contain traces of small anglesite crystals – for example at Bokkraal 344 JP in the Marico district. Some anglesite was collected at the lead-zinc mine at the Aggeneys mine in the Northern Cape.

The premier southern Africa locality for outstanding anglesite crystals is the Tsumeb mine in **Namibia**. Anglesite was relatively common at this famous mine in the Otavi mountainland (Wilson, 1977; Gebhard, 1999). The mineral formed from the oxidation of the lead ore. Some crystals are very large, over 20 cm on edge. Stunning yellow crystals (coloured by cadmium) and green and blue specimens (coloured by copper) are known from Tsumeb. Some of the crystals are transparent and gem quality, allowing them to be faceted into fancy coloured collector gemstones.

Anglesite is rare in **Zimbabwe**. It occurs in the Mutare, Belingwe and Bulawayo districts, and is found in the Copper King gossan, with pyromorphite and mimetite. The Old West mine located 2 km north-west of Penhalonga, north of Mutare, has anglesite associated with galena, cerussite and pyromorphite.

Figure 77 Platy pale blue anglesite, 5.8 cm. Tsumeb mine, Namibia. BRUCE CAIRNCROSS SPECIMEN AND PHOTO.

Figure 78 Striated, transparent anglesite crystals, 7.2 cm. Tsumeb mine, Namibia. DESMOND SACCO SPECIMEN, BRUCE CAIRNCROSS PHOTO.

Figure 79 A general view of the buildings at the Tsumeb mine, southern Africa's premier source of anglesite. The old De Wet shaft headgear can be seen in the distance. BRUCE CAIRNCROSS PHOTO, 2017.

Figure 80 Grey anglesite crystal with minor galena, 2.1 cm. Tsumeb mine, Namibia. BRUCE CAIRNCROSS SPECIMEN AND PHOTO.

Figure 81 Pale blue anglesite crystals, 4.4 cm. Tsumeb mine, Namibia. WARREN TAYLOR RAINBOW OF AFRICA COLLECTION, MARK MAUTHNER PHOTO.

Figure 82 A colourless 51.48-carat (1.73-cm) anglesite. Tsumeb mine, Namibia. WARREN TAYLOR RAINBOW OF AFRICA COLLECTION, MARK MAUTHNER PHOTO.

Figure 83 A large 59.92-carat (2.07-cm) brilliant yellow faceted anglesite. Tsumeb mine, Namibia. WARREN TAYLOR RAINBOW OF AFRICA COLLECTION, MARK MAUTHNER PHOTO.

Antimony ◆ Sb

Antimony is a metalloid and crystallizes in the hexagonal system, has a hardness of 3 to 3.5, a specific gravity of 6.7, a characteristic grey streak, and metallic lustre. It has a low melting point of 631°C. Native antimony is formed when stibnite, its main ore mineral, is chemically reduced. Specimens usually consist of large silver-white lumps displaying many cleavage surfaces.

Antimony is mixed with lead to form an alloy that is used in batteries. Antimony alloy is also used in a number of other products, including ammunition, solder and bearings. Antimony in the form of antimony trioxide is used as a flame retardant in textiles, plastics and rubber.

Platy crystals of native antimony have been found in **South Africa** at the Alpha-Gravelotte mine in the Murchison greenstone belt, associated with chlorite-rich talc-carbonate schists (Pearton and Viljoen, 1986). These schists have been metamorphically altered to a dark serpentinite by the intrusion of a dolerite dyke. On the farms Zoodorst 2IU and Nooitgezien 3IU near Steynsdorp in the Barberton district, small quantities of antimony were found in lenticular pockets of stibnite in schist at, for example, the Morning Mist antimony mine. At Malelane, an antimony deposit was exploited on the farm Amo 259 JU. About 15 tonnes of stibnite were extracted.

Native antimony has been reported from the Kwekwe area, **Zimbabwe** (Nutt and Bartholomew, 1987).

Figure 84 Coarse, platy antimony crystals with some minor quartz, 6.8 cm. Monarch mine, South Africa.
BRUCE CAIRNCROSS SPECIMEN AND PHOTO.

Figure 85 A view of the Murchison greenstone belt hills with the Consolidated Murchison mine in the foreground. Stibnite, a source of antimony, and gold, is mined here. BRUCE CAIRNCROSS PHOTO, 1992.

Aragonite ◆ CaCO₃

Aragonite crystallizes in the orthorhombic system, has a hardness of 3.5 to 4, specific gravity of 2.95, white streak, and vitreous to resinous lustre. Aragonite is a polymorph of calcite, having the same chemical composition but crystallizing in a different crystal system. Crystals of aragonite are colourless or white and are usually transparent. Trace amounts of some elements, such as copper, can add colour to aragonite. The crystals can be acicular (needle-shaped), forming very long hair-like crystals. They can also be pyramidal or tabular. Twinning is common and the classic aragonite from Spain is characteristically twinned as sixlings, forming star-like groups of crystals.

Aragonite forms under much narrower conditions than calcite and is hence not as abundant; under normal atmospheric pressure and temperature aragonite tends to transform into calcite. Speleothems (cave deposits) in limestone and dolomite-hosted caves commonly contain aragonite stalactites and stalagmites and a host of other speleothems such as anthodites or 'cave flowers'. Cave formations of aragonite usually form smooth botryoidal masses referred to as flowstone. Rounded and semi-rounded aragonite cave pearls form in agitated cave water. Many stalactites and stalagmites are composed of aragonite, which is chemically unstable and may convert to calcite. The shells of nearly all invertebrate molluscs are composed of aragonite.

Beautiful and spectacular aragonite speleothems are found in carbonate-rich limestone and dolomite caves in **South Africa**, for example, at Sterkfontein, Sudwala, Makapansgat and the Echo and Cango caves. Some of these caves have outstanding examples of aragonite frostwork, coralloids, aragonite bush, botryoidal stalagmites and stalactites, cave pearls and draperies (Cairncross et al., 2016). In the Kalahari manganese field, aragonite is occasionally found as white or colourless needle-like crystals at the N'Chwaning I and II and Wessels mines. Aragonite specimens are known from the Duke's Hill and Vaalhoek mines at Pilgrim's Rest in Mpumalanga and from the Argent lead and silver mine in Gauteng. Some outstanding crystals several centimetres long occurred as haystack-like groups at the old Riries asbestos mine in the Northern Cape.

Figure 86 Aragonite bush containing different forms, including rounded coralloids and delicate acicular crystals, 9.4 cm. Thabazimbi district, South Africa. BRUCE CAIRNCROSS SPECIMEN AND PHOTO.

Figure 87 Finely crystallized, branching aragonite comprising multicorallite bush consisting of coralloid calcite overgrown by later aragonite crystallictite, 18.5 cm. Thabazimbi district, South Africa.
BRUCE CAIRNCROSS SPECIMEN AND PHOTO.

Figure 88 *In situ* aragonite 'frost' delicately attached to the host rock dolomite, covered by white aragonite. Thabazimbi district, South Africa. Field of view approximately 1 m. BRUCE CAIRNCROSS PHOTO.

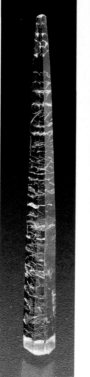

▲ **Figure 89** A stellate spray of needle-like aragonite, from the defunct Rand London quarry, Krugersdorp district, South Africa, 7.4 cm. BRUCE CAIRNCROSS SPECIMEN AND PHOTO.

▲ **Figure 90** Hollow aragonite straws from the defunct Rand London quarry, Krugersdorp district, South Africa, 13 cm. BRUCE CAIRNCROSS SPECIMENS AND PHOTO.

◄ **Figure 91** A well-formed, transparent hexagonal aragonite crystal with some internal fractures, 7.2 cm. N'Chwaning I mine, South Africa. BRUCE CAIRNCROSS SPECIMEN AND PHOTO.

Figure 92 A curtain of aragonite dripstone flowing over the exposed face in an old underground mine. Groundwater has percolated through the overlying dolomite, leaching carbonate and precipitating aragonite. Hurricane lamp for scale. Northern Cape, South Africa. BRUCE CAIRNCROSS PHOTO.

Figure 93 Yellow and white aragonite cave pearls. These speleothems form by calcium carbonate precipitating in concentric layers around a tiny nucleus. Moving water then smooths and polishes the surface, giving an appearance of a pearl. Mooinooi mine, South Africa. BRUCE CAIRNCROSS SPECIMENS AND PHOTO.

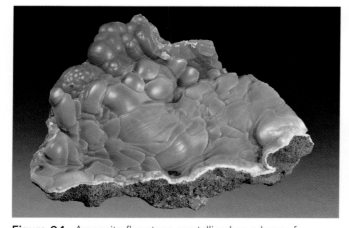

Figure 94 Aragonite flowstone crystallized on a layer of granular chromite, 13.2 cm. Montrose mine, Steelpoort district, South Africa. BRUCE CAIRNCROSS SPECIMEN AND PHOTO.

Figure 95 A 10.65-carat faceted aragonite, cut from material similar to that shown in figure 91. BRUCE CAIRNCROSS SPECIMEN AND PHOTO.

Figure 96 Two varieties of aragonite: yellow, zinc-rich nicholsonite, 5.1 cm, and white, lead-rich tarnowitzite. Tsumeb mine, Namibia. BRUCE CAIRNCROSS SPECIMENS AND PHOTO.

Beautiful aragonite crystals come from the Tsumeb and Kombat mines in **Namibia**, some of the Tsumeb specimens forming radiating sprays of crystals several centimetres long. Three chemical variations of aragonite are found at Tsumeb: green copper-rich aragonite, white tarnowitzite crystals containing traces of lead, and nicholsonite, an attractive yellow variety that contains minor quantities of zinc (these names are not officially recognized by the International Mineralogical Association [IMA], but are used extensively). Aragonite also occurs in caves and cavities in the dolomites of the Otavi mountainland. Lenses of solid aragonite are found in the Damara rocks in the Swakopmund and Karibib districts, notably near Rössing. Yellow and white-banded aragonite have been exploited as dimension stone north of Swakopmund, about a kilometre north-east of the railway bridge over the Swakop River.

In **Zimbabwe**, aragonite is found at Mangula, and at the Ethel mine on the Great Dyke in the Makonde district, which has produced some beautiful crystals. Aragonite is also found in the Chinhoyi Caves.

Figure 97 Off-white aragonite variety tarnowitzite associated with green malachite and duftite(?), 5.2 cm. DESMOND SACCO SPECIMEN, BRUCE CAIRNCROSS PHOTO.

Arsenopyrite ◆ FeAsS

Arsenopyrite crystallizes in the monoclinic system, has a hardness of 5.5 to 6, specific gravity of 6.07, a black streak, and metallic lustre. The mineral usually has a steel-grey to dull silver colour and crystals are short and prismatic. In many deposits, arsenopyrite is present as irregular lumps or masses, and it can be granular. Arsenopyrite is an important economic sulphide mineral. It has a garlic odour when struck with a hammer or metal object, which helps to distinguish it from other similar related copper and iron sulphides such as pyrite and chalcopyrite. Arsenopyrite occurs in some pegmatites and hydrothermal veins and is a commonly associated mineral in some gold deposits. It is relatively common globally, but not in vast quantities in any one locality. Arsenic is extracted from the mineral. This element is highly poisonous and is used in pesticides, herbicides and wood preservatives.

Arsenopyrite is present in many metallic ore deposits in **South Africa**, but well-formed, aesthetic crystals are rare. It was very common at the now-defunct Bushveld Complex tin mines, such as Rooiberg and Zaaiplaats (Rozendaal *et al.*, 1986). Small, silvery lustrous, well-formed crystals are known from Stavoren-Mutue Fides tin mines, Limpopo, and some small crystals were collected from the lead/silver deposits at the Argent mine near Delmas, Gauteng. Some fine specimens of crystalline arsenopyrite have been found in the Pilgrim's Rest district at the Ledouphine mine. At the Frankfort mine, crystalline arsenopyrite is found together with pyrite and quartz. Arsenopyrite is common in the ores of the gold mines in the Barberton area and in the antimony deposits of the Murchison mountain range. Other localities include Broken Hill mine and pegmatites such as Bokseputs in the Northern Cape. Arsenopyrite occurs as small, secondary metallic crystals, a few millimetres in length, in the Rietkuil uranium deposits in the southern Karoo.

As in neighbouring countries, arsenopyrite is relatively common in **Namibia** in association with base metal and gold deposits. At the Krantzberg mine in the Erongo Mountains, attractive silver-coloured crystals were found, associated with black schorl tourmaline. Other well-known sites are the Kombat, Haib, Matchless and Otjihase mines.

Arsenopyrite is found in many deposits in **Zimbabwe** associated with other common sulphides such as pyrite and chalcopyrite, notably at the Cam and Motor mine in the Karoi district and in the Mutare, Kwekwe, Gweru and Bulawayo districts and other sulphide-deposit localities. Arsenopyrite is also associated with the gold mining sector and is a common mineral in gold-stibnite-quartz vein deposits. At the Cairndhu mine, Mutare district, it occurs in talc schist together with a rare secondary arsenate, scorodite. It also occurs in talc schist at the Champion mine and in hornblende schist at the Hydra mine in the Gwanda district.

In **Eswatini**, arsenopyrite is found in the Forbes Reef areas associated with copper-iron sulphides. It occurs at the She mine in the Pigg's Peak district.

Arsenopyrite is sometimes found in association with gold deposits in the Tati schist belt in **Botswana**.

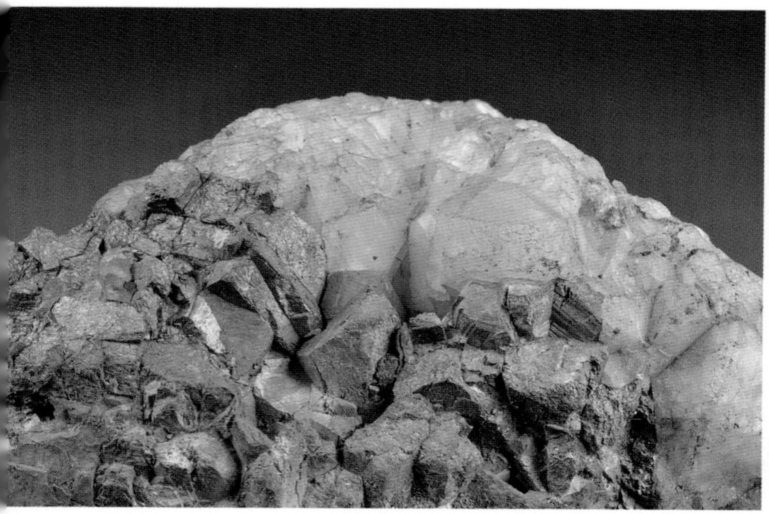

Figure 98 Arsenopyrite crystals on white quartz, 8.1 cm. Frankfort mine, Pilgrim's Rest, South Africa. PAUL MEULENBELD SPECIMEN, BRUCE CAIRNCROSS PHOTO.

Figure 99 Highly lustrous arsenopyrite crystals. Stavoren mine, South Africa. Field of view 2.1 cm. BRUCE CAIRNCROSS SPECIMEN AND PHOTO.

Figure 100 A mass of silvery arsenopyrite and a few elongate black schorl crystals. Krantzberg mine, Namibia. Field of view 6.5 cm.
BRUCE CAIRNCROSS SPECIMEN AND PHOTO.

Figure 101 Krantzberg Mountain north of the Erongo Mountains. The old mine workings and road leading up to the mine are visible on the hillside. BRUCE CAIRNCROSS PHOTO, 2014.

Asbestos ◆ see chrysotile, grunerite, riebeckite

Azurite ◆ $Cu_3(CO_3)_2(OH)_2$

Azurite crystallizes in the monoclinic system, has a hardness of 3.5 to 4, specific gravity of 3.77, a characteristic blue streak and vitreous lustre. Azurite varies from shades of pale blue to blue-black, but is characteristically a vivid blue. It can be chemically unstable and, in time, converts or alters to form green malachite.

Pseudomorphs of green malachite after azurite are common. Historically, azurite was crushed and the powder used as blue paint by artists, causing peculiar situations where the sky and ocean in old paintings now appear green. Lumps of massive azurite can be cut and polished for lapidary purposes, even though it is a relatively soft mineral. Azurite can, and does, form spectacular and beautiful crystals that are highly prized by collectors and museums. Thin crystals are usually transparent.

Azurite is a common secondary mineral associated with many copper deposits. When the primary copper sulphide ore is oxidized, it breaks down chemically to form azurite, which appears as blue smears or smudges on weathered rock surfaces. It is often associated with malachite, also a copper carbonate.

In **South Africa**, large, centimetre-long crystals of azurite were found in the nineteenth century at the defunct Willows mine east of Pretoria (Wilson Moore and Wilmer, 1893). At the Vergenoeg mine north of Pretoria, small vivid blue azurite crystals occur in association with fluorite and goethite (Crocker, 1985; Cairncross *et al.*, 2008). Rare, small (several millimetres in length) azurite crystals occurred in copper ores in the Limpopo province's

Figure 102 A view of the old De Wet shaft headgear of the Tsumeb mine, in Namibia, located close to Tsumeb town centre. This mine has been the source of some of the finest azurite crystals known. BRUCE CAIRNCROSS PHOTO, 2017.

Messina mines, while the Stavoren tin mines of the Bushveld Complex have produced some small azurite crystals that line cavities and are associated with fluorite and quartz. Several, now abandoned, mines in the Pilgrim's Rest region of Mpumalanga have also recorded crystalline azurite alongside other copper minerals. Azurite has been reported from dolerite intrusions in the Eastern Cape district, near Cradock. It is a rare accessory mineral in the copper mines in the Springbok/Okiep district.

Figure 103 A mass of bladed azurite crystals clustered together in semi-parallel arrangements. Vergenoeg fluorite mine, South Africa. Field of view 3.8 cm. BRUCE CAIRNCROSS SPECIMEN AND PHOTO.

Figure 104 A sharply terminated azurite crystal, 7 cm. Tsumeb mine, Namibia. BRUCE CAIRNCROSS SPECIMEN AND PHOTO.

Figure 105 A vug lined on the outside by massive light blue azurite while the open space inside is lined with dark blue azurite crystals, 7.2 cm. Tsumeb mine, Namibia. BRUCE CAIRNCROSS SPECIMEN AND PHOTO.

Some of the finest azurite crystals in the world – intensely blue and up to tens of centimetres – have come from the Tsumeb mine in the Otavi mountainland, **Namibia**. Top-quality azurites from Tsumeb are among the specimens most sought after by collectors (Gebhard, 1999; Von Bezing *et al.*, 2008, 2014, 2016). The azurite exists in many different forms and habits, and can occur either on its own or associated with other attractive species such as cerussite, smithsonite, calcite and wulfenite. Beautiful crystals were rarely found at the Onganja copper mine in the Windhoek district, which was also world famous for huge cuprite crystals. Some excellent drusy azurite crystals, small finger-like stalactites and platelets of an intense azure from the Tschudi deposit, about 20 km west of Tsumeb, appeared on the market in the early 1990s. Not many collectable minerals have surfaced from this deposit, which has copper-silver mineralization in sandstones. Apart from azurite in the oxidized zone, malachite is the main common mineral and occurs with cuprite.

There are hundreds of known copper deposits in Namibia – over 400 copper mines and prospects listed in *The Mineral Resources of Namibia* (1992) – many with some evidence of trace amounts of azurite. These include the old Gorob mine in the Namib Desert, Kombat mine and Sinclair mine, to name but a few.

As in Namibia, azurite is associated with many **Zimbabwe** copper deposits, usually as small crystals or in amorphous forms. Examples are: the Skipper mine, 30 km west of Kwekwe, where azurite and malachite were the main sources of copper ore and occurred in a shear zone; Copper Duke (Kadoma district); Edward (Chiredzi district), where azurite and malachite extended down to a depth of 25 m; Leonada (Kwekwe district); Luca (Hwange district) and Montana (Makonde district), where azurite is associated with malachite, chrysocolla and dioptase; and copper deposits in the Chinhoyi (Masvingo area) and on Devure farms. Other notable localities are the Lomagundi area, Bikita district, and the Scheelite King and Umkondo mines.

There are no significant occurrences of azurite in **Botswana**, except as a trace mineral associated with scattered copper prospects in the Ghanzi area, south-west of the Okavango Delta.

Traces of azurite are found in a narrow epidote vein east of Nkambeni in **Eswatini**, where it is associated with chalcopyrite, sphalerite and smithsonite.

Figure 106 A group of azurite crystals together with a twinned, reticulated cerussite, and minor green malachite, 3.1 cm. Tsumeb mine, Namibia. BRUCE CAIRNCROSS SPECIMEN AND PHOTO.

Figure 107 A mass of small azurite crystals associated with several yellow wulfenites and green malachite, 4.4 cm. Tsumeb mine, Namibia. BRUCE CAIRNCROSS SPECIMEN AND PHOTO.

Figure 108 A historic azurite specimen, circa 1920, with green arsentsumebite, 2.4 cm. Tsumeb mine, Namibia. BRUCE CAIRNCROSS SPECIMEN AND PHOTO.

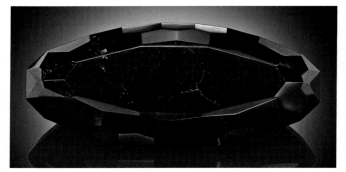

Figure 110 An unusual and difficult mineral to facet, this azurite weighs 66.85 carats and measures 3.77 cm. WARREN TAYLOR RAINBOW OF AFRICA COLLECTION, MARK MAUTHNER PHOTO.

Figure 109 An azurite crystal, 3.2 cm. Tsumeb mine, Namibia. WARREN TAYLOR RAINBOW OF AFRICA COLLECTION, MARK MAUTHNER PHOTO.

▲ **Figure 111** A drusy azurite specimen, 4.8 cm. Tschudi mine, Namibia. BRUCE CAIRNCROSS SPECIMEN AND PHOTO.

◄ **Figure 112** Bladed azurite crystals with green malachite, 3.6 cm. Tschudi mine, Namibia. BRUCE CAIRNCROSS SPECIMEN AND PHOTO.

Baddeleyite ◆ ZrO$_2$

Baddeleyite is a zirconium oxide that forms in the monoclinic system. It has a hardness of 6.5, a specific gravity of 5.74, brown-white to white streak and a greasy to vitreous lustre.

Baddeleyite is a relatively rare zirconium-bearing mineral and is of economic and academic importance, as it can be used to radiometrically date rocks using its zirconium content. It is usually found as microscopic, prismatic, shiny black crystals. Small baddeleyite crystals are often hosted in relatively unusual intrusive igneous rocks called carbonatites. These are pipe-shaped intrusive igneous rocks that have a high carbonate content. South Africa has the distinction of having produced the largest and most aesthetically pleasing baddeleyite crystals in the world.

Baddeleyite crystals up to 15 cm long were found in the early 1980s in **South Africa** at the Palabora mine, close to the Kruger National Park (Southwood and Cairncross, 2017). Many crystals over 1 cm in length were collected at the mine, which was one of the largest opencast mines in the world and is now an underground operation. Apart from copper, more than 50 minerals are known to occur in this deposit, including magnetite, uraninite-thorianite and baddeleyite as by-products, and fluorapatite and vermiculite. A variety of aesthetically pleasing zeolites were particularly plentiful and were collected as mineral specimens in the 1980s (see mesolite, fluorapophyllite). These were associated with dykes that cross-cut the orebody.

Small baddeleyite crystals occur in a dolomite occurrence south of Lüderitz in **Namibia**.

Baddeleyite is reported from **Zimbabwe** only as a rare accessory mineral in the Dorowa carbonatite.

Figure 113 A view from within the Palabora mine when the opencast operation was still running in 1990.
BRUCE CAIRNCROSS PHOTO.

Figure 114 Baddeleyite crystals partially embedded in matrix. The vertical crystal is 2.1 cm. Palabora mine, South Africa.
BRUCE CAIRNCROSS SPECIMEN AND PHOTO.

Figure 115 A terminated 10-cm baddeleyite crystal in matrix. This is reputed to be one of the largest crystals in existence.
DESMOND SACCO SPECIMEN, BRUCE CAIRNCROSS PHOTO.

Baryte ◆ BaSO$_4$

Baryte crystallizes in the orthorhombic system, has a hardness of 3 to 3.5, a specific gravity of 4.5, white streak and vitreous to resinous lustre. Baryte is a fairly common mineral that may superficially resemble calcite. It is much heavier, however, and cleaves in three directions. It is usually found as a colourless massive mineral, but can form attractive prismatic or tabular crystals, as well as fibrous masses and granular aggregates. Baryte crystals are surprisingly heavy for their size, barium being a dense element. Baryte is found in many different geological settings, but usually in hydrothermal veins in sedimentary and/or igneous rocks.

Barium has several chemical uses, notably in paint, paper, plastics, rubber, glass and ceramics. It also has medical applications, for example as a radiocontrast agent in radiography. Baryte is most widely used as an oil-drilling mud because the heavy baryte mud confines the oil and gas in the borehole during drilling. It is extremely insoluble, and does not pose an environmental problem when used in drilling operations.

The largest baryte crystals in the world are from **South Africa** and, oddly enough, from a Witwatersrand gold mine, Kusasalethu near Carletonville in Gauteng. These attractive, amber-coloured crystals were found in 1997, associated with quartz, galena, carbon and pyrite, 2,600 m below the surface. The two largest measured 83 cm and 70 cm, and weighed 76.5 kg and 64 kg, respectively (Cairncross and Rademeyer, 2001). Baryte has been found in other Witwatersrand gold mines, such as the Kopanang and Randfontein Estates mines, albeit as much smaller crystals.

Figure 117 Probably the second-largest well-formed baryte crystal known, measuring 70 cm and weighing 64 kg (see text). Attached are several galena crystals and powdery black pyrobitumen (kerogen). Kusasalethu gold mine specimen, Witwatersrand gold fields, South Africa. BRUCE CAIRNCROSS PHOTO.

▲ **Figure 116** Pale yellow, tabular baryte together with well-formed sphalerite crystals. Randfontein Estates gold mine, Witwatersrand gold fields, South Africa. BRUCE CAIRNCROSS SPECIMEN AND PHOTO.

Figure 118 ➤
Yellow baryte crystals associated with galena, quartz and pyrobitumen (kerogen), 5.6 cm. Kusasalethu gold mine, Witwatersrand gold fields, South Africa., BRUCE CAIRNCROSS SPECIMEN AND PHOTO.

Figure 119 A tabular baryte crystal showing multiple stages of crystallization. An early-formed crystal was coated by an unidentified brown mineral that was subsequently partly overgrown by layers of white baryte. Field of view 2.1 cm. N'Chwaning II mine, South Africa. BRUCE CAIRNCROSS SPECIMEN AND PHOTO.

Figure 121 A single yellow baryte crystal with two orange shigaite crystals attached, 2.1 cm. N'Chwaning II mine, South Africa. BRUCE CAIRNCROSS SPECIMEN AND PHOTO.

Figure 120 Pale blue baryte crystals, 6.5 cm, collected in 1988 at the N'Chwaning II mine, South Africa. BRUCE CAIRNCROSS SPECIMEN AND PHOTO.

Figure 122 A large terminated baryte crystal, 18.6 cm. Salpeterkop, Sutherland district, South Africa. BRUCE CAIRNCROSS SPECIMEN AND PHOTO.

Beautiful large baryte crystals, up to 15 cm, are sometimes collected in the mines of the Kalahari manganese field (Cairncross and Beukes, 2013). These tend to be white or colourless, but may be blue or pale yellow. In a volcanic carbonatite at Salpeterkop (Verwoerd, 1990), in the southern Karoo, large, spear-shaped baryte crystals, up to fist size, are found in hydrothermal veins, cavities and vugs. There is a similar geological deposit in the Goudini carbonatite, where baryte veins are up to 65 m long.

Baryte was found occasionally at the Messina copper mines as attractive, semi-transparent, pale brown, tabular crystals (Cairncross, 1991). It also occurs in the Aggeneys region in the Northern Cape, 20 km west of the Gamsberg. Several deposits with small crystals or massive baryte are known from Limpopo and the Barberton district in Mpumalanga. Prismatic orange-yellow baryte crystals up to 2 cm long have been found in the Okiep copper district, notably from the Jan Coetzee and Nababeep West mines, where the baryte crystals are often on the surfaces of large quartz crystals.

Figure 123 A spray of chisel-shaped baryte crystals, 4 cm. Rosh Pinah mine, Namibia. BRUCE CAIRNCROSS SPECIMEN AND PHOTO.

Figure 124 A large radiating cluster of orange baryte crystals, 22.5 cm. Rosh Pinah mine, Namibia. DESMOND SACCO SPECIMEN, BRUCE CAIRNCROSS PHOTO.

Baryte occurs as well-formed microscopic crystals at a few localities in KwaZulu-Natal, notably Pinetown, Westville and Sinkwazi. At Pinetown, a road-cutting on the highway passes through the farm Klaarwater in the Umhlatuzana Valley, where crystals of quartz, siderite, pyrite and baryte were found in a fault zone. The baryte occurs as small, single, transparent, yellow crystals (less than 3 mm long) arranged in attractive sprays. Beautiful micro-baryte crystals come from several deposits in the Bushveld Complex, including Argent, Boekenhouthoek, Stavoren, Vaalkop Dam (in granite), and Houtenbeck (Atansova *et al.*, 2016).

Baryte is widely distributed in **Namibia** (Schneider and Seeger, 1992a). Superb specimens, arguably the finest aesthetically pleasing baryte crystals from Namibia and possibly southern Africa, were collected in early 1989 at the Rosh Pinah lead-zinc mine in the Lüderitz district, 20 km north of the Orange River (Van Vuuren, 1986; Cairncross and Fraser, 2012). This sulphide deposit has been exploited since 1969, with mineralization occurring in volcaniclastic, volcanic and sedimentary rocks. The colour of the best baryte specimens is a rich golden-orange, and large crystals over 4 cm long are arranged in radiating sprays up to 30 cm in diameter. Some 'floater' specimens are composed of complete 360° discs, resembling Aztec suns. The finest specimens have the baryte crystals on the matrix. Other rare barium-bearing minerals found at Rosh Pinah include norsethite, benstonite, barium-rich calcite and witherite, although these have not yet been found as remarkable specimens. Galena and sphalerite crystals are occasionally collected at this mine.

Figure 125 Unusual vermiform or stalactitic group of yellow baryte, 6.5 cm. Rosh Pinah mine, Namibia. BRUCE CAIRNCROSS SPECIMEN AND PHOTO.

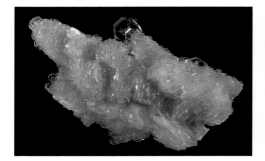

Figure 126 Gem-quality, transparent baryte from Rosh Pinah mine, Namibia, 11 cm. DESMOND SACCO SPECIMEN, BRUCE CAIRNCROSS PHOTO.

Figure 127 A colourless 29.24-carat (1.71-cm) baryte gemstone. Rosh Pinah mine, Namibia. WARREN TAYLOR RAINBOW OF AFRICA COLLECTION, MARK MAUTHNER PHOTO.

Figure 128 Brown, tabular baryte, 5 cm. Tsumeb mine, Namibia. BRUCE CAIRNCROSS SPECIMEN AND PHOTO.

Figure 129 Yellow zoned baryte, colourless cerussite and fine-grained willemite on the matrix: **A** under daylight and **B** under 365-nm long-wave ultraviolet light. The baryte fluoresces bright red. Field of view 2.2 cm. Tsumeb mine, Namibia. BRUCE CAIRNCROSS SPECIMEN AND PHOTO.

Large (up to 15 cm on edge), pale yellow baryte crystals have been found in veins outcropping at the extinct Gross Brukkaros volcano in the Keetmanshoop district. Some bright yellow and completely transparent crystals were cut into ornate faceted gemstones. Small, tabular baryte crystals occur with fluorite and quartz at the Okorusu fluorite mine. Beautiful but small, gemmy, orange baryte crystals were occasionally found associated with dioptase, quartz and malachite at the Omaue copper deposit, Kaokoveld. Baryte is one of the economic minerals found at the Otjosondu manganese deposit. Several kilometres west of Tses in Namaland, an interesting deposit of baryte and chalcedony is found. On the farm Blydskap 268 in the Outjo district, veins of baryte that cross-cut gneiss and thick layers of baryte, several kilometres long, are known in the same region, north of Tsongoari. The famous Tsumeb mine occasionally produced chocolate-brown to orange, tabular crystals, some up to 10 cm long. White to cream baryte occurs sporadically in the Goboboseb Mountain basalts together with quartz and calcite. Some of these crystals can be up to 4 cm, but most are much smaller. Some bladed crystals aggregate into tiny, attractive radiating spheres.

Baryte is fairly widespread in **Zimbabwe**, but not many deposits are economic (Morrison, 1970). The Dodge mine in the Shamva district is an exception. Here, lenses of baryte ore are found in limestones, quartzites and phyllites, and the mineralization appears to be fault controlled. Crudely crystalline baryte is found at Bumburudza in the Mwenezi district.

Small deposits of baryte are found in granites in the north-east of the central district in **Botswana**. Pockets of baryte and quartz are found at Suping, about 10 km north of Molepolole, associated with a dolerite dyke that has intruded into siltstone.

Baryte has been commercially mined in **Eswatini** in the Mbabane district, south of Oshoek, where veins of solid baryte are found in metamorphosed sedimentary rocks. There is a baryte mine just north of Oshoek near the South African border. This occurrence is in Onverwacht Group rocks, as are several other showings in various localities in the Barberton greenstone belt (Reimer, 1980). Baryte is also known to be present in veins up to 1 m thick in chert in the Pigg's Peak region on the eastern bank of the Londolozi River. Baryte is found in several stratigraphic levels in the Mapepe Formation of the Fig Tree Group in Barite Valley (Lowe et al., 2019).

Figure 130 Bright yellow baryte crystals. Tsumeb mine, Namibia. Field of view 1.5 cm. BRUCE CAIRNCROSS SPECIMEN AND PHOTO.

Figure 131 A large baryte specimen displaying more than one stage of crystal growth, 11 cm. Gross Brukkaros, Namibia. BRUCE CAIRNCROSS SPECIMEN AND PHOTO.

Figure 132 A doubly terminated baryte crystal, 7 cm. Gross Brukkaros, Namibia. BRUCE CAIRNCROSS SPECIMEN AND PHOTO.

Figure 133 A transparent, gem-quality baryte, 4.6 cm. Gross Brukkaros, Namibia. Faceted stones, such as the one shown here, have been cut from similar material. BRUCE CAIRNCROSS SPECIMEN AND PHOTO.

Figure 134 A large 54.17-carat (2-cm) faceted baryte. Gross Brukkaros, Namibia. WARREN TAYLOR RAINBOW OF AFRICA COLLECTION, MARK MAUTHNER PHOTO.

Figure 135 Yellow baryte crystals scattered on pale blue plancheite, a copper silicate. Omaue mine, Namibia. Field of view 5 cm. BRUCE CAIRNCROSS SPECIMEN AND PHOTO.

Figure 136 A radiating cluster of yellow baryte on brown calcite, on quartz. Field of view 1.7 cm. Goboboseb Mountains, Namibia. BRUCE CAIRNCROSS SPECIMEN AND PHOTO.

Figure 137 Flat, tabular yellow baryte with white quartz, 5.5 cm. Okorusu fluorite mine, Namibia. BRUCE CAIRNCROSS SPECIMEN AND PHOTO.

Beryl ◆ $Be_3Al_2Si_6O_{18}$

Beryl crystallizes in the hexagonal crystal system, has a hardness of 7.5 to 8, a specific gravity of 2.6 to 2.9, a white streak, and vitreous lustre. Beryl is well known among gemologists, lapidarists and mineral collectors. The most sought after coloured varieties are:

aquamarine	(blue-green from Fe^{2+} and Fe^{3+})
heliodor	(yellow from Fe^{3+})
morganite	(pink from Mn^{2+})
emerald	(green from Cr^{3+})
goshenite	(a colourless variety of beryl)

Even though beryl occurs in several colours, it is relatively easily identified by its well-developed hexagonal crystals with flat (pinacoid) terminations, a feature that helps distinguish it from quartz. Beryl is found almost exclusively in granitic pegmatites. Common beryl is thought to be characteristically green, but it can be white to colourless, making it difficult to recognize when surrounded, as it often is, by white quartz and feldspar.

Beryl is an important economic mineral, mined for beryllium. An alloy made of beryllium, copper and aluminium is harder than steel but lighter. Beryllium is used in certain components in nuclear reactors.

In **South Africa**, beryl is one of the most common minerals in Northern Cape pegmatites (Hugo, 1986; Cairncross, 2005b). At Middel Post pegmatite in the Keimoes district, blue to yellowish-green, smoky grey, milky white, deep rose or transparent aquamarine beryl crystals are found in different parts of the pegmatite. In lithium-rich pegmatites, the beryl is usually white to pink. It commonly occurs as very small to over a metre-long subhedral to euhedral hexagonal crystals, but instead of well-formed crystals, large lumps or masses (weighing over 10 tonnes) occur in some pegmatites. In a few of the more complexly zoned pegmatites, irregular masses of aquamarine quality are found. The Angelierspan pegmatite was the most productive beryl producer in the region, where up to 2-m-long crystals were found and a 60-tonne beryl crystal was exposed in the 1950s.

Figure 138 The open pit workings of the old Gravelotte emerald mine in Limpopo, South Africa. The steeply dipping strata are schists that form part of the Murchison greenstone belt. Remnants of white quartz are visible, and this is often associated with the emeralds. BRUCE CAIRNCROSS PHOTO, 1992.

Beryl is a common pegmatite mineral at several deposits in the Musina and Soutpansberg districts in Limpopo. It also occurs in Archaean rocks in the Letaba district. Crystal clusters of beryl up to half a metre wide, with individual crystals 8 cm in diameter, were once found in a small pegmatite on the farm Welgevonden 886 LS, about 20 km from Mooketsi Station. There are several beryl-bearing pegmatites in the Polokwane district, for example on the farm Verbrandhoek 983 LS, 65 km east of the town, and at Kalkfontein 615 LS, 25 km north of Polokwane. There are also several smaller pegmatites

Figure 139 A cluster of emerald crystals partially embedded in white vein quartz, 4.4 cm. Gravelotte emerald mine, Limpopo, South Africa. BRUCE CAIRNCROSS SPECIMEN AND PHOTO.

Figure 140 Emerald crystals in micaceous schist. Gravelotte emerald mine, Limpopo, South Africa. Field of view 4 cm. BRUCE CAIRNCROSS SPECIMEN AND PHOTO.

Figure 142 A lustrous beryl variety emerald partially embedded in quartz, 1.8 cm. Gravelotte emerald mine, Limpopo, South Africa. BRUCE CAIRNCROSS SPECIMEN AND PHOTO.

Figure 141 Beryl crystals embedded in micaceous schist, 4.2 cm, and a loose crystal from the same locality. Cobra Pit, Gravelotte emerald mine, Limpopo, South Africa. WARREN TAYLOR RAINBOW OF AFRICA COLLECTION, MARK MAUTHNER PHOTO.

Figure 143
A faceted emerald, 6.44 carats (1.28 cm). Cobra Pit, Gravelotte emerald mine, Limpopo, South Africa.

WARREN TAYLOR RAINBOW OF AFRICA COLLECTION, MARK MAUTHNER PHOTO.

south-west of the town. At Kalkfontein, two pegmatites contain beryl crystals, 15 cm in diameter, associated with quartz, cleavelandite feldspar and spodumene. Beryl was mined at the Witkop pegmatite in the eMkhondo district, Mpumalanga, and heliodor has been reported from the farm Kitchener 504 MS, located between Musina and Polokwane.

Aquamarine is found in Namaqualand in the Northern Cape, near Jakkalswater. Peach-pink to rose-pink morganite has been found in pegmatites at Steyns Puts in Namaqualand, and near Leydsdorp in Limpopo.

Emeralds have been mined at the Gravelotte emerald mine (GEM) Cobra Pit in the Murchison greenstone belt. The emeralds are found in pegmatites that cross-cut biotite, chlorite and actinolite schists, the best crystals usually occurring in biotite schist close to the point of contact with pegmatites. The crystals are mainly small and mostly opaque, and only of specimen value (Lum *et al.*, 2016a). Some, however, reach a length of 6 cm, and stones up to 10 carats have been faceted. At Gravelotte, emerald mining was more important in the past, and 664,585 carats were mined there over eight years in the 1930s, yielding £67,374 / R92 million in 2021 terms (Haughton, 1936). The emeralds were discovered in 1927, and two years later there were five companies mining the gemstones. Associated minerals are feldspar, molybdenite, pyrite, blue fluorapatite, schorl tourmaline and quartz. Emeralds have also been mined at the Phalmartin emerald mine, and were found in biotite schist at Uitvalskop, near Schweizer-Reneke.

Beryl tends to be concentrated in pegmatites in **Namibia** in a broad region from Brandberg West-Uis in the north, to Sandamap-Karibib-Usakos in the south, and in Tantalite Valley in the south of the country. Over 50 Namibian pegmatites have been worked for beryl.

Figure 144 The Erongo Mountains in Namibia, viewed looking east. Exceptional beryl variety aquamarine crystals have been collected in this mountain. BRUCE CAIRNCROSS PHOTO, 2014.

Namibia has some of the finest examples of aquamarine in the world, particularly from miarolitic pegmatites in the Karibib-Usakos district. Superb blue crystals, up to 20 cm long and 5 cm wide, have been collected from the Erongo Mountains (Jahn 2000; Jahn and Bahmann, 2000; Cairncross and Bahmann, 2006a; Lum et al., 2016b). These are associated with black schorl tourmaline, white feldspar, quartz, and variously coloured fluorite. The most sought after aquamarines are colour-zoned from pale blue at the base to darker blue towards the terminations of the crystal. This colour shift is coupled with an increase in the clarity in the crystal, from opaque at the base to transparent in the upper sections. Excellent examples of aquamarine, heliodor and goshenite have been collected in the Erongo Mountains. Cairncross and Bahmann (2006a) have provided detailed descriptions of all these varieties, colour variations and associated minerals.

Beautiful aquamarine has been found for decades at Klein Spitzkoppe, west of Erongo, associated with smoky quartz, topaz, phenakite and fluorite. The Rössing pegmatites, 5 km north of Rössing station, and the farm Donkerhoek 91 in the Karibib district, are noteworthy localities for beryl and aquamarine. Pegmatites in the Namib-Naukluft Park also produce aquamarine, for example at the farms Riet 30 and Wilsonfontein 110, about 120 km east of Swakopmund. In the Omaruru district, high-quality aquamarine came from a pegmatite on the farm Kawab 117 (Schneider and Seeger, 1992d). Some beryl crystals, up to 3 m long and 1.5 m in diameter, have come from the Etiro pegmatite.

Figure 145 Close-up view of the south-western side of the Erongo Mountains, which has produced specimens. The granite cliffs and surfaces are pockmarked with hundreds of cavities where local artisanal diggers have been searching for minerals, including beryl. For scale, just right of centre, a local digger's temporary shelter can be seen between the bushes. BRUCE CAIRNCROSS PHOTO, 2014.

Figure 146 Gem-quality beryl variety aquamarine crystal, 2 cm. Erongo Mountains, Namibia. BRUCE CAIRNCROSS SPECIMEN AND PHOTO.

▲ Figure 147 A large colour-zoned beryl variety aquamarine, 8.8 cm. Erongo Mountains, Namibia. PHILIP HITGE SPECIMEN, BRUCE CAIRNCROSS PHOTO.

◄ Figure 149 Colourless hexagonal beryl variety goshenite, 1.7 cm. Erongo Mountains, Namibia. BRUCE CAIRNCROSS SPECIMEN AND PHOTO.

Figure 148 A large specimen of stellate beryl variety aquamarine, with the top sections of crystals tending towards yellow heliodor, 13 cm. Erongo Mountains, Namibia. DESMOND SACCO SPECIMEN, BRUCE CAIRNCROSS PHOTO.

Figure 150 Beryl variety aquamarine partly overgrown by black schorl, 3.6 cm. BRUCE CAIRNCROSS SPECIMEN AND PHOTO.

Figure 151 Intergrown crystals beryl variety aquamarine, 2.5 cm. Erongo Mountains. BRUCE CAIRNCROSS SPECIMEN AND PHOTO.

Figure 152 Beryl variety heliodor, 1.4 cm. Klein Spitzkoppe, Namibia. BRUCE CAIRNCROSS SPECIMEN AND PHOTO.

Heliodor was originally discovered in 1910 in Namibia in the Hoffnungsstrahl pegmatite, close to Rössing Siding. Some of the faceted material from this deposit was given as a gift to Empress Victoria of Germany. Attractive heliodor specimens have been collected in the Erongo pegmatites and at Klein Spitzkoppe, which has produced large crystals, up to 12 cm long and 5 cm wide, associated with fluorite (Cairncross *et al.*, 1998). Heliodor is found in the Okahandja townlands. Pink morganite crystals come from the Rubikon and Helikon pegmatites in the Karibib district and from Neu Schwaben. A small emerald deposit is located in the Maltahöhe district, on the farm Neuhof 100.

Beryl is common in northern **Zimbabwe**, in the Filabusi, Masvingo, Save and Mberengwa regions, and has been mined extensively via large-scale operations and one-man artisanal diggings (Ackermann *et al.*, 1966). About 160 deposits have been worked in pegmatites in northern Zimbabwe, each producing more than 10 tonnes of beryl. Beryl may be the prime mineral, but it is usually extracted as a by-product associated with other minerals such as cassiterite, feldspar and tantalite-(Fe). Some notable beryl pegmatites are:

- Dungusha, Karoi district (gem-grade quartz, aquamarine, topaz and mica),
- St Ann's mine, Karoi district (1,268 kg aquamarine, 1,006 kg gem tourmaline, 936 kg topaz, 14 kg amethyst and 304 kg mica, 1953–1971),
- Good Days mine, Mutoko district (315 kg beryl, 2.07 tonnes tantalite-(Fe), 0.93 tonnes cassiterite, 0.07 tonnes columbite, 1,133 tonnes spodumene, 580 tonnes lepidolite and 91 tonnes of feldspar, 1953–1972), and
- Flame Lily mine, Insiza district (12.46 tonnes beryl and quartz, molybdenite, emerald, talc and powellite).

Beryl crystals, 1.5 m long and 60 cm wide, were found at Bepe Hills, 90 km east of Mutare. A solid mass of beryl measuring 6 x 6 x 3 m was uncovered in 1956 at the Rabbit Warren mine north of Mutoko. An unusual black variety of beryl has been described from the Filabusi area.

Aquamarine is found in pegmatites in the Mwami region. Stones of good quality and colour come from this area and, to a lesser extent, from the north-east of the country. Some of the finest heliodor comes from the Green Walking Stick deposit. Heliodor has also been found at the Baboon Hill and Bul Bul claims in the Mwami district. Gem-quality aquamarine and heliodor are exploited in the Karoi, Mutoko and Pfungwe areas. Morganite is found at the Pope claims, Goromonzi district.

Together with neighbouring Zambia, Zimbabwe has been an important producer of emeralds, from several localities (Metson and Taylor, 1977; Mugumbate, 1997) and primarily from Sandawana (Zwaan *et al.*, 1997; Zwaan, 2006). Emeralds were found in the Mberengwa district in the mid-1950s (Gübelin, 1958) and occur in pegmatites in the Archaean Mweza greenstone belt. Well-formed crystals are rare, and most of the gemstone rough occurs as fragments with crudely developed crystal faces. Unlike other well-known emeralds, for example from Columbia and Zambia, the Sandawana emeralds have inclusions of actinolite, cummingtonite, albite and apatite, which sets them apart from the other localities (Zwaan *et al.*, 1997).

Beryl is found in granitic pegmatites in the foothills of the Sinceni Mountains in **Eswatini**. The largest beryl crystal found here weighed 106 kg. Crystals weighing from 5 kg to 15 kg were common. The beryl is associated with quartz, muscovite, albite and microcline, as well as lesser quantities of topaz, cassiterite, magnetite, tantalite-(Fe), schorl, monazite-(Ce) and zircon.

Beryl crystals up to 30 cm are reported to occur in the Tete Province, **Mozambique** (Lächelt, 2004).

Figure 153 A large 16.5 x 15-cm beryl from Zimbabwe. BRUCE CAIRNCROSS SPECIMEN AND PHOTO.

◄ **Figure 154**
Beryl variety heliodor crystal, backlit, 2.7 cm. Green Walking Stick mine, Zimbabwe. BRUCE CAIRNCROSS SPECIMEN AND PHOTO.

Figure 155 ➤
A large 51.25-carat (2.3-cm) faceted beryl variety heliodor. Green Walking Stick mine, Zimbabwe. WARREN TAYLOR RAINBOW OF AFRICA COLLECTION, MARK MAUTHNER PHOTO.

Figure 156
A gem-quality heliodor crystal, 6.5 cm. Green Walking Stick mine, Zimbabwe. JIM AND GAIL SPANN SPECIMEN, TOM SPANN PHOTO.

Figure 157
A 1.05-carat faceted emerald. Sandawana, Zimbabwe. ROB SMITH AFRICAN GEMS AND MINERALS SPECIMEN, BRUCE CAIRNCROSS PHOTO.

Figure 158 A large heliodor crystal, 8.7 cm. Crystals of this size are exceptionally rare as most were cut to yield gemstones. Green Walking Stick mine, Zimbabwe.
JIM AND GAIL SPANN SPECIMEN, JEFF SCOVIL PHOTO.

Figure 159 A 29.89-carat (3.45-cm) beryl variety aquamarine. Hoffnungsstrahl pegmatite, Rössing Mountains area, Namibia. WARREN TAYLOR RAINBOW OF AFRICA COLLECTION, MARK MAUTHNER PHOTO.

Figure 160 An 8.84-carat (1.6-cm) faceted beryl variety heliodor. Rössing Mountains area, Namibia. WARREN TAYLOR RAINBOW OF AFRICA COLLECTION, MARK MAUTHNER PHOTO.

Figure 161 Faceted gem beryl. Top: 4.72-carat (1.3-cm) beryl variety heliodor from Klein Spitzkoppe, Namibia. Bottom: 7.47-carat (1.2-cm) beryl variety aquamarine, Masvingo district, Zimbabwe. WARREN TAYLOR RAINBOW OF AFRICA COLLECTION, MARK MAUTHNER PHOTO.

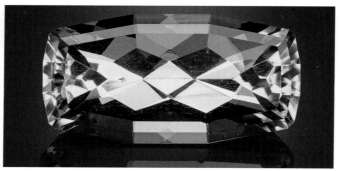

Figure 162 A 14.22-carat (2.2-cm) faceted beryl variety aquamarine. Erongo Mountains, Namibia. WARREN TAYLOR RAINBOW OF AFRICA COLLECTION, MARK MAUTHNER PHOTO.

Biotite ◆ $K(Mg,Fe^{2+})_3(Al,Fe^{3+})SiO_{10}(OH,F)_2$

Biotite crystallizes in the monoclinic system, has a hardness of 2.5 to 3, specific gravity of 2.7 to 3.4, a white streak, and vitreous lustre. Biotite is an iron-magnesium-bearing member of the mica group of minerals. Biotite per se is no longer used as a species name for the mineral. It is a solid-solution series between the iron-rich end member annite and the magnesium-rich end member phlogopite. For simplicity, most dark brown mica is referred to here as biotite. It is characteristically black, dark brown to red-brown, and the crystals form very thin paper-like layers that, as is typical of micas, are stacked one on top of each other. Where the crystals are large, their hexagonal shape may be evident. Biotite is a common rock-forming mineral that is found in southern Africa wherever there are metamorphic and igneous rocks.

Biotite is particularly common in carbonatites in **South Africa**, **Namibia** and **Zimbabwe**. It is also a common constituent in granite, gneiss and schist, and large crystals, several centimetres on edge, occur in some coarse-grained pegmatites. Aesthetic, collectable crystals are rare, in general, although large aggregates are known.

Figure 163 A 14.5-cm biotite crystal, from an unspecified pegmatite in the Karibib-Usakos region, Namibia. BRUCE CAIRNCROSS SPECIMEN AND PHOTO.

Bismuth ◆ Bi

Bismuth crystallizes in the hexagonal system, has a hardness of 2 to 2.5, specific gravity of 9.7 to 9.8, a silver-white streak, and metallic lustre. Bismuth is a native element that forms indistinct crystals or, more commonly, massive lumps. Although it is relatively soft, its heaviness in hand-specimens, caused by its high density, is distinctive. Bismuth melts at 271°C and for this reason is used in low melting point/fusible alloys. The pharmaceutical industry is a large consumer of bismuth. Bismuth occurs primarily in pegmatites, and is associated with tin, fluorite and some gold deposits.

Bismuth is sometimes found in **South Africa** as platy crystals up to a few centimetres long and massive lumps in pegmatites in the Steinkopf district, Northern Cape. It may be found with malachite, quartz and calcite. The Witkop pegmatite has produced some crystals. Bismuth was found sporadically during mining operations at the Zaaiplaats and Stavoren mines in the Bushveld Complex, usually in association with cassiterite, and with feldspar and chalcopyrite. Native bismuth, associated with silver, copper and arsenopyrite, is found in the gold mines of the Sabie-Pilgrim's Rest region.

Native bismuth is found in pegmatites in **Namibia** in the Karibib-Usakos region, at the Rubikon mine (Schneider, 1992a), and well-known subhedral crystals and massive bismuth have been collected at Etiro (Miller, 1969). It is also present in some pegmatites in Tantalite Valley. The Krantzberg mine north-east of the Erongo Mountains produced platy bismuth crystals, and specimens have been found at the farm Okarundu North West 118 in the Omaruru district. Small quantities of bismuth occur at Mesopotamia in the Khorixas district.

Bismuth is found in various pegmatites and gold-quartz veins in **Zimbabwe**. Native bismuth and bismuth minerals, such as bismutite, are found in pegmatites in these districts: Gwanda–Filabusi (Horn mine), Insiza (JCB and Sydkom mines, where 2 tonnes was mined from 1967–1974), Goromonzi (Ope mine) and Hwange (Ubique mine). Bismuth specimens were collected at the Mistress mine in the Mazowe district.

Figure 164 A mass of solid native bismuth, 4.5 cm. Etiro pegmatite, Namibia. BRUCE CAIRNCROSS SPECIMEN AND PHOTO.

Figure 165 Close-up of bismuth, showing some perfect cleavage surfaces. Etiro pegmatite, Namibia. Field of view 3.5 cm. BRUCE CAIRNCROSS SPECIMEN AND PHOTO.

Boltwoodite ◆ (K,Na)(UO$_2$)(SiO$_3$OH)•1.5H$_2$O

Boltwoodite crystallizes in the monoclinic system, with a hardness of 3.5 to 4, specific gravity of 3.6, a light yellow streak and pearly to vitreous lustre. Boltwoodite is a rare radioactive mineral containing uranium. It is very easily recognized by its vibrant, bright yellow crystals, which tend to be needle-like and arranged in sprays and bundles. Beautiful crystals are known from southern Africa, notably Namibia. Boltwoodite occurs in uranium deposits, hosted in sandstones and alaskite veins.

In **South Africa**, microscopic boltwoodite crystals have been found in Triassic sandstones in the Beaufort West district in the Karoo, where economic deposits of uranium and molybdenum were discovered in the 1970s. Although some exploration took place at the Rietkuil deposit, no commercial mining was carried out there.

In **Namibia**, superb specimens of boltwoodite come from the Goanikontes area east of Swakopmund in the Swakop River, where uranium mineralization occurs in alaskite veins (Cairncross, 2021b). This occurrence was mined specifically for mineral specimens that were first collected in 1975. All the boltwoodite comes from trenches dug on surface. The bright yellow crystals are associated with gypsum and scalenohedral 'dog's tooth' calcite crystals up to 2 cm long. Individual boltwoodite crystals are several millimetres long, but can reach 2–3 cm in length. They usually cluster in attractive radiating aggregates.

Figure 166 White veins of alaskite are visible in the country rock exposed in this dry tributary of the Swakop River, Namibia. These alaskites host the boltwoodite. BRUCE CAIRNCROSS PHOTO, 2014.

Figure 167 A large spray of boltwoodite with calcite, 6.2 cm. Goanikontes Claim, Arandis, Erongo Region, Namibia. BRUCE CAIRNCROSS SPECIMEN AND PHOTO.

▲ **Figure 168** Boltwoodite crystals on grey calcite. Goanikontes Claim, Arandis, Erongo Region, Namibia. Field of view 3.5 cm. BRUCE CAIRNCROSS SPECIMEN AND PHOTO.

◄ **Figure 169** An attractive arc of boltwoodite aggregates associated with calcite. Some of the sprays are completely coated by transparent gel-like gypsum, while others are not. Goanikontes Claim, Arandis, Erongo Region, Namibia. Field of view 3.2 cm. BRUCE CAIRNCROSS SPECIMEN AND PHOTO.

Bornite ◆ Cu_5FeS_4

Bornite crystallizes in the orthorhombic system, has a hardness of 3, specific gravity of 5.08, a grey-black streak and a metallic lustre. Bornite is one of the main copper sulphide minerals mined for copper and is therefore an important economic mineral. It is easy to identify by its distinctive colour: the surfaces of the mineral are a bright metallic purple mixed with orange, yellow and red – the reason for its common name 'peacock ore'. Although well-shaped crystals of bornite are very rare, both Zimbabwe and South Africa have produced highly sought after, large bornite crystals, some of which have found their way into collections around the world. Bornite is one of the most common copper ores and is found in many copper sulphide deposits. It is usually associated with chalcocite and chalcopyrite. Bornite, chalcocite, chalcopyrite and copper generally tend to have similar geographic distributions.

Massive bornite is found in many copper mines in **South Africa**. It is sometimes only a minor component, but can also form the bulk of the copper ore being mined. Massive bornite ore was common at the Musina copper mines in Limpopo, one of the small number of southern African localities where beautiful euhedral crystals of bornite were found. The crystal faces are somewhat dull and metallic, but the purple colour is usually still visible. Bornite was common in the Okiep copper mines and Prieska in the Northern Cape. These deposits occur in highly metamorphosed and structurally deformed rocks such as diorites, anorthosites and norites. The mineralization is relatively simple, with chalcopyrite, bornite and chalcocite the main copper ore minerals.

While massive bornite ore occurs in many copper deposits, bornite crystals are rare from **Namibia**. The Rosh Pinah, Matchless, Tsumeb and Kombat mines are a few of the localities where massive lumps of ore were found.

As in South Africa and Namibia, bornite is one of the main copper-bearing species at many copper deposits in **Zimbabwe**. One of them, the Mangula mine in the Makonde district, is world famous for its large, well-shaped, superb, large, euhedral crystals, some the size of tennis balls. They were not plentiful, but have found their way into collections around the world and are highly sought after by collectors.

In **Botswana**, at Thakadu and the adjacent areas in the Matsitama schist belt, several copper deposits contain bornite with chalcopyrite and minor chalcocite.

Bornite is found with chalcopyrite at Forbes Reef mine in **Eswatini** and at Makwanakop with chalcopyrite and malachite in an epidote-rich rock. At Kubuta, a quartz vein contains bornite, malachite, galena and cerussite.

Figure 170
Complex bornite crystal, 2.4 cm. Messina mine, South Africa.
BRUCE CAIRNCROSS SPECIMEN AND PHOTO.

▲ **Figure 171** A very large, well-formed bornite crystal, 6.4 cm. Mangula mine, Mashonaland West, Zimbabwe.
JOHANNESBURG GEOLOGY MUSEUM SPECIMEN, BRUCE CAIRNCROSS PHOTO.

◄ **Figure 172** A large, lustrous bornite crystal, with minor white calcite, 6 cm. Crystals of this size and lustre are rare. Mangula mine, Zimbabwe. DEMETRIUS POHL SPECIMEN AND PHOTO.

Brucite ◆ Mg(OH)₂

Brucite crystallizes in the trigonal system, has a hardness of 2.5, specific gravity of 2.39, a white streak, and a pearly to vitreous lustre. Brucite is a soft, micaceous mineral. Crystals are normally thin, flat and tabular, but the mineral also occurs in masses of small and scaly crystals, granules or needle-like crystals. Brucite is sectile, i.e. crystals have the interesting physical property of bending without breaking. It occurs in a variety of colours, including white, grey, pale green to vibrant green and grey-blue to sea-blue. Varieties with traces of manganese are tan to dark brown. Brucite can be found in diverse geological environments including limestones and carbonatites, and as a secondary mineral in some economic deposits, such as the Kalahari manganese field in South Africa.

The Palabora mine in Limpopo, **South Africa**, is well known for producing large plates of brucite (Southwood and Cairncross, 2017). Beautiful specimens and masses of semi-transparent to translucent deep blue brucite, tens of centimetres in diameter, occur here, as well as small hexagonal crystals. In Limpopo, in the Mokopane district, brucite was found on the farm Uitloop 3KS as small, scaly crystals together with dolomite and calcite. In the Northern Cape Kalahari manganese field, manganoan brucite is fairly common in white to brown platy crystals (Cairncross *et al.*, 2017). Attractive aggregates of brucite, shaped like small pearly spheres, are also known. Rare, small, tan-coloured, barrel-shaped crystals are occasionally found. Sky-blue brucite was found at the Wessels mine in late 2003, studded with sulphur-yellow ettringite crystals. These are stunningly colourful specimens.

Brucite is a rare mineral from **Namibia**, reported from the Kombat and Tsumeb mines.

Some of the finest specimens of brucite in the world came from the Ethel asbestos mine located on the Great Dyke, a large ultramafic/mafic intrusive complex that extends through the centre of **Zimbabwe**. The brucite forms beautiful clusters of sea-green crystals up to 2 cm long. Specimens are also reported from the Caesar chrome mine in the Lomagundi district, and the Vanguard mine, Belingwe district. Brucite is found south of Nyanga and east of Kwekwe.

  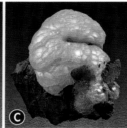

◄ **Figure 173** Brucite specimens from the Kalahari manganese field, South Africa, displaying the range of colours. **A** Pale green brucite crystals, 3.9 cm; **B** a semi-spherical cluster of terminated crystals, 2 cm; **C** semi-botryoidal aggregate of pale blue brucite, 2.4 cm. N'Chwaning II mine, South Africa. BRUCE CAIRNCROSS SPECIMENS AND PHOTOS.

◄ **Figure 174** Tan-coloured brucite together with small, white calcite and a hexagonal sturmanite crystal, 3 cm. N'Chwaning II mine, South Africa. BRUCE CAIRNCROSS SPECIMEN AND PHOTO.

Ʌ **Figure 175** A large plate of blue brucite, 29.2 cm. Palabora mine, South Africa. DEPARTMENT OF GEOLOGY, UNIVERSITY OF JOHANNESBURG, SOUTH AFRICA SPECIMEN, BRUCE CAIRNCROSS PHOTO.

Figure 176 Green gemmy brucite. Ethel mine, Zimbabwe. Field of view 8.5 cm. BRUCE CAIRNCROSS PHOTO.

Bultfonteinite ◆ $Ca_2SiO_2(OH,F)_4$

Bultfonteinite crystallizes in the triclinic system, has a hardness of 4.5, specific gravity of 2.73, white streak, and vitreous lustre. Bultfonteinite crystals are typically colourless, but may be white, pink to light brown and characteristically form as elongate acicular crystals. The type-locality is the Bultfontein diamond mine in Kimberley, where it was discovered in the early 1900s (Cairncross, 2017a).

South Africa remains the only southern African country to have produced bultfonteinite specimens, notably from the Jagersfontein, Bultfontein and Dutoitspan diamond mines, and the Wessels and N'Chwaning mines in the Kalahari manganese field. The bultfonteinite from some of the diamond mines is pink and remains extremely rare. In contrast, bultfonteinite specimens from the Kalahari manganese field are the finest in the world, are far more abundant and have been collected over a long time period. The olmiite zone in N'Chwaning II mine produced specimens associated with other species, including calcite, celestine, datolite, oyelite and manganite (Von Bezing *et al.*, 1991; Cairncross and Beukes, 2013). The bultfonteinite commonly occurred as discrete acicular crystals or attractive bundles scattered among cream-coloured euhedral olmiite crystals. Some matrix specimens are completely covered by a mat of acicular bultfonteinite. In 2009, pink and white bultfonteinite was collected in the olmiite zone at N'Chwaning II mine. Some of the bultfonteinite forms spherical aggregates several centimetres in diameter.

In May 2006, pseudomorphs of bultfonteinite after pseudocubic calcite(?) were found at N'Chwaning II mine. Most of the pseudomorphs are less than a centimetre on edge, but a few are 2–3 cm in size. Associated minerals are several generations of calcite, including water-clear, gemmy, complex scalenohedral crystals as well as brown- and black-coated calcite.

Figure 177 Bultfonteinite on calcite from Dutoitspan mine. Although not from Bultfontein mine, this specimen is historically important, as it dates back to the early part of the twentieth century, when the bultfonteinite was first found. Field of view 11 cm. MCGREGOR MUSEUM SPECIMEN, BRUCE CAIRNCROSS PHOTO.

Figure 178 Bultfonteinite pseudomorph after calcite(?), 3.6 cm. N'Chwaning II mine, South Africa. BRUCE CAIRNCROSS SPECIMEN AND PHOTO.

Figure 179 This attractive, botryoidal orange variety of bultfonteinite was found in 1985 and is different from the later discoveries shown here, 15 cm. N'Chwaning II mine, South Africa. BRUCE CAIRNCROSS SPECIMEN AND PHOTO.

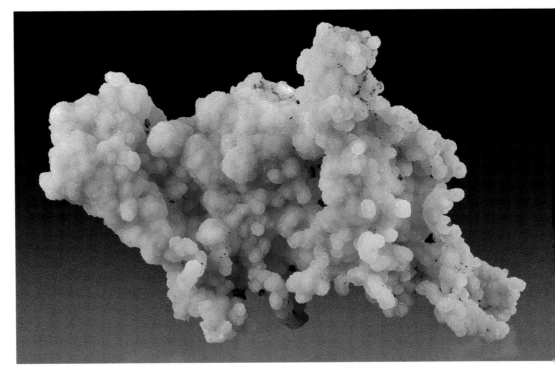

Figure 180 Aggregated spheres of bultfonteinite. N'Chwaning II mine, South Africa. Field of view 3.6 cm. BRUCE CAIRNCROSS SPECIMEN AND PHOTO.

Figure 181 Fine needles of bultfonteinite on brown lizardite. N'Chwaning II mine, South Africa. Field of view 2.5 cm. BRUCE CAIRNCROSS SPECIMEN AND PHOTO.

Figure 182 White acicular bultfonteinite crystals with caramel olmiite, 5.6 cm. N'Chwaning II mine, South Africa. BRUCE CAIRNCROSS SPECIMEN AND PHOTO.

Bustamite ◆ $CaMn^{2+}(Si_2O_6)$

Bustamite crystallizes in the triclinic system, has a hardness of 5.5 to 6.5, specific gravity of 3.3 to 3.4, white streak, and resinous, waxy to sub-vitreous lustre. Bustamite is a pink to red calcium-magnesium silicate that can be used as an attractive lapidary material.

Bustamite is rare from southern Africa. It has, however, been found in the Kalahari manganese field in **South Africa** from the N'Chwaning II mine, where it occurs as beautiful, massive pink to red specimens that can be cut and polished for ornamental purposes (Cairncross *et al.*, 2017). It does not readily occur here as well-formed crystals. In the early 2000s, specimens that were recovered from the manganese mines were initially believed to be exceptional bustamite crystals, but after being analysed were determined to be another rare mineral, schizolite (see schizolite).

Figure 183 ➤
Two specimens of bustamite: **A** a polished slice of very compact pink bustamite, 7.8 cm, and **B** rough and unpolished, 7 cm. N'Chwaning II mine, South Africa.

BRUCE CAIRNCROSS SPECIMENS AND PHOTOS.

Figure 184 A translucent sample of bustamite, cut and polished on one side to reveal its intensity of colour, 4.8 cm. N'Chwaning II mine, South Africa. BRUCE CAIRNCROSS SPECIMEN AND PHOTO.

Calcite ◆ CaCO₃

Calcite crystallizes in the hexagonal system, has a hardness of 3, specific gravity of 2.71, a white to grey streak, and vitreous, pearly lustre. Calcite is one of the most common minerals on Earth and is found in a variety of crystal forms. The most common are scalenohedrons and rhombohedrons. Calcite also occurs in massive lumps, granular masses and fibrous habits, and forms stalactites and stalagmites. Calcite occurs in almost every colour – black, white, red, yellow, green, orange, blue, brown – and it can also be colourless, transparent or opaque. Varieties of calcite include travertine and Iceland spar (transparent, optical-grade calcite). Being a carbonate of a single element (calcium), calcite forms a chemical series with magnesite (magnesium carbonate), siderite (iron carbonate), smithsonite (zinc carbonate) and rhodochrosite (manganese carbonate). It is trimorphous with aragonite and vaterite – all three minerals have exactly the same chemical composition but crystallize in different systems.

Calcite is often the chief component of limestones and marbles and is found in many other sedimentary and metamorphic rocks. It is common in numerous ore deposits, in caves, and as vug and vein fillings in lavas and dolerites. Iceland spar is found in cavities in basalt, often associated with zeolites. It polarizes light and is used in the manufacture of polarizing (Nicol) prism lenses for petrographic microscopes. During World War II, demand for optical-quality Iceland spar peaked, as it was used in gunsights.

The Kalahari manganese field in the Northern Cape has the greatest variety of calcite habits found in **South Africa** (Gutzmer and Cairncross, 2002; Cairncross *et al.*, 2017; Von Bezing *et al.*, 1991). Here, calcite is common, and beautiful specimens of rhombohedral and scalenohedral 'dog's tooth' calcite have been found at all the mines. In the Wessels and N'Chwaning I and II mines, in particular, calcite occurs as 'butterfly' twinned crystals or heart-shaped forms. The calcite is white or colourless, but may be yellow, pale pink or reddish-brown due to the presence of iron and/or manganese. Much of this calcite fluoresces pale orange when exposed to ultraviolet light (Cairncross and Beukes, 2013).

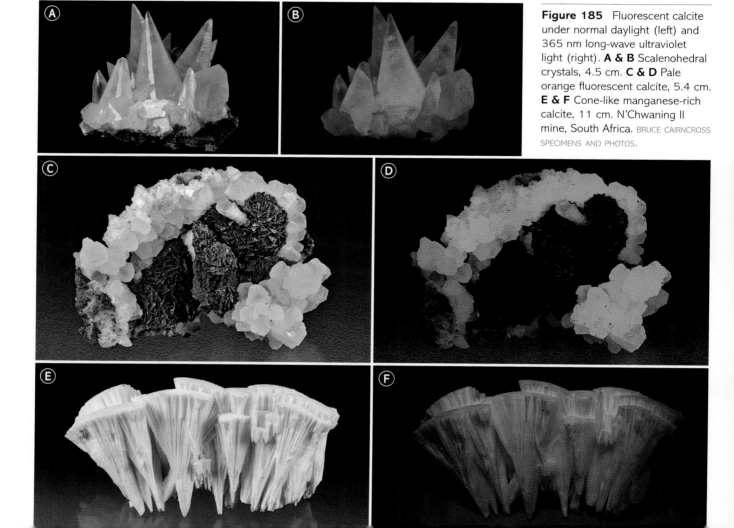

Figure 185 Fluorescent calcite under normal daylight (left) and 365 nm long-wave ultraviolet light (right). **A & B** Scalenohedral crystals, 4.5 cm. **C & D** Pale orange fluorescent calcite, 5.4 cm. **E & F** Cone-like manganese-rich calcite, 11 cm. N'Chwaning II mine, South Africa. BRUCE CAIRNCROSS SPECIMENS AND PHOTOS.

Figure 186
A doubly terminated
calcite crystal,
3.8 cm. N'Chwaning
II mine, South Africa.
BRUCE CAIRNCROSS
SPECIMEN AND PHOTO.

Figure 187 A complex, multifaceted calcite
crystal, 5.5 cm. N'Chwaning II mine, South Africa.
BRUCE CAIRNCROSS SPECIMEN AND PHOTO.

Figure 188 Scalenohedral 'dog's tooth' calcite with
orange olmiite, 4 cm. N'Chwaning II mine, South Africa.
BRUCE CAIRNCROSS SPECIMEN AND PHOTO.

Figure 189
An unusual orange
calcite composed
of multiple stacked
crystals, 4.4 cm.
N'Chwaning II mine,
South Africa. BRUCE
CAIRNCROSS SPECIMEN
AND PHOTO.

◄ **Figure 190**
Two large white
calcites on a
matrix of many
small calcite
crystals, 7.6 cm.
N'Chwaning II mine,
South Africa. BRUCE
CAIRNCROSS SPECIMEN
AND PHOTO.

Figure 191 ➤
A twinned calcite
crystal, 11.5 cm.
N'Chwaning II mine,
South Africa. BRUCE
CAIRNCROSS SPECIMEN
AND PHOTO.

Figure 192
A specimen of
calcite on hematite-
rich matrix, 22 cm.
N'Chwaning II mine,
South Africa. BRUCE
CAIRNCROSS SPECIMEN
AND PHOTO.

Iceland spar deposits are known from near Kenhardt in the Northern Cape and Calvinia in the Western Cape, in parts of Namaqualand, and at Aliwal North in the Northern Cape. At Calvinia, large masses of transparent calcite, several metres in diameter, have been found. Colourful pale blue and pink calcite is found in KwaZulu-Natal, in the quarry at Marble Delta. At the Aties mine in the Western Cape, north-west of Vanrhynsdorp, ferruginized shales of the Bokkeveld Group were mined for hematite usage in cement and pigment. This was a goethite/hematite deposit, grading downwards into pyrite stringers. Attractive calcite on dolomite specimens were collected here.

Figure 193
A faceted calcite, 14.58 carats (1.5 cm). Kalahari manganese field, South Africa.
WARREN TAYLOR RAINBOW OF AFRICA COLLECTION, MARK MAUTHNER PHOTO.

Figure 194
'Fishtail' twinned calcite studded with tiny chalcopyrite crystals. Field of view 4 cm. Buffelsfontein gold mine, Witwatersrand goldfield, South Africa. BRUCE CAIRNCROSS SPECIMEN AND PHOTO.

Figure 195 A mass of calcite crystals, 22 cm. Prieska, South Africa. BRUCE CAIRNCROSS SPECIMEN AND PHOTO.

Figure 196 Two optically clear calcite crystals; the twinned specimen on the right is 6.5 cm. Soetwater, Calvinia district, South Africa. BRUCE CAIRNCROSS SPECIMENS AND PHOTO.

Figure 197 Calcite crystals overgrown on green malachite, 4.4 cm. Messina mine, South Africa. BRUCE CAIRNCROSS SPECIMEN AND PHOTO.

Figure 198 Calcite on goethite, 4.2 cm. Aties mine, Western Cape, South Africa. BRUCE CAIRNCROSS SPECIMEN AND PHOTO.

Figure 199 Calcite on goethite, 5.4 cm. Atties mine, Western Cape, South Africa. BRUCE CAIRNCROSS SPECIMEN AND PHOTO.

Large white calcite crystals were common at the Messina copper mines. At the defunct Artonvilla mine, calcite was found together with chalcocite, native copper, chlorite and epidote. It is also associated with quartz, epidote, specularite, chlorite and sericite. At the Palabora mine, transparent, highly modified pale sherry-yellow rhombohedral crystals of calcite several centimetres in diameter were common in dykes. Small cavities in some breccia zones produce the most beautiful pale honey-coloured transparent crystals. When calcite occurs with white mesolite, green prehnite, fluorapophyllite and sulphides, very aesthetically pleasing specimens result. In many of the Witwatersrand gold mines, calcite crystals up to 25 cm are found in veins, associated with quartz and pyrite (Cairncross, 2021d). Platinum and chrome mines in the Bushveld Complex occasionally have calcite associated with pectolite, fluorapophyllite, natrolite and stilbite.

Crystalline calcite is found at other localities, including the Pilanesberg Complex, Bushveld tin mines, and the lead-zinc mines at Aggeneys in the Northern Cape. The old Okiep copper mines have also been the source of beautiful calcite specimens. Calcite crystals the size and shape of rugby balls have been collected from the Drakensberg basalt lavas in the Barkly East district.

Calcite occurs in many localities in **Namibia** and is present as a secondary mineral at most base metal mines. A vast array of calcites were found in the Tsumeb mine in the Otavi mountainland, including red, yellow, green, orange, pink and white crystals, and some transparent and colourless. Some rhombohedrons exceed 30 cm on edge, and groups of crystals weigh several kilograms. Scalenohedral calcite crystals come from the farm Kos in the Gamsberg region (Von Bezing *et al.*, 2016). The Onganja copper mine produced attractive calcite

Figure 200 Scalenohedral calcite coated by brown siderite. Coedmore quarry, Durban, South Africa. Field of view 7.5 cm. BRUCE CAIRNCROSS SPECIMEN AND PHOTO.

Figure 201 Yellow calcite crystals on a matrix of powdery white laumontite, 6.6 cm. Mooinooi mine, South Africa. BRUCE CAIRNCROSS SPECIMEN AND PHOTO.

that had inclusions of copper and chalcotrichite (Cairncross and Moir, 1996). The Navachab gold mine once produced doubly terminated, colourless calcite crystals with dark, unidentified inclusions believed to be clay. Calcite was found periodically at the Okorusu fluorite mine and included well-formed, pseudohexagonal calcites forming attractive stellate groups (Cairncross, 2018e).

High-quality Iceland spar was found on the farms Hardap 110, Dabib 112 and Narris 111 in the Mariental district, and in basalt in the Rehoboth district, sometimes as twinned crystals. Some of the cavities in this basalt are as large as 5 m in diameter. Attractive grey 'dog's tooth' calcite crystals, a few centimetres long, were found at the defunct Matchless mine, together with euhedral pyrite and water-clear, doubly

terminated quartz. The Goboboseb Mountains have produced calcite with quartz and amethyst with some calcite crystals up to several centimetres.

Calcite was mined primarily in the Hwange district, **Zimbabwe**, where veins of pure calcite, some a few metres wide, cut across Jurassic-age Karoo basalts. Most mines are located close to the Sambawiza station, west of Hwange. Some historic workings are Bumboosie and P J G, which together produced 67,738 tonnes of calcite. Calcite crystals are also found at the Dorowa and Shawa carbonatites. A manganese-rich variety of calcite, which fluoresces pink-red under ultraviolet light, is found at the Katete carbonatite and at Impala Ranch and Chiredzi. Calcite is also found in the Masvingo district.

Figure 202 An array of calcite from Tsumeb, Namibia: **A** interlocking calcite crystals, 15.5 cm; **B** bright orange and yellow calcite with arborescent copper, field of view 2.8 cm; **C** orange calcite, 3.8 cm; **D** rhombohedral calcite crystals partly coated by green malachite, 16 cm; **E** red calcite, possibly coloured by hematite or cuprite, 10.2 cm. BRUCE CAIRNCROSS SPECIMENS AND PHOTOS.

A significant deposit of calcite is found associated with a dolerite in the southern **Eswatini** lowveld area, on the slopes of Nsalitshe Hill, about 15 km south-east of Hluti. The calcite occurs in veins 1–4 m wide, individual crystals being up to 15 cm in diameter.

Calcite is relatively scarce in the vugs and geodes found in the basalts of **Lesotho**. Light brown, white to colourless rhombohedral crystals up to 10 cm on edge have been found in the Butha-Buthe district. Epimorphic casts of drusy quartz after calcite are fairly common in some of the vugs. Calcite crystals tend to be small, less than 2 cm.

Figure 203 Pseudohexagonal calcite on fluorite, 11.4 cm. Okorusu fluorite mine, Namibia. BRUCE CAIRNCROSS SPECIMEN AND PHOTO.

Figure 204 A large calcite specimen, 31 cm. Butha-Buthe, Lesotho. BRUCE CAIRNCROSS SPECIMEN AND PHOTO.

Figure 205 A cleaved and naturally etched calcite collected from an outcrop, 3 cm. Swakop River valley, Namibia. BRUCE CAIRNCROSS SPECIMEN AND PHOTO.

Figure 206 A large 212.79-carat (4.42 cm) faceted calcite. Karasburg district, Namibia. WARREN TAYLOR RAINBOW OF AFRICA COLLECTION, MARK MAUTHNER PHOTO.

Figure 207 A transparent calcite crystal overgrown on dendritic copper. Note how the copper inside the calcite is still bright and unoxidized, while the portions outside the crystal are oxidized and dull, 2.8 cm. Onganja mine, Namibia. BRUCE CAIRNCROSS SPECIMEN AND PHOTO.

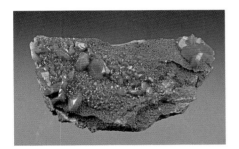

Figure 208 Red hematite(?)-rich calcite, 5.3 cm. Kombat mine, Namibia. BRUCE CAIRNCROSS SPECIMEN AND PHOTO.

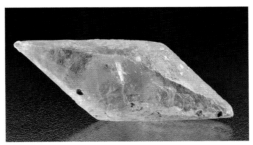

Figure 209 A doubly terminated scalenohedral calcite, 3 cm. Navachab mine, Namibia. BRUCE CAIRNCROSS SPECIMEN AND PHOTO.

Figure 210 Calcite crystals, 4.5 cm. Butha-Buthe, Lesotho. BRUCE CAIRNCROSS SPECIMEN AND PHOTO.

Cassiterite ◆ SnO$_2$

Cassiterite crystallizes in the tetragonal system, has a hardness of 6 to 7, specific gravity of 6.99, a grey-brown to white streak, and vitreous to adamantine lustre. Cassiterite is the main economic mineral mined for tin. It is brown to brown-black, but can also be grey or yellow. Crystals are opaque to translucent and may be elongate, but are usually short and prismatic. They may be grouped in botryoidal and reniform masses or finely disseminated in the host rock. Cassiterite is common in many tin-bearing granites and particularly in hydrothermal veins and pegmatites, which are often the primary host rock for this economic mineral. Crystals of cassiterite are fairly durable and heavy, hence they often occur as alluvial deposits or in eluvial material directly overlying weathered primary source rocks.

Tin is used in beneficiating the properties of other materials, for example in solder. Varying quantities of tin are used to make bronze and brass alloys and other tin-bearing materials, and in the manufacture of tinfoil and some pesticides.

Cassiterite is the major tin ore mineral exploited from several mines in the Bushveld Complex, **South Africa**. The mineralization occurs in two geological settings – in granite, and as alluvial weathered cassiterite placer deposits (Crocker, 1979; Rozendaal *et al.*, 1986). Some of the mines include Rooiberg, Neupoort, Vellefontein west of Bela-Bela, in felsites

Figure 211 Crystals of cassiterite associated with purple fluorite and small 'books' of muscovite. Rooiberg mine, South Africa. Field of view 2.5 cm. BRUCE CAIRNCROSS SPECIMEN AND PHOTO.

at Groenfontein and Zaaiplaats west of Makopane, and in granophyres at the Mute Fides and Stavoren mines east of Mookgophong. Crystals of a deep brownish-black, some over 1 cm in diameter, have been found in all the Bushveld Complex tin mines. Some very large crystals, up to 4 cm in diameter, have been recovered.

Tin deposits are known in pegmatites in schist in the Umfuli area in KwaZulu-Natal, 15 km east of Melmoth. Here, cassiterite was associated with mica, garnet, tourmaline, quartz and feldspar and was most commonly present as small crystals. However, some well-formed prismatic crystals up to 1.5 cm in diameter have been found.

Figure 212 Typical granular, disseminated cassiterite in matrix, 9 cm. Groenfontein, Zaaiplaats tin field, South Africa. BRUCE CAIRNCROSS SPECIMEN AND PHOTO.

Figure 213 Cassiterite crystals in quartz matrix, 3.5 cm. Vredehoek tin mine (an old mine that operated from 1911 to 1916), Devil's Peak, Cape Town. BRUCE CAIRNCROSS SPECIMEN AND PHOTO.

Figure 214 An aggregate of cassiterite crystals with minor pyrite (bottom right), 6.2 cm. Rooiberg mine, South Africa. BRUCE CAIRNCROSS SPECIMEN AND PHOTO.

Cassiterite is present as beautiful pale yellow-brown to dark brown crystals in scattered tin deposits in quartz veins that cut through Cape granites and metamorphosed rocks in a 40-km radius north and north-east of Cape Town (Wagner, 1909). Economically important minerals here include cassiterite, ferberite, arsenopyrite and molybdenite. These deposits occur near Kuils River on the farms Annex Langverwacht 245, Haasendal 222 and Rosendal 249 (Krige, 1921). The quartz veins extend for about 2 km in outcrops and are found in association with pale granite composed of quartz, albite and muscovite. Large crystalline grains and well-formed crystals of cassiterite and beautiful druses lined with large crystals occur in mineralized quartz veins that cut through metamorphic rocks north-west of Durbanville in Cape Town. Here, the cassiterite is associated with arsenopyrite and micro-crystalline needles of tourmaline. These Western Cape deposits are historic and were mined primarily before 1920.

Cassiterite is common in the Straussheim I and II pegmatites in the Northern Cape, where it occurs as crystals up to 20 mm, together with spessartine garnet and schorl tourmaline in a sugary albite matrix. It is also found in crystals 10–25 mm in diameter with platy albite crystals in small pockets in the Straussheim I pegmatite. In the metamorphosed stratiform deposit at Van Rooi's Vley, north of Upington, nodules and crystals of cassiterite are found with ferberite.

There are four 'tin belts' in central **Namibia** (Diehl, 1992a). They extend north-east to south-west from Brandberg West and Uis to Cape Cross, and south to west from Kohero to Nainais and Omaruru to Erongo to Sandamap (Gevers, 1929; 1969). Mindat.org lists 53 known tin deposits. The Uis mine was the largest tin-bearing pegmatite in Namibia and closed in 1991, but mining recommenced in 2018 (Maritz and Uludag, 2019). At De Rust, north of the Brandberg, cassiterite was mined from a lithium-rich pegmatite. The Molopo mine produced euhedral crystals, averaging 1 cm in diameter. The Nomgams pegmatite in the Kohero-Nainais belt contained cassiterite crystals, several centimetres in size, associated with blue-green elbaite tourmaline and pink lepidolite crystals.

Figure 215 A cassiterite crystal, 5.7 cm. Uis mine, Namibia.
WARREN TAYLOR RAINBOW OF AFRICA COLLECTION, MARK MAUTHNER PHOTO.

Figure 216
A rare faceted stone; gem-quality cassiterite, 13.22 carats (1.4 cm). Uis mine, Namibia. WARREN TAYLOR RAINBOW OF AFRICA COLLECTION, MARK MAUTHNER PHOTO.

Figure 217 Panoramic view of the opencast Uis tin mine in Namibia, at one time the largest tin mining operation in the country.
BRUCE CAIRNCROSS PHOTO, 2017.

Figure 218 The old Nainais tin mine south of Uis, Namibia. Here, finely disseminated cassiterite was extracted from the host pegmatite. BRUCE CAIRNCROSS PHOTO, 2014.

Figure 219 The entrance to an adit on the side of the Krantzberg Mountain. BRUCE CAIRNCROSS PHOTO, 2013.

Figure 220 A large cassiterite crystal, 4.5 cm. Omaruru district, Namibia. BRUCE CAIRNCROSS SPECIMEN AND PHOTO.

At the Krantzberg mine, cassiterite was associated with schorl, fluorite and beryl with crystals up to 3 cm (Haughton *et al.*, 1939). Ruby-red cassiterite crystals were found on the flank of the Krantzberg hill, about 1 km south-east of the mountain and, at Klein Spitzkoppe, in pegmatites on the western slope of the mountain.

The Brandberg West mine is a good example of the other type of cassiterite mineralization found in Namibia. Here, hydrothermal quartz veins host not only tin, but also tungsten in ferberite and scheelite. Cassiterite is found with these two ore minerals, as well as with marcasite, galena, fluorapatite, fluorite and beryl. Large cassiterite crystals, over 5 cm in diameter, were found in the Arandis tin mine, while the pegmatites in the Erongo Mountains have produced some jet-black, extremely lustrous crystals.

Figure 221 A cassiterite crystal, 3.1 cm. Krantzberg mine, Namibia. DEBBIE WOOLF SPECIMEN AND PHOTO.

Figure 222 A highly lustrous, complex cassiterite specimen, 4 cm. Erongo Mountains, Namibia. BRUCE CAIRNCROSS SPECIMEN AND PHOTO.

Figure 223 A portion of the open pit mine at Brandberg West mine, Namibia. BRUCE CAIRNCROSS PHOTO, 1999.

Many pegmatites and alluvial deposits associated with them have been exploited for cassiterite in **Zimbabwe**. Several tin-bearing pegmatites also yielded beryl and tantalite-(Fe) and ferberite (Anderson, 1979). These pegmatites are relatively small and are quickly worked out. The Karoi district, west of the Great Dyke, has a number of cassiterite-bearing pegmatites. The Nyomgoma mine has produced over 60 tonnes of cassiterite and the Mawala mine a considerable quantity of tin together with tantalite-(Fe). In north-east

Zimbabwe and the Harare district, over two dozen pegmatites have been exploited for cassiterite. The Beryl Rose mine (Shamva district), Benson and Vee Cee mines (Mudzi district), Jack mine (Harare district) and Ronspur mine (Bindwa district) were the major producers.

The Kamativi mine in the Hwange district, where the most notable cassiterite deposit in Zimbabwe was located, was the country's most important producer of tin (Rijks and Van der Veen, 1972). Over 30,000 tonnes of cassiterite was mined here since 1957, and beautiful, dipyramidal, black cassiterite crystals up to 3–4 cm were collected. Good crystals were also found at the Bikita and Mutoko pegmatites. In Hwange, eluvial cassiterite has been mined at the Gwaii and Kapata pegmatites. Cassiterite deposits are also found in the Goromonzi, Mudzi, Gutu and Mutare districts.

There are a few minor cassiterite deposits in eastern **Botswana**, none of which have been commercially exploited.

Tin was originally discovered in **Eswatini** in 1892 in gravels in the Mbabane River, near Mbabane (Hall, 1913). Forbes Reef was discovered in 1880, and was exploited for cassiterite hosted in granite and some of the surrounding schist. Alluvial deposits are concentrated close to Mbabane and south-east along the Ezulwini valley towards the defunct McReedy's alluvial tin mine. Deposits are also found at Makwanakop, near the border with Mpumalanga. Pegmatite-hosted cassiterite occurs 12 km south-west of Mbabane (Trumbull, 1995), where it is associated with a host of other pegmatite species, including magnetite, ilmenite, beryl, monazite, mauve fluorite, biotite, muscovite, pyrite and chalcopyrite. Cassiterite crystals vary in diameter from a few millimetres to as large as 4 cm.

Tin has been mined at Vila Machado, north-west of Beira, **Mozambique**.

Figure 224 A translucent cassiterite crystal, 2.2 cm. Forbes Reef, Eswatini. BRUCE CAIRNCROSS SPECIMEN AND PHOTO.

Figure 225 Three cassiterite crystals, the largest being 3 cm. Kamativi mine, Zimbabwe. BRUCE CAIRNCROSS SPECIMENS AND PHOTO.

Celestine ◆ SrSO$_4$

Celestine crystallizes in the orthorhombic system, has a hardness of 3 to 3.5, specific gravity of 3.97, a white streak, and vitreous lustre. Celestine crystals are characteristically pale blue and heavy. The high specific gravity of this mineral is caused by the presence of strontium. Crystals vary from transparent to opaque and can be prismatic, elongate and spear-shaped. Celestine is found mainly in limestones and other sedimentary rocks. It can also form by evaporation in arid areas where it is associated with gypsum and halite. It may be a secondary mineral species in some metallic ore deposits.

Celestine has been recorded from the Glenover phosphate mine and the Phalaborwa Complex in **South Africa**. Pale blue crystals up to 2 mm long have been found on calcite and limestone in the Cango Caves in the Western Cape. Celestine has been found as blocky, pale blue crystals on calcite at the N'Chwaning I mine, and as slightly curved, pearly blue crystals in association with hydroxyapophyllite-(K), calcite, pectolite and baryte at the N'Chwaning II and Wessels mines in the Kalahari manganese field.

Some of the finest Kalahari manganese field celestine to date was collected at N'Chwaning III in February 2010 (Cairncross and Beukes, 2013; Cairncross *et al.*, 2017). The crystals are pale blue, elongate and up to 15 cm. Few single crystals were collected, as most of the specimens consist of divergent, radiating clusters, some interlocking with one another. Matrix specimens were also collected, which are the best examples of celestine from southern Africa. An interesting phenomenon of some of the crystals is that they are colourless

Figure 226 Radiating spray of celestine, 6.5 cm. N'Chwaning III mine, South Africa. BRUCE CAIRNCROSS SPECIMEN AND PHOTO.

Figure 227 Elongate celestine, 8.5 cm. N'Chwaning III mine, South Africa. BRUCE CAIRNCROSS SPECIMEN AND PHOTO.

▲ **Figure 228** Somewhat unusual flattened celestine with white bultfonteinite, 3.2 cm. N'Chwaning II mine, South Africa. BRUCE CAIRNCROSS SPECIMEN AND PHOTO.

◄ **Figure 229** A mass of very pale blue celestine studded with caramel-coloured olmiite crystals. N'Chwaning II mine, South Africa. Field of view 8.5 cm. BRUCE CAIRNCROSS SPECIMEN AND PHOTO.

when collected underground but, immediately turn blue upon exposure to sunlight. The colour change is irreversible. Equally attractive celestine, but in smaller aggregates, is associated with some olmiite specimens from N'Chwaning II mine.

The Wessels mine has also yielded white to colourless, slightly curved, prismatic to nearly fibrous crystals associated with andradite, strontium-rich calcite, hausmannite and sturmanite. N'Chwaning II has produced white to colourless,

columnar celestine crystals, up to 4 cm long. Although rare, very large blue celestine crystals have been found in some kimberlite diamond mines, notably the Bultfontein mine in Kimberley.

Namibia has very few occurrences of celestine, other than rare pale blue prismatic crystals that were found at Okorusu fluorite mine (Cairncross, 2018e). These are associated with calcite and fluorite.

◄ **Figure 230** Stout celestine crystals together with drusy calcite and tan olmiite, 4.6 cm. N'Chwaning II mine, South Africa. BRUCE CAIRNCROSS SPECIMEN AND PHOTO.

▼ **Figure 231** A large celestine crystal with minor white pectolite, 7.4 cm. Bultfontein diamond mine, South Africa. MCGREGOR MUSEUM SPECIMEN, BRUCE CAIRNCROSS PHOTO.

Cerussite ◆ PbCO₃

Cerussite crystallizes in the orthorhombic system, has a hardness of 3 to 3.5, specific gravity of 6.55, a white streak, and adamantine to resinous lustre. It has two characteristic features: it is heavy and it often forms beautiful snowflake-like twinned crystals. Cerussite is usually white to grey or colourless, but inclusions of other minerals or elements can produce red, blue, green or black crystals. Cerussite forms almost exclusively from the oxidation of galena, and hence is found in association with lead deposits. Its presence in gossans is indicative of associated lead deposits.

Figure 232
A small cerussite crystal. Old Edendale mine east of Pretoria, South Africa. Field of view 1.9 mm.
WOLF WINDISCH SPECIMEN AND PHOTO.

Figure 233 'Snowflake' twinned reticulated cerussite on goethite matrix. North West, South Africa. Field of view 3.4 cm.
BRUCE CAIRNCROSS SPECIMEN AND PHOTO.

Cerussite is found in **South Africa** as small (up to 2 cm in diameter), highly reticulated, twinned crystals in some lead deposits. Tiny micro-crystals of cerussite are found in numerous deposits in most provinces, except for the Free State. Aesthetically pleasing specimens have been found at the Edendale and Leeuwbosch lead mines and in several of the old lead-zinc deposits in the North West province. At the Argent mine east of Johannesburg, cerussite is common as small, clear to dark grey crystals, which are usually well-formed simple prisms, but are sometimes twinned. Well-formed microscopic crystals have been found at Aggeneys in the Northern Cape and the Balloch mines at Niekerkshoop.

Cerussite was a common and spectacular mineral at the Tsumeb mine in the Otavi mountainland, **Namibia**. Specimens from this mine, including large (over 30 cm in diameter), multiple twinned snowflake-like crystals are to be found in collections all around the world (Von Bezing *et al.*, 2008; 2014; 2016). They are colourless or white, often transparent, and can weigh several kilograms. Rare varieties are coloured red, green or blue by the presence of copper-bearing minerals. The vanadium deposits at Berg Aukas and surrounds

Figure 234 Bladed cerussite crystals on gossan matrix, 10.5 cm. Kindergoed farm, Carolina district, South Africa.
JOHANNESBURG GEOLOGY MUSEUM SPECIMEN, BRUCE CAIRNCROSS PHOTO.

Figure 235 Highly reticulated 'snowflake' cerussite, 8.4 cm. Tsumeb mine, Namibia. BRUCE CAIRNCROSS SPECIMEN AND PHOTO.

had white to grey cerussite crystals associated with willemite (zinc silicate) and smithsonite (zinc carbonate) crystals (Cairncross, 2021c). The Kombat mine was the source of superb cerussite crystals, albeit relatively few in number. These are flawlessly transparent and were often heart-shaped twins. They acted as natural prisms when a light was passed through the crystal.

Cerussite has been found at the Ai-Ais lead mine in southern Namibia, the Uitsab vanadium mine (with galena and pyromorphite) and the Namib lead mine 7.5 km north-east of the Namib railway siding.

Figure 236 Rare red cerussite, which gets its colour from inclusions of cuprite. Tsumeb mine, Namibia. BRUCE CAIRNCROSS SPECIMEN AND PHOTO.

Figure 237 An unusual cerussite consisting of a striated silky white crystal with several yellow twinned cerussites attached, 3.6 cm. Tsumeb mine, Namibia. BRUCE CAIRNCROSS SPECIMEN AND PHOTO.

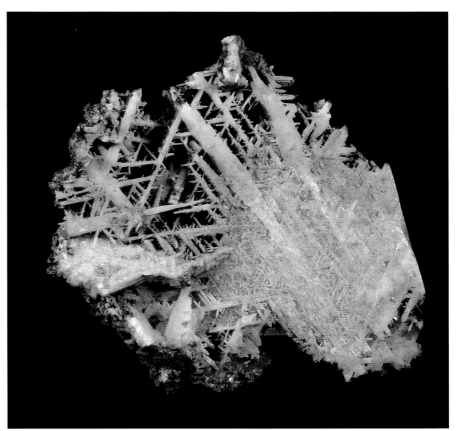

Figure 238 A twinned cerussite with partial coating of blue-green rosasite, 9.4 cm. Tsumeb mine, Namibia. DESMOND SACCO SPECIMEN, BRUCE CAIRNCROSS PHOTO.

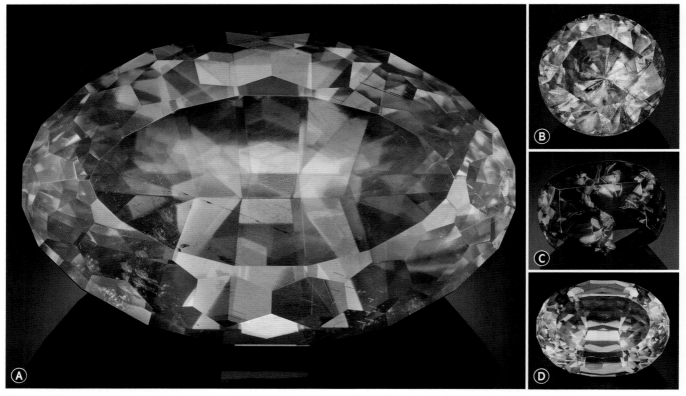

Figure 239 Cerussite is a soft mineral that cleaves and fractures easily, making it a challenge to facet. The examples shown here are outstanding for their size, weight and clarity: **A** an enormous 563.44-carat/12.58-g (5.07-cm) faceted cerussite; **B** 88.69 carats (2.22 cm); **C** 19.64 carats (1.52 cm); **D** 177.3 carats (3.11 cm). Tsumeb mine, Namibia.
WARREN TAYLOR RAINBOW OF AFRICA COLLECTION, MARK MAUTHNER PHOTOS.

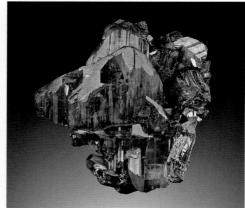

▲ **Figure 240** A grey cerussite, 3 cm. Berg Aukas mine, Namibia. BRUCE CAIRNCROSS SPECIMEN AND PHOTO.

◄ **Figure 241** A very large 430.21-carat (4.1-cm) faceted cerussite. This stone is so flawless it acts as a natural prism for light. Kombat mine, Namibia. WARREN TAYLOR RAINBOW OF AFRICA COLLECTION, MARK MAUTHNER PHOTO.

Figure 242 Cerussite and malachite, 4 cm. Christoph mine, Kunene Region, Namibia. BRUCE CAIRNCROSS SPECIMEN AND PHOTO.

▲ **Figure 243** A 'V'-twinned 1.5 cm cerussite on gossanous matrix. Rosh Pinah mine, Namibia. BRUCE CAIRNCROSS SPECIMEN AND PHOTO.

◄ **Figure 244** A 'V'-twinned cerussite, 3.9 cm. Kombat mine, Namibia. BRUCE CAIRNCROSS SPECIMEN AND PHOTO.

Figure 245 Cerussite on native copper with green malachite. Kombat mine, Namibia. Field of view 2 cm. BRUCE CAIRNCROSS SPECIMEN AND PHOTO.

Some interesting cerussite specimens have been found in the Kaokoveld. The Van der Plas mine has produced 'V'-shaped twinned crystals of 3–4 cm, as well as squat dark grey varieties (Schnaitmann and Jahn, 2010; Bowell *et al.*, 2013). Prismatic crystals have also been collected at the Christoff mine. 'V'-twinned white cerussite and more complex twinned forms have been found in gossanous matrix at the Rosh Pinah mine in southern Namibia (Cairncross and Fraser, 2012).

Small crystals of cerussite occur in some lead mines in **Zimbabwe** in the Bulawayo district (such as the OBE claims) and in the Kwekwe and Gokwe (Copper Queen mine) districts (Anderson, C.B., 1980). Cerussite occurs at the Neardy, Monte Carlo and Golden Quarry lead mines in the Mutare district and at the Hippo mine in Chipinge. It occurs as secondary enrichments in shear zones at the Elbas mine between Dett and Wankie. Cerussite is found at the Old West mine 1.8 km north-west of Penhalonga village and at the Clutha lead mine.

In **Eswatini**, cerussite occurs together with malachite, bornite and galena in a prominent quartz vein that outcrops between the Sibowe and Lubuya streams in the Kubuta district.

Chalcocite ◆ Cu$_2$S

Chalcocite crystallizes in the monoclinic system. It has a hardness of 2.5 to 3, specific gravity of 5.5 to 5.8, a black streak, and metallic lustre. Chalcocite is an important copper-bearing mineral and is most commonly found in a massive form with other copper minerals such as bornite, chalcopyrite, covellite, malachite, azurite and other metal-bearing minerals, such as galena, pyrite and sphalerite. It is closely related to, and can be confused with, another copper sulphide species, djurleite. Large, aesthetically pleasing crystals are rare, typically steel-grey and often twinned. Chalcocite is common in hydrothermal copper sulphide vein deposits. (See entries on bornite, chalcopyrite and copper for locations other than those below.)

In **South Africa**, chalcocite was one of the primary copper ore minerals in the Messina mine orebodies, where it occurred as dendritic aggregates and euhedral crystals (Cairncross, 1991). It was closely associated with bornite. The greatest concentrations were in the central cores of lodes, as discrete grains, as well as larger irregular masses and veins parallel to the banding of the host rock. Both well-formed and crude chalcocite crystals have been found in cavities in massive ore in all the mines, but mostly in Messina No. 5 Shaft. This mine yielded one of the largest known southern African chalcocite specimens (see figure 250b).

At Stavoren tin mines, chalcocite occurred in the Hillside quarry workings as solid masses up to 25 cm in diameter. Large lumps of massive chalcocite were found on the dumps along with small (1–4 mm diameter) dull grey chalcocite crystals associated with fluorite in small cavities in sulphide ore. Some of the finest crystalline chalcocite specimens in South Africa come from the Okiep copper district. Superb twinned crystals have been found in a few of the mines, notably the Nababeep West mine (Cairncross and Dixon, 1995; Cairncross, 2004b). Here, the crystals were up to 4 cm in diameter. Chalcocite is also rarely found in the Witwatersrand goldfield, the Phalaborwa Complex and the stratiform U-Mo deposits in the Beaufort West district.

Although chalcocite was one of the main copper ores at the Tsumeb mine in **Namibia**, and beautiful crystals of steel-grey chalcocite were occasionally found, good specimens were never plentiful. Some unusual, acicular hair-like chalcocite crystals, 1–2 cm long, were once found on tennantite at Tsumeb, although some of these have subsequently been identified as the rare copper-silver-lead sulphide furutobeite. Other important Otavi mountainland chalcocite localities are the Kombat and Tschudi mines. Chalcocite is found at many other copper deposits, such as the Otwane area, south of Epupa (associated with dioptase), Copper Valley in Damaraland (where several quartz-copper veins have been exploited), Henderson mine in the Karibib district and the Oamites mine, 50 km south of Windhoek. The Haib deposit in the Karasburg district and Klein Aub in the Rehoboth district contain chalcocite. Chalcocite is the main copper ore at Klein Aub, but bornite, chalcopyrite, cuprite, native copper and native silver are also present. At the Khan mine east of Swakopmund, bornite was the main copper ore, together with chalcopyrite and chalcocite. Superb euhedral silver-grey crystals came from the Khan mine.

Although not as common as bornite or chalcopyrite, chalcocite occurs in a number of economic copper deposits in **Zimbabwe**, for example at the Copper Queen, Alaska and Mangula mines. Mindat.org lists 13 known copper mines that contain chalcocite.

Figure 246 The 'Glory hole' at the Nababeep copper mine, Northern Cape, South Africa. This was one of the main copper mines in the district and produced notable mineral specimens. BRUCE CAIRNCROSS PHOTO, 1993.

Figure 247 Brilliantly lustrous, steel-grey chalcocite, 2.1 cm. Tsumeb mine, Namibia. BRUCE CAIRNCROSS SPECIMEN AND PHOTO.

Chalcocite occurs in some copper deposits in **Botswana**, usually with chalcopyrite, as at Thakadu and its satellite copper deposits. At the defunct Bushman mine, chalcocite was the main copper ore associated with minor chalcopyrite. Several copper deposits have been discovered in north-east Botswana in mudstones (similar to those at the Klein Aub mine in Namibia) at Ngwako, Ghanzi and Shinanba. These are situated in the Kalahari copper belt, which stretches from north-east Botswana to western Namibia. The Khoemacau copper-silver mine located in the Kalahari copper belt commenced operations in 2021.

Figure 248 Stacked platelets of chalcocite, 2.1 cm. Messina mine, South Africa. BRUCE CAIRNCROSS SPECIMEN AND PHOTO.

Figure 249 An 11-mm matt-grey chalcocite crystal and unidentified brown matrix. Tsumeb mine, Namibia. BRUCE CAIRNCROSS SPECIMEN AND PHOTO.

(A)

(B)

Figure 250 Two of the finest southern African chalcocite specimens. **A** Cluster of twinned steel-grey chalcocite, 7 cm. Nababeep West mine, South Africa. **B** Twinned chalcocite with minor white calcite, 9 cm. Messina mine, South Africa. NABABEEP MINE MUSEUM SPECIMEN (A), JOHANNESBURG GEOLOGY MUSEUM SPECIMEN (B), BRUCE CAIRNCROSS PHOTOS.

Chalcopyrite ◆ CuFeS$_2$

Chalcopyrite crystallizes in the tetragonal system, has a hardness of 3.5 to 4, specific gravity of 4.4, a greenish-black streak, and metallic lustre. Chalcopyrite is an important economic copper-bearing mineral. It has an attractive brass-yellow colour and may sometimes be iridescent, resembling bornite. Chalcopyrite can be confused with pyrite, although the latter commonly lacks patina or iridescence. Crystals are commonly tetrahedral, but the mineral can also have botryoidal, reniform (kidney-shaped) and massive forms. Most chalcopyrite ore is found as solid veins or masses in the host rock.

Chalcopyrite is found in many copper deposits in **South Africa**, notably the Messina copper mines where it was common, in association with bornite, pyrite, pyrrhotite, magnetite and specularite. Attractive crystals have been found attached to quartz crystals, and beautiful crystal clusters were recovered from many vugs. At the Messina mine, stalactitic chalcopyrite, up to 10 mm long and 2 mm across, came from a small stope on the sixth level and hollow stalactitic cores were filled with chalcocite and bornite. The Palabora mine is a unique copper-rich carbonatite and as such is a major source of chalcopyrite. The Maranda mine, north-west of Gravelotte, has sphalerite and chalcopyrite. At the Belfast granite quarry, where Bushveld gabbro is quarried for dimension stone, a fault was once intersected during quarrying operations and beautiful chalcopyrite crystals up to 2 cm in diameter were found together with crystalline quartz

Figure 251 The road leading up to the historic excavation of the discovery site of copper in the Okiep copper district, South Africa, by Simon van der Stel in 1685. BRUCE CAIRNCROSS PHOTO, 2011.

and hydroxyapophyllite-(K) (see figure 508B). Rare but large secondary pyrite crystals are associated with attractive quartz in some Witwatersrand gold mine specimens.

Large, aesthetic crystals of chalcopyrite are known from the Okiep copper district, notably at the Jan Coetzee and Nababeep West mines (Lombaard, 1986; Cairncross, 2004b). Chalcopyrite is one of the main ore minerals at Prieska in the Northern Cape, in addition to pyrite, galena, sphalerite and pyrrhotite (Middleton, 1976).

Figure 252 Chalcopyrite crystals and quartz. Welkom goldfield, South Africa. Field of view 7 cm. BRUCE CAIRNCROSS SPECIMEN AND PHOTO.

Figure 253 The headgear and ruins at the Nababeep mine, South Africa. In the foreground are secondary copper minerals that precipitated on the waste dump, attesting to the presence of copper. BRUCE CAIRNCROSS PHOTO, 2011.

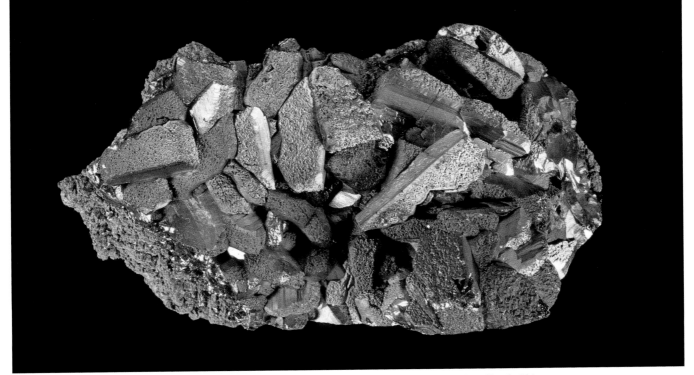

Figure 254 A cluster of bright, lustrous chalcopyrite crystals, 10.8 cm. Nababeep West mine, South Africa.
BRUCE CAIRNCROSS SPECIMEN AND PHOTO.

The distribution of chalcopyrite is very similar to that of bornite and chalcocite. These three minerals are present in varied quantities at most of over 300 documented copper deposits in **Namibia** (Schneider and Seeger, 1992b). (See entries on azurite, bornite, chalcocite and copper for some of the larger and more important chalcopyrite localities.)

Chalcopyrite is one of the main copper ores found in **Zimbabwe** at most of the copper deposits described in the entries on chalcocite, bornite, malachite and azurite in this book. Chalcopyrite is the most important source of copper at a number of mines, including Mangula, Alaska and Mistress.

There is archaeological evidence in **Botswana** that the copper deposits at Thakadu in the Matsitama schist belt were exploited in ancient times. Chalcopyrite is found with two other copper ores, bornite and chalcocite, in metamorphosed limestones here and in nearby deposits, such as Mukala and Logolo (Baldock, 1977). Copper mineralization exists in schists and dolomitic limestones north-west of the Matsitama schist belt in, for example, the Bushman deposit, where chalcopyrite is found with chalcocite. Chalcopyrite is found in amphibolite, together with pyrrhotite and pentlandite, at the Selebi-Phikwe nickel deposit and in quartz veins in the Gaborone granite at Ditshukutswane, with fluorite, covellite, galena and sphalerite. Chalcopyrite is also found in the Ghanzi area in north-west Botswana.

In **Eswatini**, copper-nickel ore is associated with the mafic Usushwana Complex rocks, for example the Mhlambanyati gabbro at Makwanakop. Chalcopyrite is associated with bornite at the Forbes Reef mine.

Figure 255 Chalcopyrite crystals with minor white calcite, 8 cm. Jan Coetzee mine, South Africa. BRUCE CAIRNCROSS SPECIMEN AND PHOTO.

Figure 256 Chalcopyrite crystals included in quartz. Griquatown district, South Africa. Field of view 9 mm. BRUCE CAIRNCROSS SPECIMEN AND PHOTO.

Chromite ◆ $Fe^{2+}Cr_2O_4$

Chromite crystallizes in the cubic system, has a hardness of 5.5, specific gravity of 4.5 to 4.8, a brown streak, and metallic lustre. It forms a solid solution with spinel, magnetite and hercynite ($FeAl_2O_4$). Chromite crystals are typically metallic grey to black, and are usually smaller than a few millimetres in size. Well-formed crystals are typically octahedral in shape. Chromite is found in mafic and ultramafic igneous rocks such as pyroxenites, gabbros and norites and with anorthosites and in some meteorites. Chromite is an important economic chromium-bearing mineral. Chrome is used for metallurgical, refractory and chemical purposes. It finds use in chrome-plating, pigments and pyrotechnics. Ferrochrome and stainless steel contain chrome because of its excellent anti-corrosive and abrasive-resistant properties.

South Africa has one of the largest concentrations of chromite in the world, in the Bushveld Complex, which covers an area of about 60,000 km². However, not all the rocks that make up the complex host chromite, only the mafic and ultramafic portions (Vermaak and Von Gruenewaldt, 1986). This enormous geological entity hosts 75% of the world's chromite reserves (Cathorn *et al.*, 2002). The chromite exists as small grains and crystals that make up rock called chromitite, found as seams that extend laterally over large distances. Elsewhere in South Africa, chromite is present in thick dolerite sills, such as those at Insizwa in the Eastern Cape. Chromite is found in heavy-mineral deposits and ancient sedimentary placer deposits such as in the Witwatersrand goldfield conglomerates.

Zimbabwe has major resources of chromite, with chromium one of the main natural resources of the country. The country has the largest reserve base of high-grade chromium ores in the world after South Africa. In 2018, Zimbabwe was the fifth-largest global producer of chromium. Most of its chromite reserves are located in

Figure 257 Granular grey chromite crystals on a matrix of white anorthosite and green pyroxene crystals, 8.4 cm. Bushveld Complex, Rustenburg district, South Africa. BRUCE CAIRNCROSS SPECIMEN AND PHOTO.

Figure 258 Layers of black chromitite (solid chromite) in the Critical Zone of the Bushveld Complex are sandwiched between white anorthosite at the Dwars River outcrop, a national heritage site in South Africa. BRUCE CAIRNCROSS PHOTO, 1998.

Figure 259 Layers of chromite separated by coarse-grained green pyroxenite, 13.4 cm. Mooinooi mine, South Africa. DEPARTMENT OF GEOLOGY, UNIVERSITY OF JOHANNESBURG SPECIMEN, BRUCE CAIRNCROSS PHOTO.

Figure 260 Grey chromitite seams, some of them fractured by natural geological processes. The green matrix is hydrogrossular garnet 'Transvaal jade' and white anorthosite, 35 cm. Bushveld Complex, Rustenburg district, South Africa. BRUCE CAIRNCROSS SPECIMEN AND PHOTO.

layers in the Great Dyke (Prendergast and Wilson, 1989), but significant deposits are also found in serpentinites of the ancient greenstone belts, for example at Selukwe and south-east of Belingwe. South Africa and Zimbabwe together hold 84% of the world's chrome ore reserves.

In **Namibia**, chromite is reported from ultramafic sills in the Kaokoveld south of Epupa. There are scattered chromite deposits in serpentinites south-east and north-east of Windhoek.

Chromite is found in some serpentinites in eastern **Botswana**, for example in chromite lenses up to 3 m thick at Majante.

Figure 261
Coarse-grained chromite crystals, 5.2 cm. North-central Zimbabwe.
BRUCE CAIRNCROSS SPECIMEN AND PHOTO.

Chrysoberyl ◆ $BeAl_2O_4$

Chrysoberyl crystallizes in the orthorhombic system, has a hardness of 8.5, specific gravity of 3.75, a white streak, and vitreous lustre. Chrysoberyl characteristically forms six-sided star-shaped crystals. Its natural colour is grey, yellow, yellow-green, olive green or dark green. Chrysoberyl also displays an optical chatoyancy, derived from the French 'cat's eye', which makes it popular with jewellers and gemologists. When transparent or translucent, it is a sought-after gemstone that can be faceted because of its hardness.

Alexandrite is a famous variety of chrysoberyl that is dichroic, that is, it displays different colours, red or green, depending on whether it is viewed in daylight or in artificial light. Gem-quality alexandrite commands high prices in the gem trade, even the synthetic, laboratory-grown alexandrite. Chrysoberyl is found almost exclusively in pegmatites and in pegmatite eluvium.

Chrysoberyl rarely comes from Northern Cape pegmatites in **South Africa**, where green to yellow-green crystals occur at Daberas, Middel Post, Wolfkop and Leeuwkop. It is found together with microcline feldspar and muscovite as crudely formed crystals and aggregates up to 10 cm in diameter. In Limpopo, chrysoberyl has also been found in pegmatites in the Letaba district, notably at Arundel 188 LT. A pale green crystal, 8 cm in diameter, was recorded on the farm Assegai 143 HT near eMkhondo in Mpumalanga.

In **Namibia**, chrysoberyl has yet to be found as well-formed crystals. It occurs in massive form at the Neu Schwaben pegmatite and, rarely, in pegmatites in Tantalite Valley in the south of the country.

The finest chrysoberyl in southern Africa comes from **Zimbabwe**. Superb specimens of alexandrite have been found at the world-famous Novello claims, close to Masvingo in the Gutu district (Schmetzer *et al.*, 2011). The alexandrite occurs in serpentinized peridotite with actinolite, chlorite, talc-carbonate schists. Small quantities of alluvial alexandrite, of much less importance, occur in the Somabula gravels. The Brentwood prospect in the Masvingo district and deposits in the Filabusi region have yielded chrysoberyl. Golden and yellow chrysoberyl displaying chatoyancy ('cat's eye' effects) is mined from the pegmatites in the Mwami area. Chrysoberyl is also found in pegmatites in the Karoi region.

Figure 263 A twinned, transparent, partly gem-quality chrysoberyl, 9 mm. Karoi district, Zimbabwe. MARTIN SLAMA SPECIMEN AND PHOTO.

Figure 262 Yellow chrysoberyl crystal, 1.7 cm. Mwami, Karoi district, Zimbabwe. BRUCE CAIRNCROSS SPECIMEN AND PHOTO.

Figure 264 A 'cat's eye' chrysoberyl cabochon, 2.2 cm. Mwami, Karoi district, Zimbabwe. ROB SMITH AFRICAN GEMS AND MINERALS SPECIMEN, BRUCE CAIRNCROSS PHOTO.

Figure 265 Sixling chrysoberyl variety alexandrite with biotite, 3.7 cm. Novello mine, Masvingo, Zimbabwe.
BRUCE CAIRNCROSS SPECIMEN AND PHOTO.

Chrysocolla ◆ $(Cu^{2+},Al)_2H_2Si_2O_5(OH)_4 \cdot nH_2O$

Chrysocolla crystallizes in the orthorhombic system, has a hardness of 2.5 to 3.5, specific gravity of 2.5, a white streak, and dull or earthy lustre. Chrysocolla is a beautiful pale blue to dark blue mineral that normally forms microscopic crystals arranged in attractive radiating bundles. It is also found as solid lumps, an opal-like form and is used for fashioning *objets d'art* such as spheres, eggs and other lapidary items. Chrysocolla is found as a secondary mineral in the oxidized zones of copper deposits.

Chrysocolla is common at many **South African** copper deposits in the Northern Cape, such as Okiep, Springbok, Nababeep and Prieska, and from the old Messina copper mines, where it forms as a bright blue mineral associated with other copper minerals.

Chrysocolla is relatively common as thin crusts or coatings in the oxidized portions of the 300+ copper deposits known in **Namibia**. Thousands of kilograms of solid chrysocolla have been mined commercially at various copper prospects in the Kaokoveld, where it is associated with other copper-

bearing minerals such as dioptase and shattuckite. Significant quantities were also found at the Onganja copper mine. Chrysocolla was common at the Gorob mine and occurred at the Matchless, Natas and Tsumeb mines.

Chrysocolla is reported from the Montana, Elephant, Chimasa and Anna copper deposits, and from a few other copper prospects in **Zimbabwe**. Economic deposits are found at the Ethel mine and at Shabani in the Shurugwi district.

Figure 266 A polished chrysocolla egg, fashioned from the rough material that forms the base, 6 cm. Sodalite mine, Orotumba, Swartbooisdrif, Namibia.
ERIC FARQUHARSON SPECIMEN, BRUCE CAIRNCROSS PHOTO.

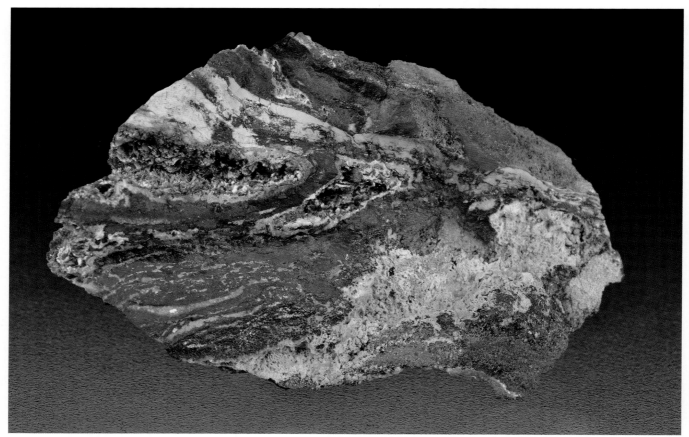

Figure 267 Blue chrysocolla associated with fibrous green malachite, 8.1 cm. Theta mine, Pilgrim's Rest, South Africa.
BRUCE CAIRNCROSS SPECIMEN AND PHOTO.

Cinnabar ◆ HgS

Cinnabar crystallizes in the trigonal system, has a hardness of 2 to 2.5, specific gravity of 8.1, a characteristic red or scarlet streak, and adamantine to dull lustre. This mineral has a very distinctive bright red to red-orange colour. It forms as rhombohedral or tabular crystals, but more often, in southern Africa, as crusts and powdery coatings. Cinnabar is an economic mineral and the main source of mercury, usually found in low-temperature hydrothermal ore deposits in quartz and calcite veins, in metamorphic schists and volcanic rocks.

The old Monarch cinnabar mine in the Murchison greenstone belt in Limpopo, **South Africa**, produced mercury from finely disseminated cinnabar dispersed in schist that imparts a red colour to this normally drab grey metamorphic rock (Pearton, 1986). Crystals of cinnabar are rare and are usually less than 1 mm. Sometimes, tiny droplets of native mercury are also present. Cinnabar is known from several places along the Inyoka fault in the eastern Barberton mountainland. The most important deposit is on the farm Kaalrug 465 JU, where the cinnabar is found in quartzites and fills cavities in white quartz veins. A small quantity of mercury was once mined from this deposit. Tiny crystals of cinnabar were found on two occasions in N'Chwaning II mine in the Kalahari manganese field – firstly as bright red crystals scattered on platy, metallic black hausmannite, and secondly

as finely dispersed wispy stringers in transparent calcite crystals, creating attractive red-included crystals.

Cinnabar is known from two deposits in **Zimbabwe**: as small crystals in lavas at Pilgrim in the Bubi district and as scattered grains in quartz-carbonate veins at Richmond in the Kadoma area (Robertson, 1972).

In **Eswatini**, cinnabar is found 6.5 km east/south-east of Havelock, disseminated in a sheared chert-rich quartzite that has been exposed by a small quarry and an adit.

Figure 268 Stringers of cinnabar enclosed in transparent calcite crystals. N'Chwaning II mine, South Africa. Field of view 5.7 cm.
ULI BAHMANN SPECIMEN, BRUCE CAIRNCROSS PHOTO.

Figure 270 Minute crystals of red cinnabar on bladed, black hausmannite. N'Chwaning II mine, South Africa. Field of view 1.7 cm. BRUCE CAIRNCROSS SPECIMEN AND PHOTO.

Figure 269 Layers of cinnabar in schist host rock. Monarch mine, Murchison greenstone belt, South Africa. Field of view 6.5 cm. BRUCE CAIRNCROSS SPECIMEN AND PHOTO.

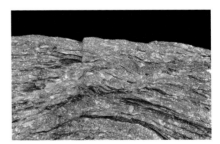

Figure 271 Red cinnabar with beads of mercury on the surface. Monarch mine, Murchison greenstone belt, South Africa. Field of view 2.6 cm. BRUCE CAIRNCROSS SPECIMEN AND PHOTO.

Chrysotile ◆ $Mg_3Si_2O_5(OH)_4$

Chrysotile crystallizes in the monoclinic system, has a hardness and specific gravity of 2.5, a white streak and silky lustre. Chrysotile is a member of the serpentine group of phyllosilicates and is the most common form of 'white asbestos'. It forms fibrous, silver to golden, silky, hair-like and small bladed crystals. This mineral has been used as a fire retardant and in fireproof textiles. It was widely used in asbestos-cement pipes, boards, roofing, paints and plaster, and in friction-proofing brake linings. Nowadays, the mineral is considered a serious carcinogenic health hazard and its use in such products no longer applies. Inhalation of fibres causes lung cancer and mesothelioma. As a result, many asbestos mines have been closed. Chrysotile is found in some greenstone belts in ultramafic serpentinites and serpentinized dolomites (Anhaeusser, 1976b).

Chrysotile was mined in **South Africa** from ultramafic serpentinites and minor dolomites at asbestos mines in the Barberton district, including the Amianthus, New Amianthus and Msauli mines (Anhaeusser, 1986b). It is found at Muldersdrift, 10 km north of Krugersdorp, in Gauteng and in serpentinites in Gordonia in the Northern Cape.

Chrysotile is contained in serpentinites in the Great Karas Mountains in the Keetmanshoop district, **Namibia**. It also occurs in the Rehoboth district and close to the old Gorob copper mine in the Namib Desert Park.

Zimbabwe contains most of the world's asbestos reserves and, before the demand declined, was the world's third-largest producer of asbestos (Laubscher, 1986; Mugumbate et al., 2001). After gold, asbestos was the country's largest income earner and a mining mainstay. The major deposits and mines are in the south of the country. About 60 deposits, scattered in the Masvingo, Insiza, Gwanda, Mberengwa and Shurugwi districts, have been worked for chrysotile, the most important asbestos species, which occurs as closely packed cross-fibres in serpentinites. The main mines, both with large reserves, were the Shabani mine in the Shurugwi district (Laubscher, 1986), and the King mine (Masvingo district), which extracted chrysotile from 1913 until the late 1980s.

Several asbestos deposits are found in south-east **Botswana**. The most important was at Moshaneng mine, where the fibres were hosted in serpentinized dolomite. In south-central Botswana, chrysotile is found in serpentinized peridotite in the western Ngaketse district at Keng Pan. It is also found at Khakhea Pan, 192 km west of Kanye. There are other deposits in banded-iron formations at Molopo River and in serpentinites in the Vukwe area 135 km west of Ramatlabama (Tati district), as well as at Christmas Kop, 10 km north/north-west of Francistown, and in serpentinized dolomite at Tautswe Hill in the Topsi area.

The Havelock asbestos mine in **Eswatini** – one of the world's largest – began production in 1939. Chrysotile fibres 1–3 cm in length occur here in veins in serpentinites. Other deposits are found in the Mbabane and Mankaiana districts.

Figure 272 Tightly packed chrysotile asbestos displaying the typical silky lustre of the mineral, 12.8 cm. Barberton district, South Africa.
BRUCE CAIRNCROSS SPECIMEN AND PHOTO.

Clinochlore ◆ $Mg_5Al(AlSi_3O_{10})(OH)_8$

Clinochlore crystallizes in the monoclinic system, has a hardness of 2 to 2.5, specific gravity of 2.6 to 3, a green to white streak and greasy, pearly lustre. Clinochlore is a phyllosilicate member of the chlorite group of minerals. It tends to form small crystals that can impart a greasy or pearly lustre to specimens. Clinochlore forms as an alteration product of mafic minerals such as pyroxenes, amphiboles and biotite.

Clinochlore is known from several **South African** localities, usually as small crystals but sometimes as larger specimens. Pale green platy crystals come from the Houtenbeck molybdenite deposit in the Bushveld granophyre, close to the well-known 'cactus quartz' locality at Boekenhouthoek. At Slipfontein in the Brits district, tightly clustered, tiny 'books' of clinochlore are found together with quartz. The Mooinooi mine, also in the Brits district, once produced some attractive, platy silver-grey clinochlore specimens. The Palabora mine has pale green clinochlore rosettes a few millimetres in diameter, associated with calcite, magnetite, pyrite, fluorapophyllite-(K) and valleriite. Some attractive specimens were found in the Witwatersrand gold mines, notably the Welkom goldfield, where clinochlore occurs as attractive green inclusions in quartz and as a partial coating on secondary quartz crystals. Other southern African occurrences are of minor importance.

Figure 274 A quartz crystal enclosing blue-green clinochlore, 4.3 cm. President Brand gold mine, Witwatersrand goldfield, South Africa. BRUCE CAIRNCROSS SPECIMEN AND PHOTO.

Figure 273
Translucent, micaceous clinochlore with many scattered magnetite crystals. Palabora mine, South Africa. Field of view 2.4 cm. BRUCE CAIRNCROSS SPECIMEN AND PHOTO.

Figure 275 ➤
Clinochlore (on the right side of specimen) with brown calcite, black dravite and quartz, 5 cm. Mooinooi mine, South Africa. BRUCE CAIRNCROSS SPECIMEN AND PHOTO.

Figure 276 A light dusting of pale green clinochlore on crystals of calcite and quartz. President Brand gold mine, Witwatersrand goldfield, South Africa. Field of view 4 cm. BRUCE CAIRNCROSS SPECIMEN AND PHOTO.

Columbite-(Fe) ◆ $FeNb_2O_6$
Columbite-(Mn) ◆ $(Mn,Fe)(Nb,Ta)_2O_6$
Columbite-(Mg) ◆ $(Mg,Fe,Mn)(Nb,Ta)_2O_6$

The columbite series general chemical formula is AB_2O_6 where 'A' represents Mg, Fe and/or Mn, and 'B' represents Nb and/or Ta. Crystals form in the orthorhombic system, with a hardness of 6, specific gravity of 6.3, brownish-black streak and submetallic to adamantine lustre.

'Columbite' per se is now no longer considered a species as such, but the columbite-(Fe)–tantalite-(Fe) series and columbite-(Fe)–columbite-(Mn) series contain the species that are characterized by the presence of iron (Fe) or manganese (Mn). There is also a related species, columbite-(Mg), that contains magnesium, an element missing in the other two iron-manganese-dominated species. Similarly, tantalite is an outdated mineral species name. Tantalite is a general name for the tantalite-(Fe)–tantalite-(Mn) series, so named because of the presence of tantalum. Mindat.org states that many specimens called tantalite actually belong to the columbite series or tapiolite series. This has led to the colloquial term 'coltan' (**col**umbite-**tan**talite) when referring to economic deposits of these metallic deposits. Columbite-(Fe) is black to brownish-black and commonly tarnishes to an iridescent sheen. These minerals are found primarily in granitic pegmatites and as rare accessory minerals in carbonatites.

Columbite-(Fe) is an economically important source of niobium (Nb). Niobium and tantalum are chemically very similar, hence the grouping of tantalite and columbite together.

Niobium, previously known as columbium, is a soft, ductile transition metal, and the main source is from the columbite series of minerals. Niobium is used primarily in special steel alloys, and minute quantities of Nb add strength to the steel. So-called superalloys containing Nb are used in jet engines and in certain components of spacecraft, such as the liquid fuel thruster nozzles. Alloys composed of niobium-germanium, niobium-tin and niobium-titanium are used in the manufacture of superconducting magnets for MRI scanners and in particle accelerators; the Large Hadron Collider uses many tonnes of superconducting wire.

Large attractive crystals of columbite-series minerals have been found in pegmatites in Mpumalanga and Limpopo in **South Africa**. Columbite has also been found as microscopic grains in the Witwatersrand goldfield and in the Pilanesberg Complex, the Glenover mine, the Murchison Range and at the Witkop pegmatite in the eMkhondo district. It has been found in the Northern Cape in pegmatites such as those on the farms Groendoorn, Jacomynspan, Kombaers Brand, Angelierspan, Stofberg, Crieff and the Groenhoekies pegmatite on the farm Steinkopf 22. Crystals in these pegmatites are usually crudely formed, but can attain a size of up to 8 cm. Some even larger crystals are known and clusters have been recovered up to half a metre in diameter.

In **Namibia**, columbite-series minerals are found in pegmatites in the Karibib-Usakos region, for example at Neu Schwaben, Davib Ost and Etiro. It is also reported from the Strathmore tin pegmatites and Sandamap.

In **Zimbabwe**, columbite, mainly columbite-(Fe) is mined from pegmatites, such as those at the Verdale tantalum deposit, the Good Days pegmatite in the Mutoko district and at Bikita. Pegmatites in the Karoi area contain columbite.

Figure 277 Columbite-(Fe), 7.5 cm. Karibib district, Namibia.
BRUCE CAIRNCROSS SPECIMEN AND PHOTO.

Copper ◆ Cu

Copper crystallizes in the cubic system, has a hardness of 2.5 to 3, specific gravity of 8.9, a light red streak, and metallic lustre. Native copper occurs in several forms, including flat sheets, wire-like fibres and, rarely, as cubic, octahedral or dodecahedral crystals. It varies in colour from copper-red in fresh, unoxidized material to brown or pale pink. Copper is an excellent conductor of electricity and is used in copper wiring, armature wiring, cables and coils. Native copper (the pure metal) is fairly common in many copper deposits in southern Africa and around the world. Such economic ore deposits also contain copper mineralization/copper ores, such as bornite, chalcocite and chalcopyrite. The distribution patterns of these minerals and native copper are often very similar.

Native copper was relatively common in **South Africa** from several economic copper deposits. At the Musina copper mines, it occurred as inclusions in quartz crystals, generally as flattened bright platelets (up to 5 mm in diameter and less than 1 mm thick), but sometimes as hair-like wires that imparted a pleasing pale cloudy copper colour to the quartz (Cairncross, 1991). Copper also occurred as irregular masses and flat sheets. Elongate, multiple hopper crystals of native copper are found on chalcocite, quartz, calcite and other silicates. Arborescent copper in the form of distorted octahedral growths, botryoidal clusters and long wires was occasionally found in cavities with quartz and epidote crystals.

Beautiful, submicroscopic copper crystals together with red cuprite are known from the Stavoren tin mines in the Bushveld Complex (Wagner, 1921). Copper occurs in a few localities in the northern regions of KwaZulu-Natal (Hatch, 1910; Thomas *et al.*, 1990). Quartz veins that contained copper were mined at Ndondo, approximately 20 km from the confluence of the Tugela and Buffalo rivers. Other minerals associated with the copper in these veins, which cut across a dyke, were calcite, cuprite, chalcopyrite, malachite and azurite. Native copper was also found at Cooper's Claim, Umhlatuzi, 12 km northeast of Nkandla. Here, native copper and malachite were present on joint planes in an intrusive syenite.

In the Okiep copper mines in the Northern Cape, copper occurs as large masses or, as at the Spektakel mine, in finely crystalline aggregates (Cairncross, 2004b). The Concordia mine is known for its mats of very fine interwoven wire copper. Similar sheets of native copper weighing several kilograms were found at the Broken Hill mine, Aggeneys.

Beautiful specimens of native copper have been found at several copper mines in **Namibia**, such as Tsumeb, Klein Aub, Otjihase and Kombat. However, the Onganja mine is arguably the best-known locality for copper specimens and has been a source of thousands, some associated with calcite, quartz and cuprite (Cairncross and Moir, 1996). The largest documented mass of natural copper found here weighed 400 kg. Crystals of native copper, which are rare in nature, and other complex forms were often found at this deposit. Beautiful arborescent groups were the most abundant form, with twinning of crystals relatively common. Although it did not produce the same quantity or quality of native copper as Onganja, Tsumeb produced a few stunning specimens, such as small dendritic groups that, with tiny brilliant red octahedrons of cuprite sprinkled around copper 'branches', resemble Christmas trees. Klein Aub mine produced native copper specimens, some associated with native silver. At the Doros Crater, native copper is found in basalts together with prehnite and chalcocite. Dendritic groups of native copper in small quartz geodes came from Copper Valley.

Copper has been mined from ancient times in **Zimbabwe**. It occurs in the Alaska, Umkondo and Mangula mines, and also from the Mkondo mine (Bikita district), Falcon mine (Chirumanzu district) and the Odzi copper mine (Mutare district).

Figure 278 Copper plus a variety of copper-bearing minerals – red cuprite variety chalcotrichite and blue copper silicates. Messina mine, South Africa. Field of view 3.2 cm.
BRUCE CAIRNCROSS SPECIMEN AND PHOTO.

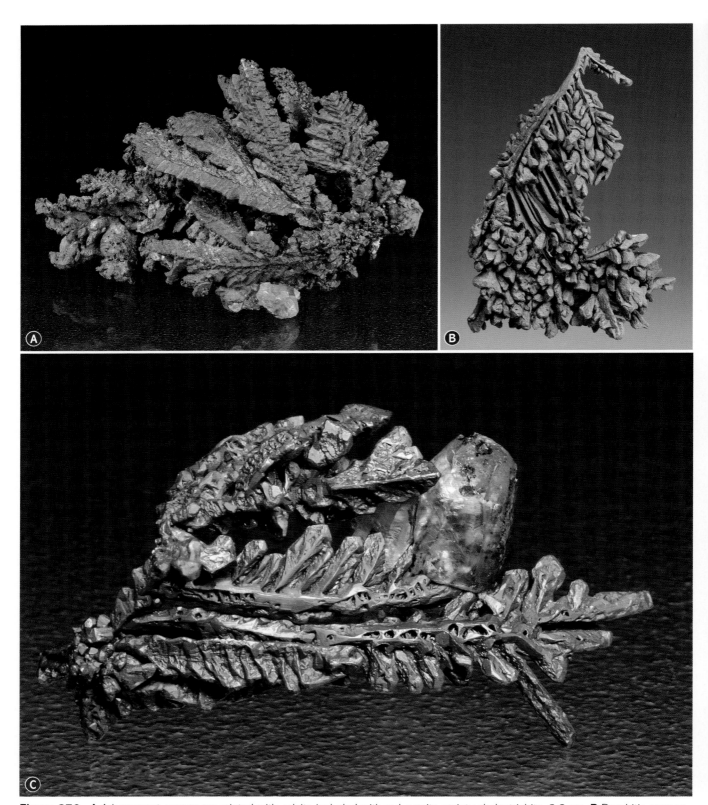

Figure 279 **A** Arborescent copper associated with calcite included with red cuprite variety chalcotrichite, 3.2 cm. **B** Dendritic copper associated with calcite, 7 cm. **C** Arborescent, twinned copper, 2.1 cm. Onganja mine, Namibia. BRUCE CAIRNCROSS SPECIMENS AND PHOTOS.

Figure 280 Flattened arborescent copper, 5 cm. Otjihase mine, Namibia. BRUCE CAIRNCROSS SPECIMEN AND PHOTO.

Figure 281 Somewhat unusual smooth, wire-like copper from the third oxidation zone, 3.8 cm. Tsumeb mine, Namibia. BRUCE CAIRNCROSS SPECIMEN AND PHOTO.

Figure 282 Platy copper interspersed in quartzite matrix. Klein Aub mine, Namibia. Field of view 4 cm. BRUCE CAIRNCROSS SPECIMEN AND PHOTO.

Figure 283 Copper with a cerussite crystal attached, 2.5 cm. Kombat mine, Namibia. BRUCE CAIRNCROSS SPECIMEN AND PHOTO.

Cordierite ◆ $Mg_2Al_4Si_5O_{18}$

Cordierite crystallizes in the orthorhombic system, has a hardness of 7 to 7.5, specific gravity of 2.5 to 2.8, a white streak, and vitreous lustre. Cordierite crystals are stubby and prismatic, but the mineral can also be granular and massive. Although most often blue, cordierite crystals can also be green, violet, grey, brown or yellow. The transparent gem variety of cordierite is called iolite or 'water sapphire', as its colour is similar to that of a blue sapphire. Iolite has an optical property called dichroism, appearing either colourless or lilac-blue when viewed from different angles. Cordierite occurs in granites and various thermally metamorphic rocks. Because it is a fairly hard mineral (7–7.5), it is found in alluvial deposits and in some countries is mined as an alluvial gemstone.

In **South Africa**, cordierite is found in association with metamorphic rocks, for example in the gneisses of the Namaqualand Metamorphic Complex, where bright blue, high-quality transparent crystals up to several centimetres in length have been found. Gem-quality cordierite is recorded from Hout Bay on the Cape Peninsula, in the contact zone between Cape granite and shales (McIver, 1966). Cordierite has also been found in the Musina district of Limpopo and at Marble Delta, KwaZulu-Natal. It is found in the Bushveld Complex contact metamorphic aureole at Steelpoort and in surrounding areas in the North West province.

Alluvial crystals of iolite are found in **Namibia** between Walvis Bay and Rooibank, north of Zebra Mountain (Schneider and Seeger, 1992d). The source of this gemstone variety of cordierite is schist and gneiss. Quartz-feldspar veins that cross-cut garnet-cordierite gneiss between Henties Bay

and Swakopmund also contain iolite. Cordierite crystals have been found at Husab in the Namib-Naukluft Park. Crystals up to several centimetres occur in schists that outcrop in the region of the Brandberg West mine. Cordierite is also reported from outcrops along the Swakop River. Alluvial deposits of blue cordierite are found north-east of the Usakos-Uis road, 8 km south-east of Nainais.

Cordierite variety iolite occurs in **Zimbabwe**, notably in the Makuti area in the Karoi district, where transparent pieces, up to 8 cm in diameter, have been found in amphibolite. Another interesting deposit is in the Beitbridge district, at the Rainbow Mite deposit, where iolite is found with apatite, tourmaline, translucent feldspar and dumortierite. Good-quality iolite also comes from the Treasure Casket prospect in the Rushinga district, which yielded about 5 kg of material, and from the Mutoko district.

Figure 284 A rough specimen of cordierite variety iolite, backlit, and viewed from two different angles showing the dichroic colour change, 2.5 cm. Karoi district, Zimbabwe. BRUCE CAIRNCROSS SPECIMEN AND PHOTOS.

Figure 285 A polished teardrop-shaped cordierite variety iolite, 3 cm. The dichroic nature is evident when the stone is viewed backlit from two different angles. Karoi district, Zimbabwe. BRUCE CAIRNCROSS SPECIMEN AND PHOTO.

Corundum ◆ Al₂O₃

Corundum crystallizes in the hexagonal system, has a hardness of 9, specific gravity of 4 to 4.1, a white streak, and vitreous lustre. Corundum is found as well-shaped crystals, varying in length from 1 cm to over 20 cm, and also as fragments referred to as 'boulder-variety' corundum. Corundum crystals are easily recognized by their hexagonal, barrel-shaped crystal form, and their hardness, second only to diamond. They may be colourless or may occur in a variety of colours, including olive-green, brown, yellow and grey. Crystals of transparent red corundum are the variety ruby; transparent blue crystals of corundum are sapphire. Sapphire may also be pink, yellow, violet, green or colourless. 'Star' sapphires display asterism – six-armed star-like reflections – when the crystal is polished *en cabochon*. Sapphire is coloured blue by the presence of titanium and iron.

Corundum was once mined and used as an abrasive. Some corundum is used in the production of high-alumina refractory bricks. Corundum occurs mainly in metamorphic rocks and is sometimes found in igneous rocks, notably pegmatites. Water-worn corundum crystals, which are hard and resistant to mechanical breakdown, can be found in the alluvium of rivers draining corundum-bearing rocks.

Corundum was mined from the early 1900s in **South Africa** in a broad area of metamorphic rocks from Polokwane to Musina to Leydsdorp (Wagner, 1918; De Villiers, 1976). Haughton (1936) states that 38,665 tonnes of corundum was mined from 1932 to 1935. In the southern part of this belt, it is found in margerite, a mica-like mineral, and in association with feldspar, mica, garnet and tourmaline. The largest corundum crystal in the world – a grey, 59-cm-long crystal weighing 151 kg – was found in Limpopo. It is on display as part of the Council for Geoscience collections in the Ditsong Museum in Pretoria. Transparent gem-quality ruby and sapphire corundum do not occur in South Africa, although some translucent red and blue varieties have been found. One such deposit of ruby-red corundum occurs in grey-white anorthosite host rock south of Aggeneys, in the Northern Cape.

◀ **Figure 286**
This 151-kg corundum crystal is reputed to be the largest in the world, 59 cm. Limpopo, South Africa. COUNCIL FOR GEOSCIENCE COLLECTION, DITSONG MUSEUM PRETORIA, SOUTH AFRICA, BRUCE CAIRNCROSS PHOTO.

◀ **Figure 287**
Hexagonal corundum crystals contained in biotite-quartz matrix. Soutpansberg, Limpopo, South Africa. Field of view 6 cm. BRUCE CAIRNCROSS SPECIMEN AND PHOTO.

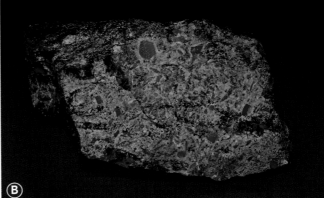

Figure 288 Ruby corundum crystals in matrix: **A** under natural daylight; **B** fluorescing in long-wave ultraviolet light, 12.8 cm. Limpopo, South Africa. BRUCE CAIRNCROSS SPECIMEN AND PHOTOS.

Figure 289 A large, terminated corundum crystal, 17.6 cm. Limpopo, South Africa. BRUCE CAIRNCROSS SPECIMEN AND PHOTO.

Figure 290 A doubly terminated corundum crystal, 8.2 cm. Undisclosed locality, Zimbabwe. BRUCE CAIRNCROSS SPECIMEN AND PHOTO.

Some corundum has been found in the Lydenburg and Barberton districts. In the pegmatite belt of the Northern Cape, corundum is found in the Namaqualand region north of Steinkopf, the Pella Mission near Pofadder, and near Kenhardt. Small blue corundum crystals are found as inclusions inside andalusite at Kakamas-Suid, Mottelsrivier and Bokvasmaak. Corundum occurs near Kranskop in KwaZulu-Natal. It has been mined from deposits south-east of eMkhondo in Mpumalanga.

In **Namibia**, grey-brown to pale pink corundum crystals, up to 4 cm long, occur in biotite schist on the farms Harib 142, Hochveld 112 and Girtis 109 in the Karasburg district (*Mineral Resources of Namibia*, 1992). Corundum is not widespread in Namibia. It occurs on the farms Bethlehem 27 and Kransnek 269 and in the Windhoek district at Kyanite Kop, a prominent hill composed of corundum, kyanite and diaspore. Alluvium on Epako 38 (Omaruru district) contains cassiterite, tourmaline, bismuth, gold and corundum. In Damaraland and the Karibib district, corundum is found in biotite schist associated with the Sandamap pegmatite. Corundum also occurs in Precambrian rocks close to Warmbad.

Zimbabwe was once the world's largest producer of corundum (Morrison, 1972). Over 160,000 tonnes was mined at the O'Brian Prospect in the Mazowe district, which was by far the most productive deposit and still contains substantial reserves of boulder corundum. The Beitbridge district has several economic deposits, such as blue corundum (sapphire) at the Andre prospect and ruby corundum at Red Mite. Gem-quality corundum has been extracted at the Devera Delta prospect in the Goromonzi district. Ruby corundum is found in alluvial gravels close to Gweru and in the Somabula diamond region, where sapphire and ruby were discovered as early as 1903, when some small stones were faceted from this material. Corundum is also found in the Chiredzi district: at the Banga prospect (Shurugwi district), where purple corundum is hosted in green schist, and at the Chiswutu prospect (Rushinga district), where pink gem corundum occurs. At Barauta (Mudzi district), small pegmatites in gneiss contain turquoise-coloured corundum, sometimes as translucent sapphire. Clear blue gem sapphire was found at the Corundum Blue mine (Buhera district), where it was associated with tantalite-(Fe), beryl and spodumene.

In **Eswatini**, corundum, hosted in mafic rocks associated with pegmatites, was mined from deposits in the Hlatikulu eluvial region, mainly as eluvial deposits. Crystals were up to 18 cm long and 10 cm in diameter.

In **Botswana**, sapphire-blue corundum crystals up to 3 cm long occur in alluvial deposits north-east of the Lotsane River in the Tuli Block.

Crocoite ◆ PbCr⁶⁺O₄

Crocoite crystallizes in the monoclinic system, has a hardness of 2.5 to 3, specific gravity of 5.97 to 6.02, yellow-orange streak, and sub-adamantine, resinous to waxy lustre. Crocoite forms characteristic elongate bright orange crystals. It forms a secondary mineral in lead deposits where there are also associated chromium-bearing rocks.

The Bushveld Complex in **South Africa** holds large reserves of chromite ore, which in turn is the source of chromium in the secondary crocoite. Such a deposit is located at Argent, Mpumalanga (Southwood and Viljoen, 1986; Atanasova *et al.*, 2016). Beautiful small crocoite crystals have been known here since the late 1800s, when the deposit was discovered and mined for lead.

The King's Daughter mine in **Zimbabwe** is reported to contain crocoite (Bartholomew, 1990a).

Although perhaps out of the geographic domain of southern Africa, noteworthy small crocoite crystals have been found in the Monarch gold mine in the Manica Province of **Mozambique**, and are worthy of mention for this rare but vibrantly coloured southern African species.

Figure 291 Bright red crocoite crystals with yellow pyromorphite. Argent mine, South Africa. Field of view 3.8 mm.
WOLF WINDISCH SPECIMEN AND PHOTO.

Figure 292 Crocoite crystals on talc schist. Monarch mine, Manica Province, Mozambique. Field of view 4.2 cm.
BRUCE CAIRNCROSS SPECIMEN AND PHOTO.

Cuprite ♦ CuO

Cuprite crystallizes in the cubic system, has a hardness of 3.5 to 4, specific gravity of 6.14, a red-brown streak, and adamantine to earthy lustre. Crystals of cuprite are typically octahedral, cubic or dodecahedral. Chalcotrichite is a variety that forms hair-like, fibrous red crystals. The vibrant blood-red to purple-red colour of cuprite crystals is striking and distinctive. However, if exposed to sunlight for extended periods, the crystals alter to a metallic black colour. Cuprite usually forms from the oxidation of primary copper sulphides, and as such can be an important source of copper ore.

Aesthetically pleasing cuprite is not very common in **South Africa** and, where present, is usually found with native copper. Small bright red crystals were very rarely found in the Messina mines. Small, shiny, red cubes and hair-like chalcotrichite have been found in oxidized ore at Stavoren. Massive cuprite was found in some pegmatites in the Northern Cape, notably the Noumas pegmatite. There is an unusual occurrence of small quantities in sandstone at Clocolan in the Free State.

Onganja copper mine in **Namibia** (Moore, 2010)became famous in 1972 for the enormous malachite-coated cuprite crystals that were discovered there (Cairncross and Moir, 1996). Crystals 6–8 cm, enormous for the species, were not uncommon; the largest crystals of cuprite in the world, up to 14 cm in diameter and 2.1 kg in weight, came from the Onganja copper mine in Namibia. Below the thin green malachite coating, most of this cuprite is transparent. This has allowed some remarkable faceted cuprite gemstones to be cut from this material. One disadvantage of these cut gems is that they tarnish very easily, leaving permanent fingerprints if improperly handled and turning silver when exposed to light. Matrix specimens of cuprite are uncommon. Chalcotrichite was occasionally found at the Onganja mine either on matrix or as inclusions in clusters of clear, euhedral calcite crystals.

Excellent specimens of cuprite, with crystals up to 1 cm, came from Otjizonjati, 8 km west of the Onganja mine. The Tsumeb mine produced blood-red crystals of cuprite, often associated with other colourful, secondary minerals – such as red cuprite on green smithsonite. Chalcotrichite was common at this mine, either as free-standing sprays or inclusions in other minerals such as cerussite and calcite. Transparent, brilliant red octahedrons of cuprite were found on white

(A)

Figure 293
Cuboctahedral cuprite crystals:
A coated by green malachite, 6 cm;
B with some of the coating removed, 3.7 cm. Onganja mine, Namibia.

DESMOND SACCO SPECIMEN (A), BRUCE CAIRNCROSS SPECIMEN (B) AND PHOTOS.

(B)

Figure 294 The old workings at the Onganja mine, Namibia, the source of outstanding cuprite and copper specimens.
GEORGE HENRY PHOTO, 2012.

calcite matrix at the Kombat mine, and small quantities of cuprite are sometimes found in other copper deposits, for example at Klein Aub, Gorob, Copper Valley and the Natas mine.

Cuprite was found in some of the oxidized portions of copper prospects in **Zimbabwe**, such as Canadians (Shurugwi district), Edward (Chiredzi district), Elephant (Bikita district), Luca (Hwange district as massive lumps), Nkai Copper (Nkayi district), Odzi Copper (Mutare district) and Zimbabwe Copper (Masvingo district).

Figure 295 Three large cuprite crystals with a thin coating of malachite, 9 cm. Onganja mine, Namibia.
DESMOND SACCO SPECIMEN, BRUCE CAIRNCROSS PHOTO.

Figure 296 Dodecahedral cuprite crystals partially coated by malachite. Onganja mine, Namibia. Field of view 19 mm.
BRUCE CAIRNCROSS SPECIMEN AND PHOTO.

Figure 297 Cuprite crystals associated with yellow mimetite and calcite. Field of view 4.2 cm. Tsumeb mine, Namibia.
BRUCE CAIRNCROSS SPECIMEN AND PHOTO.

Figure 299 Semi-skeletal cuprite crystal showing multiple growth features, on calcite, 1 cm. Tsumeb mine, Namibia.
BRUCE CAIRNCROSS SPECIMEN AND PHOTO.

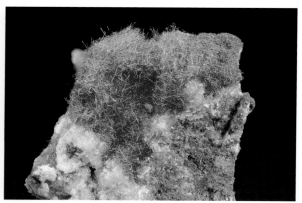

◄ **Figure 298** Cuprite variety chalcotrichite. Field of view 1.4 cm. Tsumeb mine, Namibia. BRUCE CAIRNCROSS SPECIMEN AND PHOTO.

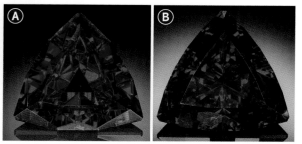

Figure 300 Extraordinary large gem-quality cuprite crystals: **A** 149.36 carats (3.22 cm); **B** 293.6 carats (3.96 cm). These require careful handling and storage; if exposed to direct light, the surfaces will tarnish and turn an opaque silver. Onganja mine, Namibia. WARREN TAYLOR RAINBOW OF AFRICA COLLECTION, MARK MAUTHNER PHOTOS.

Descloizite ◆ PbZn(VO$_4$)(OH)

Descloizite crystallizes in the orthorhombic system, has a hardness of 3 to 3.5, specific gravity of 6.24 to 6.26, a brown-red to orange streak, and greasy lustre. Descloizite forms attractive dark orange-red to brown or black crystals of a variety of shapes and habits, including prismatic, stalactitic, granular and tabular. The crystals are often pyramidal and clustered together in attractive groups. Descloizite is an important ore of vanadium, used in the manufacture of steel alloys, catalysts, blue and yellow paint and ceramic pigments, electrodes in fuel cells and as a drying agent in ink and paint. A vanadium and gallium compound, V$_3$Ga, is used as a superconductor.

In southern Africa, descloizite occurs in sedimentary rocks, either as a primary vanadium ore mineral, or as a secondary mineral. Descloizite forms a chemical series with another vanadium-bearing mineral, mottramite (see page 193).

In **South Africa**, descloizite and vanadinite occur as small, orange or light to dark brown crystals at the Argent mine, Edendale lead mine and on Kindergoed 332 JT in Mpumalanga (Atanasova *et al.*, 2016). Small (under 5 mm) crystals are known from veins in dolomite at Shingwedzi, near Sibasa, in Limpopo.

The largest deposits of vanadium in southern Africa are found in the Otavi mountainland in **Namibia** (Wartha and Schreuder, 1992). This coincides with the world's largest concentration of descloizite, particularly from the Berg Aukas and Abenab mines, but also from some of the other vanadium deposits in the region, for example Baltika and Abenab West (Cairncross, 1997). Berg Aukas descloizite is unrivalled in its variety of forms and crystal habits, and specimens can be seen in collections around the world (Cairncross, 2021c). The most common form is pyramidal, dark brown crystals. Associations of different minerals are rare. Descloizite was sometimes found with calcite and/or smithsonite, but was more commonly found on a hard, pink, clay-rich matrix. Colour often varies according to crystal size; small, thin translucent crystals are red, orange or yellow, while the more stout crystals are opaque and silver, black, brown, red-brown to dark olive-green.

The Abenab mine, which is somewhat similar in geology and mineralogy to Berg Aukas, was worked from 1922–1958 and also produced superb specimens of descloizite. The orebody was a steeply dipping, pipe-like structure that was filled with brecciated country rock. One form of descloizite

Figure 301 A scanning electron image of descloizite crystals. Edendale mine, South Africa. Field of view 173 microns. WOLF WINDISCH SPECIMEN, MARIA ATANASOVA IMAGE.

Figure 302 The mine dump at Berg Aukas mine, source of the finest descloizite specimens known. BRUCE CAIRNCROSS PHOTO, 2017.

Figure 303 Metallic descloizite with red vanadinite crystals. Namib lead mine, Namibia. Field of view 2.1 cm. BRUCE CAIRNCROSS SPECIMEN AND PHOTO.

that came from the Abenab mine is pseudomorphing descloizite after vanadinite, ranging from thin descloizite coatings around stout vanadinite crystals up to 12 cm in length, to complete replacement by descloizite, producing hollow pseudomorphs after vanadinite.

Highly lustrous, millimetre-sized metallic and dark amber crystals associated with red vanadinite and white calcite come from the Namib lead mine. Descloizite is also reported from the Kaokoveld, 20 km north of Opuwo, where it occurs with galena in dolomite.

Figure 304 Descloizite from Berg Aukas, Namibia, comes in a variety of colours and habits. **A** Elongate descloizite displaying two colours, 5.7 cm. **B** Crystals of descloizite on matrix, such as in this specimen, are relatively rare, 11.5 cm. **C** An arborescent group of descloizite crystals, 12.8 cm. **D** Large plate of semi-parallel aligned descloizite, 18.5 cm. **E** Typical pyramidal highly lustrous descloizite crystals, 6.5 cm. **F** Descloizite rimmed by a thin film of dolomite, 4.6 cm. BRUCE CAIRNCROSS SPECIMENS AND PHOTOS.

Diamond ◆ C

Diamond crystallizes in the cubic system, has a hardness of 10, specific gravity of 3.51, a white streak, and adamantine to greasy lustre. Although diamonds are usually colourless – as seen in most diamond jewellery – many colour variations are known, including yellow, pink, brown, red, blue, grey, green and black. Diamonds with colour are referred to as 'fancy'

stones. 'Bort' is the name for a grey-black, opaque variety. Diamonds are known in many crystal forms in addition to the typical octahedron. They include dodecahedrons, spherical stones, flattened 'mackles' and combinations of forms. Synthetic diamonds have been manufactured for some time, for both industrial as well as jewellery purposes, and coloured

Figure 305 A set of replica rough and faceted diamonds from Africa, manufactured from cubic zirconia. With four exceptions, all are from South Africa. Their names, weight, country of origin and date of discovery are listed. WARREN TAYLOR RAINBOW OF AFRICA COLLECTION, MARK MAUTHNER PHOTO.

Faceted diamond replicas (anticlockwise from the centre of the spiral):
- Eureka (11 carats) Kimberley, South Africa, 1866
- Heart of Eternity (27.64 carats) Premier Diamond mine, South Africa, ~2000
- Williamson Pink (23.6 carats) Williamson Diamond mine, Tanzania, 1947
- Blue Lili (30.06 carats) Premier mine, South Africa, year unknown

- Pink Star (formerly known as Steinmetz Pink) (59.60 carats), South Africa, 1999
- Tiffany (129 carats) Kimberley, South Africa, 1878
- Allnat (101 carats) Cullinan, South Africa, ~1950
- Red Cross (205 carats) Kimberley, South Africa, 1901
- Centenary (274 carats) Cullinan, South Africa, 1988
- Golden Jubilee (545.67 carats) Premier mine, South Africa, 1985

- Cullinan I (530 carats) Cullinan, South Africa, 1905
- Cullinan II (317 carats) Cullinan, South Africa, 1905
- Incomparable (407 carats) Mbuji Mayi, Congo, ~1980
- Jonker I (126 carats) Cullinan, South Africa, 1934
- Light of Peace (130 carats) Sierra Leone, 1969
- Premier Rose (137 carats) Cullinan, South Africa, 1978

- Cullinan III (94 carats) Cullinan, South Africa, 1905
- Jubilee (245 carats) Kimberley, South Africa, 1897
- Niarchos (128 carats) Cullinan, South Africa, 1954
- Millennium Star (203 carats) Mbuji Mayi, Congo, 1990

Rough diamond replicas, from largest, bottom left:
- Cullinan (3,106 carats) Cullinan mine, South Africa, 1905

- Jonker (726 carats) Elandsfontein mine, South Africa, 1934
- Centenary (599 carats) Premier mine, South Africa, 1986
- Cullinan Heritage (507 carats) Cullinan mine, South Africa, 2009
- Premier Rose (353 carats) Premier mine, South Africa, 1978
- Niarchos (426 carats) Premier mine, South Africa, 1954

stones can now be synthesized in the laboratory. In 2019, a Swedish company produced the first 3D-printed diamond for industrial use.

Southern Africa has produced some of the largest and most famous diamonds in the world and, for more than a century, diamonds and gold have sustained the economic development of the region. Diamonds were originally the mainstay of South Africa's mining industry. Although the hardest of all minerals on the Mohs scale, diamonds can be broken mechanically. Diamonds are used in numerous tools, including drilling bits, diamond-edged saw blades and very sharp surgical scalpels, and are consumed in quantity by the jewellery industry.

SOME OF THE WORLD'S LARGEST DIAMONDS FROM SOUTHERN AFRICA

Significant Lesotho diamonds

The Lesotho Legend	910 carats	2018
Lesotho Brown	601 carats	1967
Lesotho Promise	603 carats	2006
Letšeng Legacy	493.27 carats	2007
Leseli La Letšeng	478 carats	2008
Letšeng Star	550 carats	2011
Letšeng Dynasty	314 carats	2015
Letšeng Destiny	357 carats	2015
Unnamed	442 carats	2020

Significant South African diamonds

Cullinan	3,106.75 carats	1905
The Excelsior	975 carats	1893
The Golden Jubilee	755 carats	1985
The Jonker	726 carats	1934
The Jubilee	650.8 carats	1895
Unnamed octahedron	616 carats	1964
Centenary	599 carats	1986
Cullinan Heritage	507 carats	2009
The Kimberley	490 carats	1921

Significant Botswana diamonds

The Sewelô	1,758 carats	2019
The Lesedi La Rona	1,109 carats	2015
Unnamed	1,098 crats	2021
The Constellation	813 carats	2015
Unnamed	549 carats	2020
Unnamed	472 carats	2018

South Africa no longer ranks highest in the world in diamond production, but diamonds are inextricably woven into its social, economic and political fabric and that of southern Africa.

Geologically, diamonds are mined in South Africa from igneous host rock called kimberlite and from alluvial river gravels and marine deposits off the Atlantic coast (Gurney, 1990; Gurney et al., 1991). Namibia exploits marine and alluvial diamonds, while Botswana mines two main kimberlite provinces. Most of Botswana's kimberlites are Cretaceous in age (about 90 million years old), their geology and age similar to the Kimberley pipe in South Africa. Zimbabwe exploits kimberlitic and alluvial diamonds, as does Lesotho, although primary kimberlite is the main source.

South Africa exploits three different geological sources of diamonds: kimberlite-hosted, alluvial and marine. Five major *kimberlite* diamond mines – Bultfontein, Dutoitspan, De Beers, Wesselton and Kimberley – are associated with Kimberley in South Africa (Lynn et al., 1998; Cairncross, 2014). Some of these mines have reached maturity or have closed, although reworking of the old mine dumps might prolong diamond mining in the region. Venetia mine is located several hundred kilometres north of Kimberley in Limpopo, approximately 50 km west of Musina. This is the largest diamond producer in South Africa, averaging 8 million carats per annum (https://www.mineralscouncil.org.za/sa-mining/diamonds). The Finsch mine in the Northern Cape was the second-largest producer in 2019, with 1.8 million carats, and the famous Cullinan mine east of Pretoria was ranked third in 2019, with 1.7 million carats.

Alluvial diamonds are mined from ancient gravel deposits originating from eroded kimberlites. The Vaal and Orange rivers are the two main river systems that transport and deposit diamondiferous gravel. Approximately 60 km upstream from Port Nolloth on the west coast, alluvial diamond operations are active at Baken and Bloedrif. Most operations are of a relatively small scale compared to the large kimberlite mines, but nonetheless produce significant volumes of gemstones. Large diamonds, for example the 511-carat Venter diamond, can potentially be discovered in alluvial deposits. The 726-carat Jonker stone, the largest alluvial diamond found in South Africa, came from a stream flowing near the Premier kimberlite pipe close to Pretoria.

Offshore *marine* diamonds are almost invariably high-quality gemstones, as most flawed or fractured stones are destroyed in the process of transportation in the Vaal and Orange rivers. Reserves of west coast offshore diamonds are large, calculated at 1.5 billion carats.

Another rather unusual source of South African diamonds occurs in the gold-bearing conglomerates of the Archaean Witwatersrand goldfield, first discovered in 1893, in the Klerksdorp area (Denny, 1897; Cairncross, 2021d). These are typically less than several carats in weight and invariably pale to dark green. In all cases, this coloration is surficial and exists as an outer coloured rind surrounding colourless interiors.

Figure 306 Green diamonds from the Witwatersrand goldfield. The green colour is only a thin outer coating and does not penetrate the crystals, 3.5 mm (left) and 4 mm (right). JOHANNESBURG GEOLOGY MUSEUM COLLECTION, BRUCE CAIRNCROSS PHOTO.

Raal (1969) investigated 38 green Witwatersrand goldfield diamonds loaned to him by the Anglo American Corporation of South Africa Ltd and deduced from his analyses that the green colour is caused by irradiation, presumed to emanate from the uranium present in the ores. These are not considered to be economically important, but their presence as detrital gemstones in such ancient sedimentary rocks proves that a source of diamonds already existed very early on in Earth's history (Smart *et al.*, 2016).

In 2018, South Africa produced 9.9 million carats, making it the sixth-largest producer in the world. In 2019, the figure dropped to 7.2 million carats, somewhat less than the previous year.

In 2019, **Botswana** produced 23.7 million carats, making it the world's third-largest producer of diamonds. Unlike neighbouring Namibia and South Africa, Botswana produces diamonds exclusively from kimberlite pipes, at Orapa and Jwaneng. Botswana relies heavily on diamond sales for its national income.

There are more than 50 kimberlites in Botswana. Mines are operating at Damtshaa, Jwaneng, Letlhakane and Orapa. The kimberlites are Cretaceous in age, about 90 million years old, their age and geological formation thus similar to the famous 'Big Hole' kimberlite in South Africa. The Jwaneng kimberlite to the south of Orapa is different from the other pipes: it is about 250 million years old and three kimberlite pipes have coalesced to form the present surface exposure (Richardson *et al.*, 1999). The Jwaneng diamond mine in south-central Botswana is considered the world's richest diamond mine in terms of value. On 1 June 2021, a diamond weighing 1,098 carats was found at the Jwaneng mine, making it the fourth-largest diamond discovered, after the 3,106-carat Cullinan, the 1,758-carat Sewelô and the 1,109-carat Lesedi la Rona diamonds.

Although **Lesotho** is a small landlocked country with limited natural resources, the Letšeng-la-Terae diamond mine produces some of the largest diamonds ever found (Shor *et al.*, 2015; see table). In addition to their large size, the diamonds produced are type 11a diamonds, ones that have a very low nitrogen content, which also adds to their value. Letšeng is not the only diamondiferous kimberlite in Lesotho. Others include Kao, Mothae, Liqhobong and Khaphamali. There is some artisanal diamond mining in Lesotho from alluvial deposits.

Figure 307 The Big Hole in Kimberley, South Africa. This man-made excavation produced 2,722 kg (14,504,566 carats) of diamonds by the time it ceased production in 1914. BRUCE CAIRNCROSS PHOTO, 2013.

Figure 308 The Bells Bank kimberlite, a diamond-bearing dyke exploited for diamonds. The depth of the excavation is approximately 25 m. BRUCE CAIRNCROSS PHOTO, 2005.

Figure 309 Two parcels of alluvial diamonds totalling approximately 100 carats. These multicoloured gems are from the alluvial diggings on the Vaal River. DIAMONDS COURTESY OF BENNIE CLOETE, CS DIAMONDS, KIMBERLEY, BRUCE CAIRNCROSS PHOTOS.

Figure 310 An octahedral diamond, in kimberlite matrix, showing natural etch patterns on the crystal surface, 6 mm. INSTITUTIONAL COLLECTION, BRUCE CAIRNCROSS PHOTO.

Figure 311 Several diamonds in eclogite matrix. The dark orange crystals are garnet. South Africa. Field of view 12 mm. INSTITUTIONAL COLLECTION, BRUCE CAIRNCROSS PHOTO.

Figure 312 A selection of various coloured diamonds from South Africa. Left to right: pink 1.31 carats, yellow octahedron 1.4 carats, green octahedron 0.46 carats, amber 0.32 carats and green 0.14 carats. INSTITUTIONAL COLLECTION, BRUCE CAIRNCROSS PHOTO.

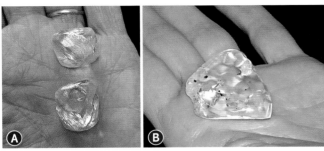

Figure 313 A Two yellow diamonds from alluvial diggings, 47 carats (top) and 49 carats (bottom). **B** A 138-carat near-flawless diamond, Vaal River alluvial diggings, South Africa. DIAMONDS COURTESY OF BENNIE CLOETE, CS DIAMONDS, KIMBERLEY. BRUCE CAIRNCROSS PHOTOS.

Figure 314 The 507-carat Cullinan Heritage diamond. Cullinan diamond mine, South Africa. BRUCE CAIRNCROSS PHOTO.

Kimberlites are known from the Gibeon district in **Namibia**, but these do not appear to host economic concentrations of diamonds. Economic deposits of diamonds in Namibia occur in offshore marine deposits along the Atlantic seaboard and Orange River (Schneider and Miller, 1992). In the Namib Desert, famous historical finds of diamonds led to the establishment of places such as Kolmanskop, outside Lüderitz. As in South Africa, some ancient alluvial diamonds are mined along the Orange River, which borders the two countries.

In 2019, Namibia produced US$914 million worth of rough diamonds.

Zimbabwe has a number of kimberlites and several of these contain diamonds. In addition, there are also alluvial diamond deposits (Stocklmayer, 1981). The Marange alluvial diamond fields in Manicaland were discovered in 2001 by De Beers and are the largest alluvial diamond deposits in

the world (Smit *et al.*, 2018). These have been producing significant quantities of diamonds during the twenty-first century, not without social and political problems. The diamonds are typically several carats in size and have smooth, resorbed surfaces, rather than sharp, well-formed crystal faces. The colours are very commonly dark green to brown and black, the dark colour formed by inclusions of graphite and brown radiation discoloration. These inclusions have produced interesting internal star-shaped features (Rakovan *et al.*, 2014).

Other localities are Colossus in the Bubi district, where the largest diamond recovered weighed 8.75 carats, River Ranch (Beitbridge district), Wessels (Bubi district), Sebungwe (Binga district), Shingwize (Mwenezi district) and Clare (Gweru district). A new diamond mine known as Murowa is planned near Zvishavane. Diamond-bearing gravels have been known for 100 years at Somabula, 20 km south-west of Gweru (MacGregor, 1921). These are found in an ancient river conglomerate deposit with an exotic mix of other gemstones, including white and blue topaz, chrysoberyl, sapphire, ruby and emerald. In 2018, Zimbabwe produced 3.2 million carats of diamonds.

Eswatini's only diamond mine, the DDM mine, ceased operations in the mid-1990s. It exploited a kimberlite in the Dvokolwako area, approximately 50 km north-east of Manzini. Some exploration for alluvial diamonds continues in the area.

Figure 315 A replica of the 3,106.75-carat (circa 621 g) Cullinan diamond. BRUCE CAIRNCROSS SPECIMEN AND PHOTO.

Diaspore ◆ AlO(OH)

Diaspore crystallizes in the orthorhombic system, has a hardness of 6.5 to 7, specific gravity of 3.3 to 3.5, a white streak and vitreous to pearly lustre. Diaspore usually forms flat, platy crystals that are stacked one on top of another. Prismatic or needle-like (acicular) crystals do occur, but are relatively rare. Pure diaspore is colourless to white, but when chemical impurities are present, it can be yellow, green, pink or red. Diaspore is often associated with other metamorphic minerals, or is found in altered orebodies.

A famous manganese-bearing variety of diaspore occurs in the Postmasburg district of **South Africa** (Cairncross *et al.*, 1997). The best crystals were found at the farm Paling M87, north of Postmasburg and from Lohathla, where intergrown laths and rosettes of platy rose-red crystals, up to 10 cm long, are found in manganese ore. Somewhat similar manganese-rich diaspore has been found in the Kalahari manganese field north of Kuruman. Although a rare discovery, these crystals are an attractive pink and are up to 10 cm long and 3 cm wide. Small, flaky diaspore crystals are known from west of Bela-Bela and in the Middelburg district (Rogers, 1924).

In **Namibia**, diaspore is found at Kyanite Kop on the farm Bethlehem 27 in the Windhoek district. Diaspore is associated with corundum deposits in the southern region of **Zimbabwe.** An attractive green chrome-diaspore is found at the Selukwe Peak mine.

Lenses of diaspore up to 10 cm diameter occur with andalusite, pyrophyllite and quartz at Sicunusa in **Eswatini**.

Figure 316
Diaspore from Droogekloof, Bela-Bela, South Africa. Field of view 1.8 cm.
PAUL MEULENBEELD SPECIMEN, BRUCE CAIRNCROSS PHOTO.

Figure 317 A vug partially filled with diaspore crystals, 11.6 cm. Inset: Interlocking platy crystals of diaspore, 6 cm. Postmasburg, South Africa. BRUCE CAIRNCROSS SPECIMENS AND PHOTOS.

Diopside ◆ CaMgSi$_2$O$_6$

Diopside crystallizes in the monoclinic system, has a hardness of 5.5 to 6.5, specific gravity of 3.22 to 3.38, a white to grey streak, and vitreous to dull lustre. Diopside crystals are prismatic, columnar or massive. They are colourless, white or grey, but may be brown, green or black. Chrome diopside is a vivid green variety, which is sometimes found as transparent crystals. Diopside commonly occurs in carbonatites and numerous kimberlites in southern Africa.

Diopside is found in large crystalline masses in **South Africa** at Marble Delta, KwaZulu-Natal. Pale lime-green crystals, some over a metre long, occur at the Palabora mine in Limpopo as large cleavages and fractured crystals (Southwood and Cairncross, 2017). Diopside is fairly common in some of the rocks of the Bushveld Complex and has been found in the Northern Cape. Small bright green crystals of transparent to translucent chrome diopside are found in kimberlites at the Jagersfontein, Roberts Victor, Monastery and De Beers mines. A rare pink diopside has been found at the Wessels mine in the Kalahari manganese field.

In **Namibia**, chrome diopside is common in a cluster of 60 kimberlites (none are diamond-bearing) in the Gibeon-Brukkaros area. Diopside is common in skarn deposits at Ais dome and in the Otjiwarongo district.

Chrome diopside occurs in kimberlites found in **Zimbabwe**.

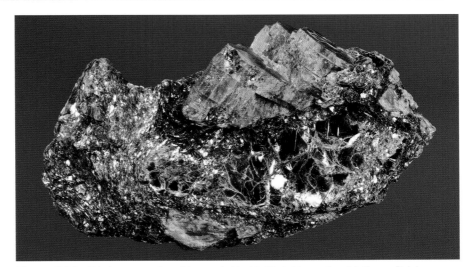

Figure 318 Bright green diopside crystal in biotite/phlogopite, 11.6 cm. Palabora mine, South Africa. BRUCE CAIRNCROSS SPECIMEN AND PHOTO.

Figure 319 A terminated 4.5-cm diopside crystal collected from outcrop exposed along the Swakop River, Namibia. BRUCE CAIRNCROSS SPECIMEN AND PHOTO.

Dioptase ◆ $Cu^{2+}SiO_2(OH)_2$

Dioptase crystallizes in the trigonal system, has a hardness of 5, specific gravity of 3.28 to 3.35, a pale blue-green streak, and vitreous lustre. The vibrant, bright green to blue-green of dioptase crystals may cause it to be confused with emerald. Dioptase is highly prized in mineral collections. Its crystals are well formed, prismatic and, in exceptional cases, can reach several centimetres in length. Small specimens are wrapped in silver or gold wire and used to make pendants. Dioptase is a secondary copper mineral that forms at the expense of primary copper ores such as bornite or chalcopyrite.

Dioptase is very rarely found in **South Africa**. Small, bright green crystals up to 10 mm long were found near Christiana, in the North West province, and attractive crystals a few millimetres in length occurred in weathered gossans that outcrop at Broken Hill mine, Aggeneys.

Namibia is world-famous for dioptase, notably from the well-known Tsumeb mine, and most important mineral collections have Tsumeb dioptase. The Tsumeb mine has produced some of the finest dioptase crystals in the world, the largest measuring 5 cm on edge (Wilson, 1977; Gebhard, 1999; Von Bezing et al., 2014), with crystals over 1 cm being common. When these brilliant green crystals occurred on snow-white calcite, they made stunning specimens. Other spectacular examples were composed of transparent twinned cerussite that are sprinkled with drusy dioptase. Good dioptase crystals were found at the Guchab mine in the Otavi mountainland to the south of the Tsumeb mine, and on the dumps at the old Rodgerberg mine.

Some copper occurrences in the Kaokoveld have been the source of fine dioptase specimens since the early 1980s.

Beautiful, bright green, transparent dioptase crystals, typically elongated on the c-axis, were dug from the surface outcrop of a small copper prospect close to the village of Omaue in Kaokoland in northern Namibia that has been known for decades (Schnaitmann and Jahn, 2010; Bowell et al., 2013; von Bezing et al., 2014). Magnificent dioptase crystals, up to 5 cm long, were collected in late 2003 at a copper prospect close to Omaue. Dioptase is also found in the Kaokoveld, south of Epupa Falls in the Otwane area, where it occurs with massive chalcocite, and at Okagwa, 10 km south-west of the Schwarze Kuppen, where it occurs with cuprite, chalcocite, malachite and azurite. The dioptase occurs in quartz veins. Although the deposit contains copper mineralization – including chalcocite, plancheite, malachite, azurite and chrysocolla – the digging was opened primarily to extract specimens of dioptase. Colourless quartz and calcite are common associated species, and euhedral, bright orange baryte (a rare accessory mineral) occurred with some dioptase specimens. Malachite is also present as acicular crystals and sprays that were sometimes coated with a thin film of pale blue drusy quartz.

Dioptase is reported from **Zimbabwe** at the Montana copper prospect in the Chiredzi district, with malachite, chrysocolla and azurite. Large, perfect crystals have been described from near Kariba. Dioptase is also found at the Alaska copper mine, west of Chinhoyi, the Copper Queen zinc-copper-lead mine approximately 100 km west of Chinhoyi and the Nevada, Old Mint and Cedric mines in the Makonde district.

Figure 320 Dioptase crystals scattered on dolomite matrix, 8.2 cm. Tsumeb mine, Namibia. BRUCE CAIRNCROSS SPECIMEN AND PHOTO.

Figure 321 Dioptase crystals, 6.5 cm. Tsumeb mine, Namibia.
BRUCE CAIRNCROSS SPECIMEN AND PHOTO.

Figure 322 A large vug lined with masses of crystalline dioptase, 16.5 cm. Tsumeb mine, Namibia.
BRUCE CAIRNCROSS SPECIMEN AND PHOTO.

Figure 324 Dioptase on blue shattuckite. Omaue mine, Kaokoveld, Namibia. Field of view 4.8 cm. BRUCE CAIRNCROSS SPECIMEN AND PHOTO.

Figure 323 Bright green dioptase crystals scattered on white calcite, 3.7 cm. Tsumeb mine, Namibia. BRUCE CAIRNCROSS SPECIMEN AND PHOTO.

Figure 325 Rare faceted dioptase, 1.22 carats (7 mm). Tsumeb mine, Namibia. WARREN TAYLOR RAINBOW OF AFRICA COLLECTION, MARK MAUTHNER PHOTO.

Figure 326 Dioptase with pale blue-green plancheite, 4.6 cm. Omaue, Kaokoveld, Namibia.
BRUCE CAIRNCROSS SPECIMEN AND PHOTO.

Dolomite ◆ CaMg(CO₃)₂

Dolomite crystallizes in the trigonal system, has a hardness of 3.5 to 4, specific gravity of 2.85, a white streak, and pearly to vitreous lustre. Dolomite exists as both a mineral and a rock, but is described here in its mineralogical form. Dolomite crystals are rhombohedral and are distinguished by the curved saddle shapes that crystal faces may form. This is useful in differentiating dolomite from calcite crystals, which they closely resemble. Most dolomite is white or colourless, but trace elements of metal, such as cobalt or copper, may impart pink and green to the mineral. Brown crystals are also known. Dolomite forms a chemical series with two other carbonate minerals: ankerite (iron-magnesium carbonate) and kutnohorite (calcium-manganese carbonate). Dolomite is quarried for lime used in the cement industry. Dolomite rock is composed of massive dolomite crystals that form underground cavities and caves (such as the Sterkfontein Caves in Gauteng and the Cango Caves near Oudtshoorn, South Africa) when dissolved in acidic water.

In **South Africa**, dolomite crystals are occasionally found as secondary vein fillings in the Witwatersrand gold mines – as at the Kloof gold mine – in the form of aesthetic pink crystals,

up to 5 mm in length, usually in association with quartz and pyrite (Cairncross, 2021d). Small crystals have been found at the Palabora mine, Limpopo. Grey-white saddle-shaped crystals, up to 2 cm in length, are found at the opencast lead-zinc mine at Pering between Vryburg and Kuruman in the Northern Cape. At the Mamatwan mine in the Kalahari manganese field, where dolomite rock hosts the economic manganese ore, small (up to 5 mm long) white dolomite

Figure 327
Pink dolomite and minor quartz, 4.2 cm. Kloof gold mine, South Africa.
BRUCE CAIRNCROSS SPECIMEN AND PHOTO.

Figure 328 Dolomite and amber-coloured sphalerite. Field of view 12.5 cm. Pering mine, South Africa. BRUCE CAIRNCROSS SPECIMEN AND PHOTO.

rhombohedrons are found. Some economic deposits of lead (galena), zinc (sphalerite) and fluorine (fluorite) are found in cavities in dolomite as at the Pering mine. Some of the caves found in dolomite formations may have secondary dolomite crystals present.

Saddle-shaped crystals and groups of crystals were common at the Tsumeb mine in **Namibia**. They were usually grey-white, but some interesting chocolate-brown varieties are known. Tsumeb has produced plates of thin dolomite casts formed when dolomitic solutions coat other minerals, usually calcite (Southwood and Robison, 2016). When the underlying calcite dissolved, it left its imprint in the dolomite. Pink-mauve cobaltian dolomite and green copper-rich dolomite casts are sought-after specimens. Some quartz from the Erongo Mountains have clusters of dolomite crystals attached. Solution cavities in the extensive dolomite formations in Namibia are often lined with drusy dolomite crystals.

Dolomite is common in **Zimbabwe** in the Lomagundi Group carbonate rocks, around Darwin and Mutoko.

Figure 329 Pink cobalt-rich dolomite, 5.9 cm. Tsumeb mine, Namibia. BRUCE CAIRNCROSS SPECIMEN AND PHOTO.

Figure 330
Dolomite cast showing residual shapes of calcite crystals that were beneath the dolomite and dissolved away, 12 cm. Tsumeb mine, Namibia. BRUCE CAIRNCROSS SPECIMEN AND PHOTO.

Figure 331
Pale green copper-rich dolomite, 14.5 cm. Tsumeb mine, Namibia. BRUCE CAIRNCROSS SPECIMEN AND PHOTO.

Figure 332 Yellow-stained dolomite on descloizite, 6.4 cm. Berg Aukas mine, Namibia. BRUCE CAIRNCROSS SPECIMEN AND PHOTO.

Dumortierite ◆ $Al_7(BO_3)(SiO_4)_3O_3$

Dumortierite crystallizes in the orthorhombic system, has a hardness of 8.5, specific gravity of 3.3 to 3.7, a white streak, and vitreous to dull lustre. Dumortierite does not usually form large or distinctive crystals. It is better known in southern Africa as an attractive, massive, blue-violet-purple mineral that is cut and polished for ornamental purposes. Dumortierite is found in specific igneous rocks, including pegmatites and granites that are enriched in aluminium.

In **South Africa**, dumortierite is found in Namaqualand in high-grade aluminium-rich gneisses (Beukes *et al.*, 1987). Outcrops of solid, blue, lapidary-grade titanium dumortierite occurs in the Northern Cape on the farm N'Rougas Suid, north of Kenhardt.

Dumortierite is found in **Namibia** north of the Erongo Mountains, on the farm Etemba 135, west of Omaruru. It is also reported from the farm Carolina 99 in the Keetmanshoop district (Schneider and Seeger, 1992d).

Dumortierite is found near the Save River in **Zimbabwe** in the Lowveld region and also in the Karoi district. At the Rainbow Mite cordierite deposit in the Beitbridge district, dumortierite occurs with iolite, tourmaline and apatite.

There is an occurrence of dumortierite in **Botswana** from Tsodilo Hills where it is found in small, attractive bowtie sprays of acicular crystals (www.mindat.org/photo-482511.html). Dumortierite is reported in minor quantities from the kyanite deposit at Halfway Kop, 15 km south-east of Francistown.

Excellent lapidary-quality dumortierite comes from a commercially worked deposit in Tete Province of **Mozambique**.

Figure 333 A cut and polished specimen of dumortierite, 10.4 cm. Mazowe, Changara, Tete Province, Mozambique. BRUCE CAIRNCROSS SPECIMEN AND PHOTO.

Figure 334 Dumortierite, 15.3 cm. Northern Cape, South Africa. BRUCE CAIRNCROSS SPECIMEN AND PHOTO.

Figure 335 Rough (left) and polished (right, 6.8 cm) dumortierite. Etemba, Namibia. BRUCE CAIRNCROSS SPECIMENS AND PHOTO.

Elbaite ◆ $Na(Li,Al)_3Al_6(BO_3)_3Si_6O_{18}(OH)_4$

Elbaite crystallizes in the trigonal system, has a hardness of 7, specific gravity of 3.03 to 3.1, white streak, and vitreous lustre. Elbaite is a member of the tourmaline group of minerals. It is a very popular gemstone and is used extensively in jewellery. Its popularity is due to its hardness, transparency and beautiful colours, ranging from green to blue, red, pink, orange, peach, mauve, yellow, white, or even colourless. It is common for crystals to be colour-zoned with a dark green outer layer surrounding a red inner core, giving rise to the popular name 'watermelon tourmaline'. Elbaite has typical striations or grooves on the surfaces, which run parallel to the long axis of the crystal. Indicolite is the blue variety of elbaite, rubellite the red. These two names are not approved by the IMA but rather used informally. Elbaite and other tourmalines are common in granitic pegmatites.

Gem-quality elbaite is rare but not unknown in **South Africa**. Bright blue, green and pink opaque elbaite crystals up to 10 cm long have been found in the Straussheim and Angelierspan pegmatites in the Kenhardt district in the Northern Cape, associated with albite, lepidolite and spodumene. They are mostly translucent single-colour red and green crystals, although some 'watermelon' crystals occur in the Straussheim pegmatite. Concentrically zoned green and pink elbaites have been found at Norrabees, north of Steinkopf.

Apart from diamonds, tourmaline is arguably the most important gemstone in **Namibia**. Forty five mines, claims and pegmatites containing elbaite are located in the Erongo Region (*The Mineral Resources of Namibia,* 1992). They are concentrated in tin-lithium-beryllium pegmatites in the Karibib-Usakos and Sandamap-Erongo areas and close to the Brandberg and Uis. Notable multicoloured and colour-zoned crystals come from Otjua mine and Neu Schwaben farm (Von Bezing *et al.*, 2014; 2016).

Figure 337 Tourmaline is one of the most important gemstones mined in Namibia. Some of the mines include: **A** Otjua mine, where gem-quality tourmaline and associated minerals are found. Some giant quartz crystal clusters were once found here (see figure 640a, page 221); **B** Omapyu mine, south-east of Omaruru; **C** Rubikon mine, south-east of Karibib. This pegmatite was mined for lithium. BRUCE CAIRNCROSS PHOTOS, 2014.

Figure 336 A selection of elbaite crystals from the Karibib-Usakos region, Namibia. **A** 11.8 cm; **B** 7.4 cm, with lepidolite; **C** 10 cm. BRUCE CAIRNCROSS SPECIMENS AND PHOTOS.

Figure 338 Doubly terminated elbaite on a 'floater' quartz, 12.5 cm. Otjua mine, Karibib district, Namibia. BRUCE CAIRNCROSS SPECIMEN AND PHOTO.

Figure 340 Green elbaite crystals and a stout hexagonal lepidolite crystal (right) overgrown by quartz, 8.6 cm. Karibib district, Namibia. BRUCE CAIRNCROSS SPECIMEN AND PHOTO.

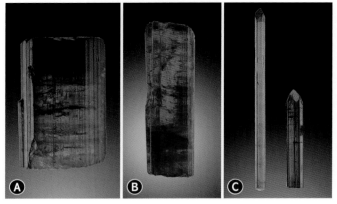

Figure 342 Backlit elbaite crystals from Namibia. **A** 3 cm, Neu Schwaben; **B** 3.1 cm, Otjua mine; **C** green elbaite, larger crystal is 11.2 cm, Omaruru district. BRUCE CAIRNCROSS SPECIMENS AND PHOTOS.

◄ **Figure 339** Backlit colour-zoned elbaite, 3 cm. Karibib-Usakos district, Namibia. BRUCE CAIRNCROSS SPECIMEN AND PHOTO.

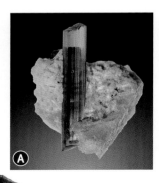

Figure 341
Two gem-quality indicolite crystals, one in massive quartz, the other attached to well-formed quartz crystals. **A** 3 cm, Neu Schwaben; **B** 4 cm, Johann Albrechtshöhe Farm 149, Karibib district, Namibia. BRUCE CAIRNCROSS SPECIMENS AND PHOTOS.

Figure 343 A slice of 'watermelon' colour-zoned elbaite cut perpendicular to the crystal's main axis, 4 cm. Karibib district, Namibia. BRUCE CAIRNCROSS SPECIMEN AND PHOTO.

Figure 344 Faceted tourmaline gemstones from Namibia. **A** 13.17 carats (1.5 cm), Karibib; **B** 14.66 carats (1.6 cm) Usakos; **C** 15.59 carats (1.7 cm) Omapyu mine. WARREN TAYLOR RAINBOW OF AFRICA COLLECTION, MARK MAUTHNER PHOTOS.

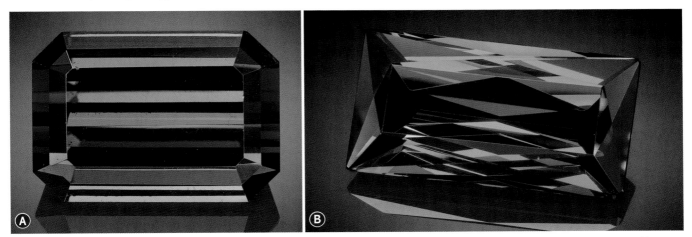

Figure 345 Two outstanding tourmaline variety indicolite gemstones from Namibia, cut from rough material found in 1966 at Neu Schwaben, which is famous for its blue tourmalines. **A** 16.23 carats (1.6 cm); **B** 35.58 carats (2.7 cm). WARREN TAYLOR RAINBOW OF AFRICA COLLECTION, MARK MAUTHNER PHOTOS.

Figure 346 A 16.81-carat elbaite variety rubellite (1.72 cm), Karibib district, Namibia. WARREN TAYLOR RAINBOW OF AFRICA COLLECTION, MARK MAUTHNER PHOTO.

Figure 347 Achroite is a virtually colourless variety of elbaite tourmaline. This 13.42-carat achroite is from Wilhelmstal, Karibib district. WARREN TAYLOR RAINBOW OF AFRICA COLLECTION, MARK MAUTHNER PHOTO.

Rubellite crystals from the Mwami pegmatites in **Zimbabwe** tend to be translucent and hence are not transparent gemstone material. Green tourmaline is more common in the Mwami region and pegmatites in the north-east. Beautiful green and red elbaite has come from the famous St Ann's pegmatite, which has produced half of Zimbabwe's gem tourmaline. This pegmatite was also an important source of blue topaz. Gem-quality elbaite was extracted from the Somabula gravels and the Fungwe Gem mine in the Rushinga district.

A pegmatite near Kubuta in **Eswatini** contains rubellite tourmaline with lepidolite.

Figure 348 A large quartz crystal with dark green elbaite penetrating the quartz, 17.5 cm. St Ann's mine, Zimbabwe. SIMON HARRISON SPECIMEN AND PHOTO.

Ephesite ◆ $NaLiAl_2(Al_2Si_2O_{10})(OH)_2$

Ephesite crystallizes in the triclinic system, has a hardness of 3.5 to 4.5, specific gravity of 2.9, a white streak, and vitreous to pearly lustre. Ephesite is a relatively rare mineral but can form attractive crystals. It is typically found in geological settings that have high concentrations of aluminium and an absence of quartz. Being a member of the mica group, it forms barrel-shaped hexagonal crystals, 'books' or flakes that cleave easily.

Beautiful pink crystals come from the Postmasburg manganese field in **South Africa**. Crystals have been found at the Bishop, Glosam, Lohathla and Paling mines in the region, some associated with diaspore and bixbyite (Cairncross *et al.*, 1997).

Figure 349 Bright red ephesite crystals with black manganese ore, 7.5 cm. Lohathla mine, Postmasburg, South Africa. BRUCE CAIRNCROSS SPECIMEN AND PHOTO.

Figure 350 **A** Masses of pink ephesite crystals in manganese ore, 9.2 cm. Lohathla mine, Postmasburg, South Africa. **B** Close-up showing details of the crystals. Field of view 2.1 cm. BRUCE CAIRNCROSS SPECIMEN AND PHOTOS.

Epidote ◆ $Ca_2(Fe^{3+},Al)_3(SiO_4)_3(OH)$

Epidote crystallizes in the monoclinic system, has a hardness of 6 to 7, specific gravity of 3.35 to 3.5, a white-grey streak and vitreous lustre. Epidote is typically found as light green to dark black-green crystals. As with some other minerals, chemical impurities can cause it to take on different colours. Manganese in the crystal structure produces a red-orange variety of epidote often incorrectly referred to as piemontite, which is a distinct, separate species to manganese-rich epidote. A rock known as unakite, which is popular with lapidarists, is a mixture of green epidote and pink orthoclase feldspar. Epidote ranges from transparent (in thin crystals) to opaque (in larger crystals). Most common habits are smooth, striated, elongate and prismatic, but can also be needle-like clusters and fibrous hair-like groups, or massive lumps in igneous and metamorphic rocks. Epidote is found in greenstone belts and many metamorphic rocks, such as gneiss and schist, and is hence fairly widespread. It also occurs in pegmatites, alpine cleft veins and hydrothermal deposits.

A particularly large deposit of epidote occurs near Neilersdrif between Kakamas and Keimoes in the Northern Cape, **South Africa**. This material can be fashioned into cabochons. Dark green epidote as well as unakite comes from pegmatites in the Onseepkans area in the Northern Cape. Good-quality unakite is collected in Limpopo in the Limpopo River region close to Musina. At the Musina copper mines, epidote was a common hydrothermal mineral in the ore, occurring as emerald-green to yellow-green crystals (up to 15 cm long, but generally under 1 cm in length) that fill cavities or adhere to quartz crystals. Small manganese-bearing epidote comes from the Musina mines. These specimens are an attractive, transparent, orange to burnt-orange colour and average 7 mm in length.

The copper mines of the Okiep district have produced dark green epidote, commonly associated with quartz and chalcopyrite. Beautiful large crystals of brass-coloured chalcopyrite are sometimes intergrown with pencil-like crystals of dark green epidote. Some pegmatites in the Onseepkans area on the Orange River, Northern Cape, contain epidote. Epidote is occasionally found in the Witwatersrand gold mines, together with other secondary minerals such as quartz and euhedral pyrite crystals.

Figure 351 Two different types of epidote from the Messina mine, Limpopo, South Africa. **A** Green epidote and quartz, 11.4 cm; **B** orange-green manganese-rich epidote on a quartz crystal. Field of view 2.8 cm. BRUCE CAIRNCROSS SPECIMENS AND PHOTOS.

Figure 352 Polished slab of unakite, an intergrowth of green epidote and orange orthoclase feldspar, 8.7 cm. Limpopo, South Africa. BRUCE CAIRNCROSS SPECIMEN AND PHOTO.

◄ **Figure 353** Epidote crystals, 5.7 cm. Free State Geduld mine, Welkom goldfield, South Africa. BRUCE CAIRNCROSS SPECIMEN AND PHOTO.

At Naugas in the Gamsberg region of the Rehoboth district in **Namibia**, epidote crystals hosted in granite veins rival some of the finest known (Von Bezing *et al.*, 2016). Namibian crystals are dark green to black, and extremely lustrous. Some are semi-transparent, and over 10-cm-long groups or clusters of crystals are often found together. Epidote crystals are also found with orange-red hematite quartz from pegmatites in the southern extremity of Namibia. Small, dark green crystals are occasionally found in basalt cavities in the Tafelkop amethyst deposits, west of Brandberg. Highly lustrous epidote crystals scattered on pink quartz come from close to the Schakalskruppe Siding, Berseba, 35 km east of Aus.

Figure 354
Two epidote specimens, **A** 5.2 cm, **B** 5 cm. Rehoboth rural area, Namibia. JOHANNESBURG GEOLOGICAL MUSEUM SPECIMEN (A), BRUCE CAIRNCROSS SPECIMEN (B) AND PHOTOS.

Figure 355 Tiny dark green epidote associated with quartz and pale green prehnite. Goboboseb Mountains, Namibia. Field of view 3.2 cm. BRUCE CAIRNCROSS SPECIMEN AND PHOTO.

Epidote is common in calc-silicate rocks, altered greenstone belt mafic and ultramafic rocks, granites and metamorphic rocks in **Zimbabwe**. Lapidary-grade unakite is exploited in the Beitbridge district. Crystals of epidote are found at the Ganyanhewe workings in the Rushinga district. The mineral also occurs in the Masvingo and Karoi districts.

Epidote is fairly widespread in mafic schists in the Tati schist belt, **Botswana**. It also occurs with orthoclase, quartz and hornblende in a prominent ridge of syenite west of Francistown.

Epidote is found in the mafic and ultramafic greenstone belt rocks in north-west **Eswatini**. At Makwanakop, an epidote-rich rock contains malachite, bornite and chalcopyrite.

Figure 356 Epidote on albite and quartz, 7.9 cm. Amis Gorge, Brandberg, Namibia. GERHARD LOUW SPECIMEN AND PHOTO.

Erythrite ◆ $Co_3(AsO_4)_2 \cdot 8H_2O$

Erythrite crystallizes in the monoclinic system, has a hardness of 1.5 to 2.5, specific gravity of 3.06, pale red to pink streak, and waxy, pearly to earthy dull lustre. Erythrite crystals are typically bright red, to pink or crimson. It forms as a secondary mineral in some cobalt-nickel-arsenic deposits. The crystals tend to be small, flattened and prismatic and frequently form stellate and radiating clusters.

Beautiful specimens of erythrite come from the old Kruisrivier cobalt mine located in the Groblersdal district, approximately 180 km north-west of Johannesburg, **South Africa** (Mellor, 1907). The crystals are of secondary origin and occur in fractures and vugs in Bushveld Complex felsite and gabbro (Reeks, 1996; Atanasova *et al.*, 2016). The erythrite varies from dark and pale purple, peach, cream, pink, pale pink, pale rose to white. Thin prismatic and acicular crystals form spectacular radiating aggregates most often found in cracks and vugs close to the cobalt sulphide ore vein.

Figure 357 The old workings at the Kruisrivier cobalt deposit, South Africa. BRUCE CAIRNCROSS PHOTO, 2016.

Figure 358 A myriad pink erythrite 'puffballs' on matrix. Kruisrivier cobalt mine, South Africa. Field of view 10 mm. BRUCE CAIRNCROSS SPECIMEN AND PHOTO.

Figure 359 Elongate erythrite crystals. Kruisrivier cobalt mine, South Africa. Field of view 2 cm. BRUCE CAIRNCROSS SPECIMEN AND PHOTO.

Figure 360 Close-up of radiating erythrite crystals. Kruisrivier cobalt mine, South Africa. Field of view 3.1 mm. WOLF WINDISCH SPECIMEN AND PHOTO.

Ettringite ◆ $Ca_6Al_2(SO_4)_3(OH)_{12} \cdot 26H_2O$

Ettringite crystallizes in the trigonal system, has a hardness of 2 to 2.5, specific gravity of 1.77, a white streak, and vitreous lustre. Ettringite is a rare mineral, but is included here because the finest crystals come from South Africa. The mineral forms hexagonal prismatic crystals of a beautiful yellow, with shades varying from amber to brown to orange-yellow, canary-yellow and white. It is a soft mineral that is lightweight because of its low specific gravity. Being highly hydrated (note the 26 water molecules at the end of the chemical formula), it tends to be somewhat unstable and will dehydrate, alter and change colour over time. It occurs as a secondary alteration mineral in South Africa's Kalahari manganese mines.

Before the discoveries of ettringite in the early 1980s, notably from the N'Chwaning II mine in the Kalahari manganese field, ettringite was known mainly as microscopic crystals. Individual crystals among the **South African** specimens were over 15 cm long. The crystals were an attractive canary-yellow colour, but also light brown, white or colourless, and were associated with manganite, calcite and hematite (Von Bezing et al., 1991). Ettringite may be confused with other minerals such as thaumasite, jouravskite and sturmanite, which are also found in the Kalahari manganese field, and single crystals may be mineralogically zoned by more than one species (Cairncross and Beukes, 2013).

Figure 361
A large ettringite crystal on calcite with hematite, 11 cm. N'Chwaning II mine, South Africa.
ULI BAHMANN SPECIMEN, BRUCE CAIRNCROSS PHOTO.

Figure 362
Ettringite on gaudefroyite, 6.1 cm. N'Chwaning II mine, South Africa.
DESMOND SACCO SPECIMEN, BRUCE CAIRNCROSS PHOTO.

◄ **Figure 363**
Ettringite, calcite and black metallic manganite, 7.2 cm. N'Chwaning II mine, South Africa.
DESMOND SACCO SPECIMEN, BRUCE CAIRNCROSS PHOTO.

Figure 364 ➤
Bright yellow ettringite with glassy white calcite, 5 cm. N'Chwaning II mine, South Africa.
BRUCE CAIRNCROSS SPECIMEN AND PHOTO.

Euclase ◆ BeAlSiO$_4$(OH)

Euclase crystallizes in the monoclinic system, has a hardness of 7.5, specific gravity of 3.05 to 3.1, a white streak, and vitreous lustre. Euclase is a relatively rare mineral. It is included here because Zimbabwe produces fine blue crystals that may be faceted into gemstones. Crystals are prismatic and can be either long or short and stubby. Some euclase is colourless to pale green, but the most desired variety is blue. Individual crystals may be colour-zoned, with blue terminations grading to white towards the base. Euclase, found in granitic pegmatites, is usually formed by the chemical replacement of beryl.

Gem-quality, intensely blue euclase crystals are found in some pegmatites in the Mwami area, such as The Falls, MWM, Mishek and Trim, all in the Karoi district, **Zimbabwe** (Anderson, 1979b). Yellow, green and colourless varieties are more common. The Last Hope pegmatite mine, 13 km north/north-east of Karoi, was mined for muscovite mica and is the premier locality for excellent, dark blue, gemmy euclase crystals. Transparent, dark indigo or cobalt-blue to peacock-blue euclase crystals were found here in a 2.5-m-wide pegmatite together with beryl. Gem-quality specimens also came from the St Ann's mine. At the Rattis mine, also in the Karoi district, gem-quality euclase was associated with aquamarine, chrysoberyl and gem-quality almandine garnet. In mid-2003, excellent pseudomorphs of euclase after beryl were collected from a pegmatite in the Mwami area. The euclase had partially or wholly replaced the beryl, some crystals of which were over 5 cm in diameter and had maintained their original hexagonal shape.

Euclase was found in the Stiepelmann pegmatite, Klein Spitzkoppe, **Namibia**. Some euclase has been found in the miarolitic cavities in the Erongo Mountains. These are colourless, translucent and up to 2 cm.

Figure 365 Blue euclase, 4.4 cm. Lost Hope mine, Karoi district, Zimbabwe. JIM AND GAIL SPANN SPECIMEN, JEFF SCOVIL PHOTO.

Figure 366 Two well-terminated euclase crystals, 3.5 cm. Lost Hope mine, Karoi district, Zimbabwe. BRUCE CAIRNCROSS SPECIMEN AND PHOTO.

Figure 367 A well-terminated euclase crystal, 1.5 cm. Lost Hope mine, Karoi district, Zimbabwe. BRUCE CAIRNCROSS SPECIMEN AND PHOTO.

Figure 368 A 3.35-carat (9-mm) faceted euclase from the Lost Hope mine, Karoi district, Zimbabwe. WARREN TAYLOR RAINBOW OF AFRICA COLLECTION, MARK MAUTHNER PHOTO.

Feldspar Group ◆ see albite, microcline, orthoclase

Ferberite ◆ $Fe^{2+}WO_4$

Ferberite crystallizes in the monoclinic system, has a hardness of 4 to 4.5, specific gravity of 7.51, black to brown-black streak, and adamantine to submetallic lustre. Ferberite (an iron species) is one of two tungstates that form the 'wolframite group', the other being hübnerite (manganese species). These minerals form heavy, prismatic crystals that can be black, brown or red-brown to metallic grey. These two minerals are found in granitic pegmatites.

Tungsten, which has the highest melting point (3,410°C) of all metals, is extremely hard yet tensile. Tungsten carbide is used in drill bits and abrasives and in tool and die apparatus. Electric filaments in light bulbs are made of tungsten.

A large number of small tin-tungsten deposits are known from the Northern Cape, **South Africa**. In the Springbok-Okiep-Nababeep area, ferberite occurs in quartz veins in schist that contain minor quantities of molybdenite, chalcopyrite and bismuth-bearing minerals (Söhnge, 1950; Bowles, 1988). Many of these deposits were mined sporadically on a small scale, notably in the vicinity of Okiep. The most important tungsten deposit occurs at Van Rooi's Vley near Upington, and consists of several tungsten- and tin-bearing quartz-tourmaline veins.

Numerous small tungsten deposits are found in a broad belt along the Orange River between Vioolsdrif and Upington. In the western part of this belt the deposits are mainly quartz-tourmaline veins containing only scheelite as the main tungsten ore mineral. In the eastern sections, however, the mineralization is more complex, and ferberite, scheelite, cassiterite and molybdenite are found in many places. There is also an unusual occurrence of tin-zinc-tungsten-topaz at Rhenosterkop near Augrabies.

Ferberite comes from the Bushveld Complex, notably in the Mutue Fides and Stavoren deposits. In the Western Cape, within a 40-km radius of Cape Town, small economic deposits of ferberite occur with cassiterite, arsenopyrite and molybdenite in veins that cut through granite and metamorphosed rocks.

Ferberite is found in pegmatites and quartz veins in **Namibia**, often associated with scheelite (Pirajno and Jacob, 1987; Diehl, 1992b). A substantial quantity of ferberite, including well-formed crystals, occurred at the Krantzberg tungsten mine north-east of Erongo Mountains (Pirajno and Schlögl, 1987). The mineral was associated with titanite, tourmaline, topaz, quartz and fluorite. Some of the Erongo pegmatites that contain ferberite outcrop at Davib Ost and Ameib (Cairncross and Bahmann, 2006a). The Brandberg West mine and the Paukuab pegmatites provided further ferberite mineralization.

Figure 369 A cleaved ferberite crystal, 9.2 cm, from Namaqualand, South Africa, collected in 1965. BRUCE CAIRNCROSS SPECIMEN AND PHOTO.

Figure 370 Parallel aligned ferberite crystals, 3.4 cm. Northern Cape, South Africa. BRUCE CAIRNCROSS SPECIMEN AND PHOTO.

Figure 371 Ferberite crystal, 4.2 cm. Karibib district, Namibia.
BRUCE CAIRNCROSS SPECIMEN AND PHOTO.

Figure 372 Metallic ferberite crystal perched on fluorite, 2 cm. Erongo Mountains, Namibia. ULI BAHMANN SPECIMEN, BRUCE CAIRNCROSS PHOTO.

Figure 373 An unusual mineral to facet, this is a rare cut ferberite crystal of 84.03 carats (3.4 cm). Kunene Region, Namibia.
WARREN TAYLOR RAINBOW OF AFRICA COLLECTION, MARK MAUTHNER PHOTO.

Ferberite is found in granites and in metamorphic rocks in **Zimbabwe** (Anderson, 1979). It has been mined at several deposits in the Karoi district in the northern and north-western parts of the country. It occurred in quartz veins at Helen's Hope mine and in garnet schist at the Makashi deposit. At the Honey mine, approximately 110 tonnes of ferberite was mined from quartz veins that were hosted in a quartz-mica schist. This deposit had well-formed, large ferberite crystals. The Union Jack mine produced tungsten

from ferberite, while at the Richardson mine, ferberite was found associated with calcite, fluorapatite, fluorite, galena, pyrrhotite and sphalerite. Both these deposits are in the Umzingwane district. Ferberite occurs in tourmaline quartz veins in the Hwange area, at the RHA mine, where the veins cut through muscovite schists.

In **Eswatini**, ferberite is reputed to have come from Mtshengu's Drift on the Mkondo River near Goedgegun and the Forbes Reef tin-tungsten-nickel occurrence (Davies, 1964).

Fluorapatite ◆ Ca$_5$(PO$_4$)$_3$F

Fluorapatite is a member of the apatite group of minerals, crystallizes in the hexagonal system, has a hardness of 5, specific gravity of 3.1 to 3.2, a white streak, and vitreous to resinous lustre. Colour can be highly variable, from colourless to yellow, mauve or purple and crystals can be transparent to opaque. Fluorapatite is common and widespread in igneous and metamorphic rocks and in hydrothermal vein deposits. It is found in alkaline complexes, carbonatites and pegmatites.

Phosphorous is derived from fluorapatite and finds a wide range of applications in mineral feedstock, detergents and various pharmaceutical products. Most of the phosphate produced from fluorapatite concentrate is used in the manufacture of fertilizers.

Fluorapatite has been reported from many places in **South Africa**, including Marble Delta in KwaZulu-Natal, the Palabora mine in Limpopo (where it is an ore mineral at the FOSKOR pit) and the Glenover phosphate mine. Other carbonatite complexes that contain fluorapatite are Spitskop, north-east of Groblersdal, the Schiel Complex between Louis Trichardt and Giyani, and several smaller deposits (Verwoerd, 1986). Fluorapatite is found in the Pilanesberg at Wydhoek 92 JQ, as massive lumps associated with purple fluorite and as large blue crystals (up to 10 cm on edge) in feldspar at Mica Siding, Limpopo, together with corundum. In the Murchison mountain range at Gravelotte, gem-quality blue-green crystals are found in quartz. In the Soutpansberg, at Barend 523 MS and at Redhill 103 LS, blue 0.5-cm crystals occur with corundum and titanite crystals. At Horst 89 LS, fluorapatite occurs as massive green aggregates with pink baryte crystals and monazite and at Bandelierskop, green crystals are found with yellow monazite-(Ce). It is also recorded in a pegmatite from Palakop in the Letaba district with garnet, muscovite, feldspar and beryl, and at De Hoek 547 LT and Doreen KU in the Tzaneen district. Furthermore, fluorapatite is a relatively common mineral associated with the tin deposits of the Bushveld Complex.

The best South African fluorapatite crystal specimens have come from pegmatites in the Northern Cape and Limpopo. In the Northern Cape, large blue crystals up to 6 cm on edge were found in quartz at Jakkalswater, 2-cm white crystals in quartz from Upington, and green prismatic crystals from Uranoop and Groendoorn. Some jewellery-quality fluorapatite comes from the farm Zeekoe-Streek 9 in the Kakamas district. Fluorapatite is found at Steenkampskraal, together with chalcopyrite and monazite.

Fluorapatite crystals are found in many pegmatites and carbonatites in **Namibia**. Some of the finest purple-violet to green-

Figure 374
Calcium-rich fluorapatite, 8.7 cm. Wondergat, Richtersveld, South Africa. BRUCE CAIRNCROSS SPECIMEN AND PHOTO.

blue crystals have come from the Clementine II pegmatite in the Karibib district, while colour-zoned pink-green crystals were found in the Otjua mine (Von Bezing *et al.*, 2016). At the now-defunct Uis tin mine, pale lavender barrel-shaped crystals up to 2 cm long occur in quartz veins. Deep blue crystals were collected at the De Rust lithium-tin-tungsten pegmatite, and large crystals have come from the Rubikon pegmatite. Carbonatites and other alkali-rich rocks commonly have fluorapatite as an accessory mineral. In the Osongombe Complex, crystals form aggregates up to 30 cm in diameter. The rocks west of Epembe in the Kaokoveld have fluorapatite associated with minerals such as aegirine, arfvedsonite and biotite. The Otjisazu (alkaline) Complex, 20 km north-east of Okahandja, contains fluorapatite in mafic pegmatites in association with aegirine, augite and titanite. Three different habits occur: finely disseminated grains, pale green massive nodules up to 50 cm in diameter, and crystals up to 5 cm.

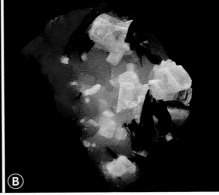

Figure 375 Colourless, transparent fluorapatite crystals with quartz and muscovite, 3.6 cm. **A** is under normal daylight and **B** under 365 nm long-wave ultraviolet light, showing the vivid yellow fluorescence of the fluorapatite crystals. Blesberg mine, Northern Cape, South Africa. BRUCE CAIRNCROSS SPECIMEN AND PHOTOS.

Figure 376 A scattering of colour-zoned blue fluorapatite crystals on matrix. Erongo Mountains, Namibia. Field of view 1.8 cm. BRUCE CAIRNCROSS SPECIMEN AND PHOTO.

Figure 377 Typical fluorapatite crystal, 3.2 cm. Erongo Mountains, Namibia. BRUCE CAIRNCROSS SPECIMEN AND PHOTO.

Figure 378 Pale lavender fluorapatite with quartz. Uis mine, Namibia. Field of view 2.2 cm. BRUCE CAIRNCROSS SPECIMEN AND PHOTO.

Figure 380 Cascading fluorapatite crystals and cream-white cookeite, 5.3 cm. Otjua mine, Karibib district, Namibia. JOHN RAKOVAN SPECIMEN AND PHOTO.

Figure 379 A colour-zoned fluorapatite crystal, 2.2 cm. Omaruru district, Namibia. BRUCE CAIRNCROSS SPECIMEN AND PHOTO.

Since it is found in igneous rocks and is a relatively common species in granites, fluorapatite is widespread in **Zimbabwe**. It occurs in what are termed alkaline-ring carbonatite complexes. These are usually circular intrusive bodies and may have an exotic array of rock types and minerals. Beautiful blue crystals come from the Karoi district. At Dorowa in the Buhera district, fluorapatite was exploited from a pipe-like alkaline-ring complex in association with magnetite, vermiculite and serpentinite. Two other deposits, Shawa and Chishaya, contain large, potentially economic, deposits of fluorapatite.

Fluorapatite is also known from pegmatites in the Karoi area, where it often forms attractive, barrel-shaped, blue, green or brown to white crystals. These pegmatites include the Turning Point claims (where muscovite and beryl were also mined) and the Miami and Pope mines (associated with lepidolite, beryl and topaz). In some pegmatites, fluorapatite is associated with other rare phosphate species, such as brazilianite, triplite and lazulite. Good crystals have been collected near the Marondera and Pfungwe areas. At the Benson pegmatites, north of Mutoko, large fluorapatite crystals, up to 15 cm long, are known. There is a manganese-rich variety of fluorapatite at the Vee Cee Claims, Makaha, and the Al Hayat deposit in the Bikita district, which forms attractive, translucent, green crystals that fluoresce yellow-white.

In **Botswana**, fluorapatite is found in anorthosite, 6 km east of Mmadinare, and in pegmatites at the Lose quarry.

Some pegmatites in **Eswatini** contain fluorapatite.

Figure 382. Two faceted fluorapatite gemstones cut from rough material found in the Gamsberg area, Namibia. **A** 10.32 carats (1.7 cm); **B** 25.61 carats (2.6 cm). WARREN TAYLOR RAINBOW OF AFRICA COLLECTION, MARK MAUTHNER PHOTOS.

Figure 381
Fluorapatite together with quartz, mica and tourmaline, 6.8 cm. Clementine II pegmatite, Okatjimukuju Farm 55, Karibib district, Erongo Region, Namibia. JOHN RAKOVAN SPECIMEN AND PHOTO.

Figure 383 Blue-capped fluorapatite crystals, 2.4 cm. Karoi district, Zimbabwe. JOHN RAKOVAN SPECIMEN AND PHOTO.

Figure 384 A fluorapatite crystal, 1.7 cm. Karoi district, Zimbabwe. JOHN RAKOVAN SPECIMEN AND PHOTO.

Fluorapophyllite ◆ $KCa_4(Si_8O_{20})(F,OH)\cdot 8H_2O$

Fluorapophyllite crystallizes in the tetragonal system, has a hardness of 4.5 to 5, specific gravity of 2.37, white streak, and pearly to vitreous lustre. Fluorapophyllite is the fluorine-bearing member of the apophyllite group (see hydroxyapophyllite-(K)). It has a confusing variety of shapes, often forming crystals that are pseudocubic, tabular, pyramidal or elongate prisms. However, it is almost always white, cream or colourless, although some varieties (uncommon in southern Africa) may be green or pink. Fluorapophyllite forms in cavities and fissures in basalt and hydrothermal veins, and is often associated with zeolite minerals.

Well-formed crystals of fluorapophyllite were common in intrusive dykes at the Palabora mine, **South Africa** (Southwood and Cairncross, 2017). The largest crystals are 2 cm and were found in cavities, and were associated with prehnite, calcite and mesolite. Large pseudocubic crystals have come from the Penge asbestos mine. Some public collections, such as the McGregor Museum in Kimberley and the Council for Geoscience in Ditsong Museum, Pretoria, have large plates of colourless fluorapophyllite that came from the Kimberley, De Beers, Bultfontein, Jagersfontein and Premier diamond mines. Individual colourless to yellow crystals are several centimetres in size. Some fluorapophyllite was collected in the Witwatersrand gold mines.

Cavities in the Aris phonolites south of Windhoek, **Namibia**, have produced beautiful, clear fluorapophyllite crystals.

The geodes and vugs in the basalts at Butha-Buthe, **Lesotho**, have yielded apophyllite, although analyses have not been conducted to determine the exact species (Cairncross and Du Plessis, 2018). Crystals are tetragonal, and measure from less than 5 mm to 4 cm. The crystals are typically white to off-white or cream-coloured. Some pockets contain abundant apophyllite, whereas others do not. Associated species include stilbite, quartz and laumontite. Apophyllite is relatively rare compared with the abundance of stilbite and quartz encountered in the mineralized cavities.

Fluorapophyllite occurs as well-formed crystals at the Ayshire, Jeppe and Beatrice mines, and in basalts in the Victoria Falls region, **Zimbabwe**.

Beautiful tabular fluorapophyllite crystals were found in **Botswana** in a fissure at the Selebi-Phikwe mine. The crystals are up to 2 cm on edge, and some enclose minute red hematite crystals that impart an attractive speckled red hue.

Figure 385 A large slab of cream-coloured fluorapophyllite crystals with some yellow calcite, 35 cm. Palabora mine, South Africa. COUNCIL FOR GEOSCIENCE SPECIMEN, BRUCE CAIRNCROSS PHOTO.

Figure 388 Fluorapophyllite crystals with minor yellow calcite, 25 cm. Palabora mine, South Africa. BRUCE CAIRNCROSS SPECIMEN AND PHOTO.

Figure 386
Fluorapophyllite crystals on matrix, 4.5 cm Letšeng-la-Terae mine, Lesotho. BRUCE CAIRNCROSS SPECIMEN AND PHOTO.

Figure 387
Fluorapophyllite crystal, 2.1 cm. Butha-Buthe district, Lesotho. HERMAN DU PLESSIS SPECIMEN, BRUCE CAIRNCROSS PHOTO.

Figure 389
Fluorapophyllite crystals on drusy quartz and associated with white laumontite, 6.4 cm. Butha-Buthe district, Lesotho. HERMAN DU PLESSIS SPECIMEN, BRUCE CAIRNCROSS PHOTO.

Figure 390 A mass of in situ white apophyllite/fluorapophyllite crystals with beige stilbite in basalt. Butha-Buthe district, Lesotho. HERMAN DU PLESSIS PHOTO, 2016.

Figure 391 A mass of fluorapophyllite crystals included by red hematite and black goethite, 15 cm. Selebi-Phikwe, Botswana. BRUCE CAIRNCROSS SPECIMEN AND PHOTO.

Fluorite ◆ CaF$_2$

Fluorite crystallizes in the cubic system, has a hardness of 4, specific gravity of 3.18, a white streak, and vitreous lustre. Fluorite is one of the chameleons of the mineral kingdom, and is found in virtually all shades of the spectrum – white, yellow, orange, green, brown, red, pink, blue, purple, black – and may even be colourless. Crystals of fluorite commonly form simple cubes, although octahedral and dodecahedral forms also exist and combinations thereof. The mineral often fluoresces under ultraviolet light. Because of its very good octahedral cleavage and relative softness, fluorite is notoriously difficult to facet into fancy gemstones, but when done successfully, the results are stunning.

Fluorite is a major economic mineral and is mined for fluorine, which is used in the metallurgical and chemical industries. One of its main uses is in the production of hydrofluoric acid. It is also used as a flux. Fluorine is added, somewhat controversially, to drinking water to safeguard teeth. Fluorite occurs in association with many lead-zinc deposits worldwide, often hosted in limestones and/or dolomites. It is common in granite pegmatites.

Fluorite is relatively common in **South Africa**. The best known locality, for quality and quantity of specimens, is Riemvasmaak, close to the Orange River in the Northern Cape (Cairncross, 2009). Vibrant green octahedral and modified cuboctahedral crystals up to 15 cm on edge have been collected. Many are transparent and gem quality. Most are coated by a thin film of white, drusy quartz that needs to be removed to reveal the fluorite. Apart from the ubiquitous green crystals, yellow, yellow-green, and purple crystals are found. Apart from the Riemvasmaak locality, fluorite crystals come

Figure 392 Octahedral fluorite crystals with quartz, 7.2 cm. Riemvasmaak, South Africa. BRUCE CAIRNCROSS SPECIMEN AND PHOTO.

Figure 393 A large cuboctahedral fluorite crystal with a smaller octahedron displaying modified cube corners, 13.5 cm. Riemvasmaak, South Africa. DESMOND SACCO SPECIMEN, BRUCE CAIRNCROSS PHOTO.

Figure 394 A large group of octahedral fluorite crystals, 23.4 cm. Riemvasmaak, South Africa. RONNIE MCKENZIE SPECIMEN, BRUCE CAIRNCROSS PHOTO.

Figure 395 A large modified cuboctahedral fluorite on quartz, 10-cm diameter. Riemvasmaak, South Africa. DESMOND SACCO SPECIMEN, BRUCE CAIRNCROSS PHOTO.

Figure 396 **A** Fluorite specimen, 4.6 cm, together with **B** an 18.04-carat faceted fluorite. Riemvasmaak, South Africa. MASSIMO LEONE SPECIMEN (B), BRUCE CAIRNCROSS SPECIMEN (A) AND PHOTO.

Figure 397 A 21.22-carat (1.4-cm) green faceted fluorite. Riemvasmaak, South Africa. WARREN TAYLOR RAINBOW OF AFRICA COLLECTION, MARK MAUTHNER PHOTO.

from various unnamed localities generally labelled as 'Orange River'. These vary in colour, as do those from Riemvasmaak, and are often associated with red, hematite-included quartz.

Multicoloured transparent fluorite is found at Ottoshoop in the North West. In the western part of the Pilanesberg, peculiar feather-like habits of fluorite are found intergrown with other minerals. Fluorite was plentiful in the tin mines of the Bushveld Complex (Crocker, 1979), as at Zaaiplaats, where yellow, violet, pale green, colourless or colour-zoned fluorite occurred, some crystals as large as bricks. A number

of economic fluorite deposits occur in the Rooiberg-Bela-Bela-Mookgophong (Naboomspruit) area as veins and irregular bodies in the granite, granophyre and rhyolite host rocks. One of these deposits is the Buffalo fluorite mine close to Mookgophong. Here, mineralization consists of several large fluorite bodies of low-grade ore and, at times, minor monazite-(Ce). Small quantities of green and purple fluorite crystals have also been recovered, and bright green octahedral crystals, up to 20 cm across, were found in cavities (Atanasova *et al.*, 2016).

Figure 398 Panoramic view of the Buffalo fluorite mine north of Pretoria, South Africa. Veins in the Bushveld granite were exploited for fluorite. BRUCE CAIRNCROSS PHOTO, 2010.

Figure 399 Very dark purple octahedral fluorite crystals in quartz-lined vugs. The orange granite matrix that hosts the mineralization is visible, 7.8 cm. Buffalo fluorite mine, South Africa. BRUCE CAIRNCROSS SPECIMEN AND PHOTO.

Figure 400 Close-up of purple fluorite cubes on drusy quartz. Buffalo fluorite mine, South Africa. Field of view 3.3 cm. BRUCE CAIRNCROSS SPECIMEN AND PHOTO.

Figure 401 Very dark purple cubes of fluorite displaying complex surface growth features, with quartz. Field of view 2.1 cm. Buffalo fluorite mine, South Africa. BRUCE CAIRNCROSS SPECIMEN AND PHOTO.

Figure 402 A cut and polished specimen with the fluorite present in the rock as twinned, feather-like growths, 12.8 cm. Pilanesberg, South Africa. BRUCE CAIRNCROSS SPECIMEN AND PHOTO.

At the Vergenoeg fluorspar mine, north-east of Pretoria, large economic deposits of fluorite are mined (Crocker, 1985; Cairncross *et al.*, 2008). These occur in an ancient volcanic pipe that measures 700 x 900 m at the surface but tapers off with depth. The oxidized ore consists of hematite and fluorite. An interesting feature of the deposit is the presence of numerous cavities and caves in the ore that have unusual fluorite and goethite stalactites, associated with rare minerals such as prosopite, gearksutite and nitrocalcite as white, powdery encrustations. Minor azurite and discrete balls of white kaolinite are also associated with the stalactitic goethite.

In KwaZulu-Natal, veins of fluorite, up to 65 cm wide, are found north and south of Hlabisa (Snyman, 1998). At the old Rhebok workings a branching network of chalcedony veins cuts through coarsely crystalline, sugary encrustations of fluorite,

often with colour banding. Very good purple banding is present at the old Rabe's Claims. Where veins have not completely filled, gemmy, euhedral fluorite crystals up to 10 cm on edge, are sometimes found with quartz crystals up to 1 cm. Fluorite crystals, up to 4 cm on edge, are common, and the fluorite shows colour zoning from outer layers of pale green to mauve cores. Some of this material is transparent and has been faceted. Other KwaZulu-Natal localities are the Aladdin fluorite mine, Sinkwazi on the north coast, and the area south of Nongoma.

In the copper mines of the Okiep district, notably the Jan Coetzee and Nababeep West mines, fluorite was found as bright purple to colourless crystals associated with quartz, chalcopyrite, epidote, calcite, pumpellyite and axinite. Crystals are usually small, but some specimens have cubes several centimetres on edge (Cairncross, 2004b). Important deposits of low-grade

Figure 403 The opencast Vergenoeg fluorite mine, Rust de Winter, South Africa. Although a fluorite mine, the locality is better known for its goethite specimens. BRUCE CAIRNCROSS PHOTO, 2013.

Figure 405 Unusual glassy, vermiform (worm-like) fluorite on goethite. Vergenoeg fluorite mine, South Africa. Field of view 1.7 cm. BRUCE CAIRNCROSS SPECIMEN AND PHOTO.

Figure 404 Goethite partly coated by fluorite, 5.2 cm. Vergenoeg fluorite mine, South Africa. BRUCE CAIRNCROSS SPECIMEN AND PHOTO.

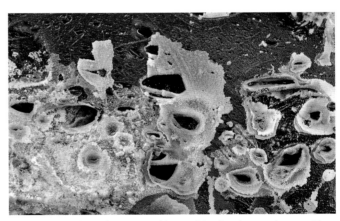

Figure 406 An anthropomorphic goethite-fluorite specimen, with portions resembling faces. The matrix consists of brown goethite punctured by irregularly shaped cavities. This has been partially coated by sugary white and purple fluorite. Vergenoeg fluorite mine, South Africa. Field of view 4 cm. BRUCE CAIRNCROSS SPECIMEN AND PHOTO.

fluorite, associated with very minor quantities of lead and zinc ore, occur in the top section of the dolomites south of Zeerust, North West province. Most of the fluorite forms lenses.

The Okorusu deposit, 48 km north of Otjiwarongo in **Namibia**, is the largest deposit of fluorite in Namibia and produces outstanding specimens (Schneider and Seeger, 1992c; Cairncross, 2018e). This deposit has been known since the early German colonial era and was mined on a small scale in the 1920s. The main orebody consists of sugary fluorite, vein fillings in fissures, in breccia and as replacement of limestone. Green and purple colours predominate, although yellow, grey and colourless varieties are sometimes encountered. Large cavities, often over a metre long, are lined and filled with clusters of euhedral cubic fluorite crystals. Most are simple cubes, the larger crystals attaining dimensions of over 10 cm on edge. The best specimens consist of groups of lustrous cubes stacked on one another. Internal colour zoning of deep green, sea-green and purple layers is common and produces striking specimens. Associated minerals include baryte, chalcedony, hematite and quartz.

During the first two decades of the 2000s, the Erongo Mountains produced some extraordinary fluorite specimens (Jahn and Bahmann, 2000; Cairncross and Bahmann, 2006a). Vibrant green cubic and cuboctahedral crystals associated with quartz, feldspar or muscovite typify the specimens. The colour varies from bright green to almost black, with colour zoning in many crystals. The latter are typified by what became referred to as 'lizard eyes' or 'snake's eyes' fluorite; bizarre internal colour zones alternate between an inner, dark opaque core with translucent green cubic faces that appear as eyes when backlit. This is then covered by transparent, colourless fluorite. Spinel twinned light purple fluorite up to 15 cm has also been found.

Figure 407 Panoramic view of the Okorusu opencast mine, Namibia. This large open-pit operation produced innumerable multicoloured fluorite during its period of operation. BRUCE CAIRNCROSS PHOTO, 2017.

Figure 408 Translucent green fluorite, 7.1 cm. Okorusu mine, Namibia. BRUCE CAIRNCROSS SPECIMEN AND PHOTO.

Figure 409 Purple fluorite rimmed by quartz, 13 cm. Okorusu mine, Namibia. BRUCE CAIRNCROSS SPECIMEN AND PHOTO.

Figure 410 Backlit fluorite specimen showing internal colour zoning. This variety has been referred to as 'harlequin fluorite', 7.2 cm. Okorusu mine, Namibia. PHILIP HITGE SPECIMEN, BRUCE CAIRNCROSS PHOTO.

Figure 411 Cubic yellow fluorite crystals with purple corners, 11 cm. Okorusu mine, Namibia. BRUCE CAIRNCROSS SPECIMEN AND PHOTO.

Figure 412 Cubes of fluorite coated by a thin film of iridescent goethite and capped by white calcite, 6 cm. Okorusu mine, Namibia. BRUCE CAIRNCROSS SPECIMEN AND PHOTO.

Figure 413 Coloured-zoned fluorite with yellow muscovite, 16 cm. Erongo Mountains, Namibia. DESMOND SACCO SPECIMEN, BRUCE CAIRNCROSS PHOTO.

Figure 414 A view of one of the granite cliff faces on the south-western side of the Erongo Mountains, Namibia. The rock is pockmarked by myriad cavities dug by artisanal miners searching for mineral specimens such as fluorite. BRUCE CAIRNCROSS PHOTO, 2014.

Figure 415 Coloured-zoned fluorite crystal, 3.4 cm. Erongo Mountains, Namibia. BRUCE CAIRNCROSS SPECIMEN AND PHOTO.

Figure 416 From the Erongo Mountains of Namibia, a fluorite, backlit to show the complex internal colour zoning, 8 cm. DESMOND SACCO SPECIMEN, BRUCE CAIRNCROSS PHOTO.

Figure 417
A pale green 126.72-carat (3-cm) faceted fluorite. Klein Spitzkoppe, Namibia.
WARREN TAYLOR RAINBOW OF AFRICA COLLECTION, MARK MAUTHNER PHOTO.

Figure 418
A pale blue 32.46-carat (2.6-cm) faceted fluorite. Klein Spitzkoppe, Namibia.
WARREN TAYLOR RAINBOW OF AFRICA COLLECTION, MARK MAUTHNER PHOTO.

Figure 419 A parcel of faceted Namibia fluorites totalling 318.7 carats. Top left to right: Okorusu mine 50.86 carats (2.4 cm); Bergsig farm, Omaruru district 22.55 carats (1.7 cm); Okorusu mine 22.67 carats (1.9 cm). Bottom left to right: Okorusu mine 68.02 carats (2.8 cm); Karasburg district 76.91 carats (2.5 cm); Karasburg district 77.69 carats (2.5 cm).
WARREN TAYLOR RAINBOW OF AFRICA COLLECTION, MARK MAUTHNER PHOTO.

Figure 420 Two vivid green faceted fluorites: **A** 23.33 carats (1.9 cm); **B** 24.9 carats (1.9 cm). Bergsig farm, Erongo Region, Namibia. WARREN TAYLOR RAINBOW OF AFRICA COLLECTION, MARK MAUTHNER PHOTOS.

Green fluorite occurs sparsely in the Klein Spitzkoppe pegmatites (Cairncross, 2005a). Drusy purple fluorite occurs in cassiterite-bearing pegmatites in southern Namibia, such as those west of the farm Girtis. Beautiful cubic green fluorite crystals, up to 15 cm on edge, were found at the Berg Aukas mine during early production. Fluorite is reported from the Namib lead mine, and a breccia pipe at the Garub mine, 100 km north of Karasburg, contains a core zone of fluorite in granite-gneiss (Schneider and Seeger, 1992c). The fluorite crystals are blue, green and violet and are associated with other interesting minerals, such as calcite, quartz, galena, azurite, cerussite, chrysocolla and anglesite. Quartz veins containing honey-coloured fluorite are found in granite, 25 km north-east of Uis. Similar quartz veins are found on the farm Rhinelands 18 and Blydskap 268 in the Outjo district, as well as on the farm Platrand 154 in the Karasburg district. The Seven Pillars mine in the extreme south of Namibia, close to the Orange River, has produced attractive sea-green cubic fluorite crystals, some sprinkled with small grains of pyrite. Pale green fluorite was also found in small quantities at the Navachab gold mine.

In **Zimbabwe**, fluorite is found in the Hwange district, in veins as fracture fillings contained in Karoo rocks, and in the Guruve district, in igneous and metamorphic rocks such as granite gneiss. Pegmatites in the Karoi district and other areas often contain green, pink or purple fluorite, either as granular masses or crystallized cubic crystals. The P & O mine in the Masvingo district and the Tinde deposit close to Hwange have produced choice, transparent, colourful fluorite crystals. At the Marion mine, Hwange district, pale green and deep purple fluorite is found in quartz veins in gneiss. Several other fluorite mines in the Hwange district, such as Betlyn, Fluor, Entuba, Jab, Luso, Tinde, and P J G East, all have fluorite and quartz in veins and lenses in granite. Good specimens have come from the Rusape municipal quarry. At the Fluorite The First mine, brown cubic crystals were found in veins that infilled shear zones and fault surfaces. The fluorite from the Kamota, Marion and Jab prospects is usually pale green.

Fluorite is found in quartz veins at the Bushman mine, 60 km south-east of Nata in **Botswana**, together with galena and chalcopyrite. Quartz veins in Gaborone granite at Ditshukutswane, between Kika and Manyana, also contain fluorite. The fluorite is purple, white or colourless and is associated with sphalerite, galena, chalcopyrite and covellite. Another fluorite locality is known from west of the Taupse River.

Two deposits of fluorite are found in the Hlatikulu district, **Eswatini**. One is at Mhlosheni, 29 km from Goedgegun near the Hluti road, where veins of fluorite are hosted in granite and the fluorite is sometimes covered by interesting chalcedony casts. The other deposit is in quartz veins 2–60 cm wide, which are located 2.5 km south/south-east of Hluti. Here, fluorite is sometimes found with quartz crystals up to 5 cm long.

Fluorite has been exploited in **Mozambique** at Serra da Gorongosa in the Sofala Province at two localities close to the north-eastern border with Zimbabwe. The fluorite is attractive and ranges in colour from yellow to blue-green or green (Lächelt, 2004).

Figure 421 Amber-coloured cleaved fluorite octahedron, 7 cm. Undisclosed locality in Zimbabwe. BRUCE CAIRNCROSS SPECIMEN AND PHOTO.

Galena ◆ PbS

Galena crystallizes in the cubic system, has a hardness of 2.5, specific gravity of 7.58, grey streak, and metallic lustre. It is the main economic ore mineral of lead. It is steel-grey and is characteristically very heavy due to the lead component. Although the most common crystal form is cubic, octahedrons and combination forms exist. Galena is a common mineral in base metal, gold and other metamorphic and hydrothermal deposits, and is typically found in low-temperature metamorphic rocks and hydrothermal veins. It is most commonly concentrated in limestone/dolomites in so-called 'MVT' (Mississippi Valley Type) lead-zinc deposits, associated with sphalerite. These MVT deposits occur not only in southern Africa, but in other parts of the world as well. Lead was used as a petrol additive, but the atmospheric pollution it causes has led to a shift to unleaded petrol. Lead is used to make ammunition, brass and bronze alloys, and as an additive in pesticides.

In **South Africa**, galena is found in many localities in the Malmani dolomite, often associated with palaeokarst (Hammerbeck, 1970; Philpott and Ainslie, 1986). The most significant deposit was at the Pering mine in the North West province, which ceased operations in 2003 (Southwood, 1986; Wheatley *et al.*, 1986b). The mine consisted of ring-shaped bodies of collapsed breccia in dolomite similar to the Bushy Park deposit in the Griquatown district (Wheatley *et al.*, 1986a). In Limpopo, the Leeuwbosch lead deposit 16 km north of Thabazimbi produced good galena specimens. Similar deposits are found at Bokkraal 344 JP, Rhenosterhoek 343 JP and the old Doornhoek lead mine.

At the Nababeep West mine at Nababeep, Northern Cape, rare, euhedral octahedral crystals 3 cm on edge are associated with chalcopyrite. In the Witwatersrand gold mines, secondary cubic galena crystals up to 5 cm on edge have (rarely) been found in quartz veins. At the Broken Hill mine at Aggeneys, large subhedral and euhedral galena crystals up to 15 cm on edge are found.

Figure 422 Elongate and partially skeletal galena crystals on drusy rhodochrosite. N'Chwaning I mine, South Africa. Field of view 1.6 cm. BRUCE CAIRNCROSS SPECIMEN AND PHOTO.

Figure 423 Galena crystals with minor quartz, 5.7 cm. Kusasalethu gold mine, Carletonville district, South Africa. BRUCE CAIRNCROSS SPECIMEN AND PHOTO.

Figure 424 Many small galena crystals scattered on olive-green sphalerite that is coating calcite, 4.3 cm. Wessels mine, South Africa. BRUCE CAIRNCROSS SPECIMEN AND PHOTO.

Galena is rare at the N'Chwaning II mine in the Kalahari manganese field, where small, very attractive cubes up to 5 mm on edge were found with calcite in 1991. Small, frequently distorted galena crystals were found on calcite and celestine in the 1990s. The Wessels mine produced a few unusual specimens consisting of elongate calcite 'fingers' coated by olive-green sphalerite and this studded by many cubic galena crystals. A handful of highly lustrous, distorted and elongate galena cubes associated with small pink rhodochrosite were collected in N'Chwaning I in 2016.

Namibia has many small lead-zinc/base metal deposits that contain galena (Wartha and Genis, 1992). At the Ai-Ais mine in southern Namibia, veins of solid galena up to 30 cm thick were exploited. Good-quality galena crystals made up of cubes and composite crystal shapes have been collected at the Rosh Pinah mine, as well as the Namib lead mine (Van Vuuren, 1986; Cairncross and Fraser, 2012). Abenab West had significant quantities of the mineral. Lead mineralization occurs 20 km north-west of Grootfontein on the farm Olifantsfontein. Galena was one of the main ore minerals at the Tsumeb mine. Although not plentiful, crystals of galena were found at this mine, many with characteristic etched and striated crystal faces and some with a peculiar flattened habit.

In **Zimbabwe**, galena is mostly associated with gold deposits in ancient greenstone belts, lead being obtained as a by-product (Anderson, C.B., 1980). Exceptions are the Copper Queen zinc mine and the Elbas mine in Hwange district. At the latter, quartz veins with galena, pyrite, chalcopyrite and sphalerite occur in a variety of metamorphic rocks, well-formed galena crystallizing in cavities and brecciated zones in the quartz

veins. At the King's Daughter mine, galena is found with an interesting suite of minerals such as pyrite, crocoite, sphalerite, chalcopyrite, arsenopyrite and gold.

In eastern **Botswana**, galena occurs at the Lady Mary mine and, with sphalerite, at Dihudi. The Bushman mine south-east of Nata has galena with fluorite, quartz and chalcopyrite. Quartz veins with galena and fluorite are found in Gaborone granite at Ditshukutswane, at the headwaters of the Kolobeng River.

Galena and a secondary lead mineral, cerussite, are found in quartz veins at Kubuta, **Eswatini**, with malachite and bornite. A mineralized quartz vein contains galena, pyrite, chalcopyrite and arsenopyrite in outcrops in the Ngwavuma valley, east of Hlatikulu.

Figure 425 Galena, 3 cm. Rosh Pinah mine, Namibia. BRUCE CAIRNCROSS SPECIMEN AND PHOTO.

Figure 426 Striated galena crystals, 4.7 cm. Tsumeb mine, Namibia. BRUCE CAIRNCROSS SPECIMEN AND PHOTO.

Garnet ◆ See almandine, andradite, demantoid, grossular, pyrope and spessartine

Goethite ◆ α-Fe³⁺O(OH)

Goethite crystallizes in the orthorhombic system, has a hardness of 5 to 5.5, specific gravity of 3.3 to 4.3, a brown-orange streak and dull to submetallic lustre. Goethite forms from the oxidation of iron-bearing minerals such as hematite, pyrite and other iron oxides and sulphides. Its colour is typically ochre to orange-red. Earthy to massive goethite is the most common iron mineral in many orebodies, where it imparts a reddish coloration in hand specimens. Goethite is normally found as soft, earthy coatings on rocks and on other minerals, but it does form aesthetially pleasing specimens of botryoidal masses and tiny golden crystals. Goethite is common throughout southern Africa, and only important examples are given below.

In **South Africa**, goethite is found in abundance at the Vergenoeg fluorspar mine, where it occurs as black or iridescent stalactitic masses in vugs and cave fillings (Cairncross *et al.*, 2008). Goethite occasionally forms minute golden crystals that are included in quartz crystals, as at the Musina copper mines. Goethite pseudomorphs after pyrite are found in the suburb of Monument Park in Pretoria as

◀ **Figure 427** Goethite from Gamsberg mine, South Africa. Field of view 4 cm. BRUCE CAIRNCROSS SPECIMEN AND PHOTO.

▼ **Figure 428** Goethite after pyrite, 2 cm. The crystal form is a trapezohedron. Mount Anderson, Lydenburg district, South Africa. BRUCE CAIRNCROSS SPECIMEN AND PHOTO.

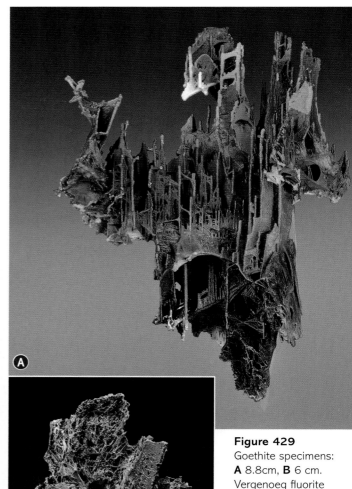

(A)

(B)

Figure 429
Goethite specimens:
A 8.8cm, **B** 6 cm.
Vergenoeg fluorite mine, South Africa.
BRUCE CAIRNCROSS SPECIMENS AND PHOTOS.

Figure 430 Goethite pseudomorphs after pyrite. Crystal shapes include dodecahedron, and other pyrite forms. **A** 4.4 cm, **B** 5.5 cm; **C** 9.7 cm. Monument Park, Pretoria, South Africa. BRUCE CAIRNCROSS SPECIMENS AND PHOTOS.

crystals 5 cm on edge that exhibit several forms, for example cube, dodecahedron and penetration twins. This locality was discovered when houses were built, and excavations in the dolomite revealed hundreds of these crystals. They are also found as groups of crystals. Similar large pseudomorphs are found in the Lydenburg district of Mpumalanga, especially near Mount Anderson. Very attractive ochre-red goethite/hematite pseudomorphs after pyrite were found, cementing pieces of brecciated rock, at the N'Chwaning II mine in the Kalahari manganese field in 2003.

Goethite is found in large amounts in banded iron formations on the farm Eisenberg 78 in the Otjiwarongo district, **Namibia**. Because of the irregular weathering and oxidation of iron ore, goethite is usually present in varying amounts in regions that contain banded iron formation rocks. Goethite was abundant at the Berg Aukas mine, and it is associated with fluorite at Okorusu fluorite mine. It is widespread in iron-rich rocks in the Otjosondu manganese field. Pseudomorphs of goethite after siderite are encountered in pegmatites at Klein Spitzkoppe, Erongo Mountains and the Onganja mine.

Figure 431 Yellow calcite associated with brown goethite pseudomorphs after pyrite, 5.8 cm. N'Chwaning II mine, South Africa. BRUCE CAIRNCROSS SPECIMEN AND PHOTO.

Figure 432 Many tiny black goethite globules with white dolomite and yellow mimetite. Tsumeb mine, Namibia. Field of view 2 cm. BRUCE CAIRNCROSS SPECIMEN AND PHOTO.

Figure 433 Orange goethite after pyrite pseudomorphs cementing hematitic iron ore breccia, 10.1 cm. N'Chwaning II mine, South Africa. BRUCE CAIRNCROSS SPECIMEN AND PHOTO.

Figure 434 Goethite pseudomorph after siderite with topaz, 5 cm. Erongo Mountains, Namibia. BRUCE CAIRNCROSS SPECIMEN AND PHOTO.

Figure 435 Shiny black goethite coating calcite crystals. This specimen contains white calcite and green dioptase that formed on the goethite. Tsumeb mine, Namibia. Field of view 3.5 cm. BRUCE CAIRNCROSS SPECIMEN AND PHOTO.

Figure 436 Goethite pseudomorph after pyrite, 2.6 cm. Krantzberg mine, Namibia. DEBBIE WOOLF SPECIMEN AND PHOTO.

Goethite has formed from the weathering of banded iron formations in the Mwanesi range and elsewhere in **Zimbabwe**.

Beautiful golden goethite micro-crystals included in quartz occur at the Devil's Reef gold deposit in the Pigg's Peak district, **Eswatini** (Jones, 1962). These minute crystals sometimes aggregate into circular haloes, scattered inside the quartz. Goethite is a very common constituent of the iron ore deposits at Ngwenya.

In **Lesotho**, small amounts of black botryoidal goethite are found on some of the quartz and stilbite specimens from the Butha-Buthe area.

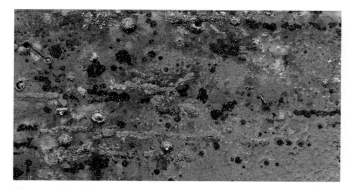

Figure 437 Close-up of the surface of a quartz crystal showing the underlying included yellow-orange goethite and some silver hematite. Devils Reef, Eswatini. Field of view 1.3 cm. BRUCE CAIRNCROSS SPECIMEN AND PHOTO.

Figure 438 Tiny spheres of black goethite on drusy quartz, some of which is bent and distorted. Butha-Buthe, Lesotho. Field of view 2.2 cm. BRUCE CAIRNCROSS SPECIMEN AND PHOTO.

Gold ◆ Au

Gold crystallizes in the cubic system, has a hardness of 2.5 to 3, specific gravity of 19.3, a gold streak, and metallic lustre. Probably the most famous precious metal known, gold has had high monetary value from antiquity. Its metallurgical properties also make it ideally suited for jewellery, as it is highly malleable and can be hammered, stretched and rolled without breaking. Crystals are extremely rare and are highly prized by collectors. Gold is very durable and when it is eroded and transported away in rivers as grains and nuggets, economic concentrations can be formed. This is particularly applicable to source rocks such as greenstone rocks, gold-rich quartz veins and other gold-bearing rocks.

The best crystalline gold specimens come from so-called vein-type gold deposits or hydrothermal deposits. Here, gold usually crystallizes as coarse aggregates and can form beautiful specimens.

Gold is synonymous with southern Africa and, in particular, South Africa, where it has been mined for centuries. Gold occurs mostly in three geological environments in southern Africa:

● placer deposits (gold weathered out of host rocks and transported and deposited elsewhere), both modern and ancient
● associated with shear zones and quartz veins, and
● in pegmatites.

In **South Africa**, beautiful gold artefacts such as beads, wire gold, amulets and gold animals have been exhumed from Mapungubwe in Limpopo, west of Musina, and at Thulamela in the Kruger National Park. The University of Pretoria has many of the gold artefacts found at the Mapungubwe site, including the famous gold-plated rhinoceros excavated in the 1930s. There are also similar archaeological sites in Zimbabwe and in Botswana.

The famous Witwatersrand goldfield accounts for most of South Africa's gold production. Much of the gold mined from these metamorphosed sedimentary deposits is microscopic, but some beautiful large specimens have been found, these having formed much later than the sediments (Kershaw, *et al.*, 2003; Cairncross, 2021d). The Witwatersrand goldfield contains, and has yielded, more gold than any other deposit on Earth and has historically produced more than ten times the gold of all other known deposits. To date (2021), over 52,000 metric tonnes of gold has been extracted from the deposit, and there is still an estimated 30,000 tonnes of inferred resources remaining. There is a strong association between pyrobitumen (carbon) and gold in some layers. Where the carbonaceous material is present, gold particles are often visible and the gold grades increase prodigiously. Virtually all the gold is present as tiny, microscopic particles. However, when these ancient sedimentary rocks were buried, metamorphosed and exposed to hydrothermal fluids, some gold became remobilized and subsequently recrystallized as large gold specimens. For over a century, the origin of gold has been debated, with three main theories holding sway: i) the placer origin, whereby gold was eroded from a source rock and entered the sedimentary basin together with the clastic debris; ii) hydrothermal gold that was introduced later into the already existing sedimentary strata, and iii) a combination of the two processes (Cairncross, 2021d).

Figure 439 The Main Reef outcrop at the heritage site at Langlaagte, south of Johannesburg. The strata are dipping steeply to the right (south) and the old adit is visible below the green bush (centre). This photograph was taken in 1986, before subsequent vandalism and degradation of the site.
BRUCE CAIRNCROSS PHOTO.

When the Witwatersrand goldfield is discussed, what is sometimes overlooked is that gold was discovered and mined in several other regions of South Africa prior to the bonanza discovery of the Witwatersrand deposit and, more importantly, even in areas close to the famous discovery site at Langlaagte. The earliest (European) discoveries took place decades before in the old Transvaal Republic and, later, at Pilgrim's Rest, Barberton and in the former Cape Colony at places such as Millwood close to Knysna (Duff, 2020), Cradock, and Prince Albert (Cairncross and Anhaeusser, 1992). In Prince Albert, gold was discovered in 1870, initiated by a two-ounce (67-g) alluvial nugget found in the weathered soil.

Figure 440 Old dumps on the Klein Waterval farm, Prince Albert, South Africa. Gold was discovered at the historic Prince Albert goldfield in 1870 and was mined from alluvial deposits. BRUCE CAIRNCROSS PHOTO, 2016.

Figure 441 ➤
Sheared conglomerate reef with visible gold, 12.8 cm. Rietfontein mine, Witwatersrand goldfield South Africa. BRUCE CAIRNCROSS PHOTO.

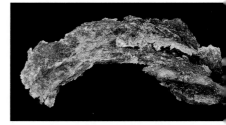

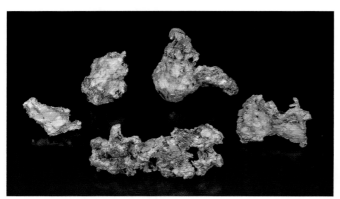

Figure 442 A selection of historic gold specimens from the Witwatersrand goldfield, South Africa (clockwise from far left): Nigel Reef Robinson Deep mine, 3.5 cm; 5.2 cm; West Driefontein mine 8.4 cm; 6.5 cm; Nigel Reef Vogelstruisbult mine 10.5 cm. BRUCE CAIRNCROSS PHOTO.

Figure 443
Gold in white quartz and pyrite, 8 cm. Robinson Deep mine, Witwatersrand goldfield, South Africa. BRUCE CAIRNCROSS PHOTO.

Figure 444
Gold, 2 cm. Witwatersrand goldfield, South Africa. BRUCE CAIRNCROSS PHOTO.

Figure 445 Skeletal octahedral gold crystals, 2.5 cm, with quartz. Van Dyk mine, Witwatersrand goldfield, South Africa. BRUCE CAIRNCROSS PHOTO.

Figure 446 Gold, smeared on the upper surface of thick, black pyrobitumen, 13 cm. Basal Reef, Lorraine gold mine, Witwatersrand goldfield, South Africa. BRUCE CAIRNCROSS PHOTO.

Figure 447 Gold and quartz, 15.6 cm. Witwatersrand goldfield, South Africa. BRUCE CAIRNCROSS PHOTO.

Figure 448 Gold collected in 1895, 15 cm. Rose Deep Ltd mine, Witwatersrand goldfield, South Africa. BRUCE CAIRNCROSS PHOTO.

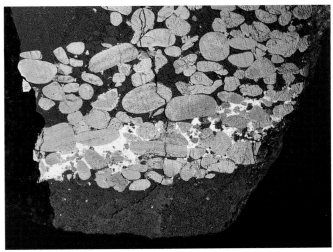

Figure 449 A cut and polished slice of Witwatersrand goldfield conglomerate reef seen under reflected light, 4.5 cm. The smooth, rounded grains are pyrite ('fool's gold'), while the yellow infilling between the pyrite at the base is solid gold. Welkom goldfield, South Africa. BRUCE CAIRNCROSS PHOTO.

Figure 450 Gold, with galena and minor sphalerite, 6.5 cm. St Helena mine, Witwatersrand goldfield, South Africa. GAIL AND JIM SPANN SPECIMEN, TOM SPANN PHOTO.

The Eersteling goldfield in the Polokwane district was the site of the first European gold discovery in the Transvaal Republic. After that, further discoveries occurred in the Pilgrim's Rest and Barberton districts. Some fabulously rich deposits, such as the 'Golden Quarry' at the Sheba mine (Wagener and Wiegand, 1986), were once worked at the Barberton goldfield (Anhaeusser, 1986a), the second-most important gold-producing region in South Africa. In 1886, the discovery of the Witwatersrand gold-bearing conglomerate occurred, and the development of the largest gold deposit on Earth forever changed the economic, social and political history of South Africa. The volume of literature on the discovery, history, development, mining, geology, mineralogy, environmental issues and socio-political influence of the Witwatersrand goldfield is, perhaps, enough to fill a small library.

Figure 453 Gold crystals like these are rare. Barberton goldfield, South Africa. Field of view 2 cm.
BRUCE CAIRNCROSS PHOTO.

Figure 454 Small sheet of 8 mm gold on matrix. Golden Osprey mine, Giyani, Limpopo, South Africa. BRUCE CAIRNCROSS PHOTO.

Figure 455 Gold in massive quartz, from the historical Eersteling mine, Limpopo, South Africa. Field of view 4 cm. BRUCE CAIRNCROSS PHOTO.

Figure 451 Spongiform gold etched out from the enclosing quartz, 7 cm. Sheba mine, Barberton, South Africa. BRUCE CAIRNCROSS PHOTO.

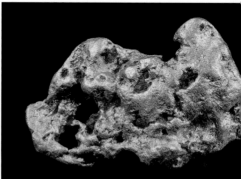

Figure 452 Alluvial gold nugget, 3.1 cm. Pilgrim's Rest, South Africa. BRUCE CAIRNCROSS PHOTO.

Although native gold is found in **Namibia**, most of the country's gold comes from trace amounts associated with sulphide minerals (Hirsch and Genis, 1992a). Gold was first discovered in Namibia in 1899, close to Rehoboth. From 1917–1963, the main gold mining district was the Ondundu goldfield in Damaraland. Currently (2021), gold is mined at the Navachab mine in the Karibib district, the largest gold mine in Namibia that extracts gold from marble, a somewhat unusual host rock. The Otjikoto mine, 48 km south-west of Otavi, had gold hosted in marble and various schists. Many

Figure 456 Two views of old workings at the Ondundu goldfield, Namibia: **A** a trench that exploited an auriferous pegmatite; **B** a small decline shaft. GERHARD LOUW PHOTOS, 2012.

Figure 457
Gold and quartz, 2.5 cm. Ondundu goldfield, Namibia. GERHARD LOUW PHOTO.

Figure 458 Gold associated with malachite on quartz. Kaokoveld, Namibia. Field of view 20 mm. BRUCE CAIRNCROSS PHOTO.

Figure 459
Alluvial gold nugget, 2.5 cm. Undisclosed locality, Namibia. BRUCE CAIRNCROSS PHOTO.

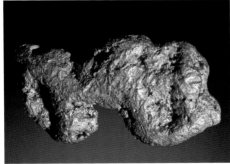

Figure 460
Two gold samples from north of Opuwa, Kaokoveld, Namibia. The gold is associated with blue chrysocolla and shattuckite; **A** 3.3 cm; **B** 3.1 cm. GERHARD LOUW PHOTOS.

Figure 461
Gold on quartz, 3.7 cm. Natas mine, Namibia. GERHARD LOUW PHOTO.

small prospects have been worked, some in gold-bearing quartz veins, for example on the farms Korechas 381 and Navarre 383, while others are associated with metamorphic rocks (Hirsch and Genis, 1992a). Dormant gold mines are located in the Windhoek, Karibib, Rehoboth, Swakopmund and Lüderitz districts. Most of these had gold associated with metals such as copper, lead, zinc and molybdenum. Gold was found at the copper-molybdenum mine at Onganja, and the Natas tungsten mine produced some native gold.

There are also alluvial gold deposits in several locations, including the Erongo region, where nuggets up to 100 g were panned with another placer mineral, cassiterite. In the Omaruru district, 43 kg of alluvial gold were extracted from 1937–1943, with some gravels containing rich grades of up to 80 gm of gold per ton of gravel. The gold occurred as flakes, rounded nuggets and porous, spongy nuggets. Some rare gold associated with chrysocolla and malachite was found in the copper deposits in the Kaokoveld.

Gold is the most important source of revenue in the mining sector in **Zimbabwe**. The country has thousands of gold deposits, many of them very small (Bartholomew, 1990b). The most important mines are Cam and Motor (Kadoma district), Globe and Phoenix (Kwekwe district) (Cairncross, 2020a), Rezende (Mutare district), Shamva (Shamva district) and Dalny (Chegutu district). Almost all Zimbabwe's gold has been extracted from ancient Archaean rocks, in particular greenstone belts (Foster *et al.*, 1986). Some gold is extracted as a by-product from a few copper mines. Geologically, about 80% of the gold occurs in veins and shear zones; the remaining 20% is associated with ancient sedimentary rocks, i.e. stratabound deposits. The Wanderer and Connemara mines produced gold from banded iron formations. Some alluvial gold nuggets were extracted from the Somabula diamondiferous gravels, an ancient alluvial deposit. The following references, although decades old, contain valuable information on Zimbabwe's gold deposits: Maufe (1920), MacGregor (1928), Keep (1929), Morgan (1929), Tyndale-Biscoe (1933), Ferguson and Wilson (1937), Phaup (1937), Phaup and Dobell (1938), and Amm (1946).

Figure 463 Two polished samples of quartz, pyrite and gold: **A** 3 cm; **B** 3 cm. Wanderer mine, zimbabwe. BRUCE CAIRNCROSS PHOTOS.

Figure 464 Gold and quartz, 4.2 cm. Globe & Phoenix mine, Zimbabwe. BRUCE CAIRNCROSS PHOTO.

Figure 462 Trenches dug in north-eastern Botswana as part of a gold exploration project. The semi-arid bush with mopani trees is typical of this region. BRUCE CAIRNCROSS PHOTO, 1985.

Gold deposits in **Botswana** are restricted to the eastern section of the country, in particular the Tati schist belt (Chatupa, 1999). Gold is frequently found here in quartz veins, for example at the Monarch mine, and in shear zones as at the Bonanza mine. The Tati Concession, where the German explorer Carl Mauch discovered ancient African workings in 1866, was the site of the first European discovery of gold in Botswana. There are now approximately 60 abandoned gold mines and diggings here, all located in greenstone belt rocks. Locations include the Vumba schist belt and Gaborone granite west of Gaborone.

Gold deposits and mines in **Eswatini** are located along the north-western border, in a zone extending between Oshoek and Horo (Sims, undated). Gold occurs here in rocks of the Barberton greenstone belt. Several mines operated in the past in two districts: the Forbes Reef area (where James Forbes first discovered gold in the late 1870s) and Pigg's Peak, the site of the country's largest mine. Gold was mined at the Daisy, Lomati and Wyldsdale mines and at Devil's Reef, where grades of up to 400 oz per tonne of rock were mined, surpassing even the fabled 'Golden Quarry' in neighbouring South Africa. Alluvial gold has been exploited from gravels in the Malolotsha River and Forbes Reef areas.

Southern **Mozambique** has gold resources and these occur in hard rock as well as in weathered eluvial and alluvial deposits (Lächelt, 2004). Alluvial gold and primary gold in quartz-carbonate veins has been exploited close to Manica at Bragança, in the Mavita region, and at Monarch mine in the Manica Province. The gold from Manica has been mined since prehistoric times, while 'modern' mining has taken place since the late 1800s. Gold-bearing gravels were dredged in the 1960s until they were exhausted. Lächelt (2004) describes further gold occurrences in southern Mozambique in detail, and these include gold-quartz veins in the Inchope-Doeroi area in Manica and Sofala provinces, alluvial gold in the Luenha and Mazowe rivers, Manica and Tete provinces, and several more in Tete Province, Missale, Chifumbaze, Mulolera, Cazula and the Fíngoè Zone.

Graphite ◆ C

Graphite crystallizes in the hexagonal system, has a hardness of 1 to 2, specific gravity of 2.1 to 2.33, a black streak and earthy, dull to metallic lustre. Graphite is characteristically soft and ductile, and crystals can be bent without breaking. It usually forms black to silvery-black platy crystals and flakes. Its softness and black streak are eminently suitable for pencil lead and it has further uses in the manufacture of brake linings, carbon brushes, batteries, crucibles and refractory bricks. Graphite is most commonly found in graphitic schists, limestones and gneisses and in South Africa in some areas where Karoo coals have been metamorphosed.

In **South Africa**, massive graphite was exploited from a metamorphosed coal seam at the Mutale mine 80 km east of Musina. At the Gumbu mine in the same district, graphite flakes up to 4 mm were extracted from metasediments. Some excellent graphite specimens come from Appingedam in the Zaaiplaats tin field, Mokopane district. Graphitic schists are found at Kwa-Xolo in KwaZulu-Natal. Graphite crystals occur at Umzimkulu in white marble. In the Northern Cape, in Namaqualand, a graphite deposit is known from the farm Oup 80.

Small flakes of graphite are found in metamorphic rocks 20 km south-west of Otjiwarongo, **Namibia**. Approximately 60 km east of the town of Aus, graphite mineralization occurred as pods and lenses in granite. Between Swakopmund and Wlotzkasbaken on the west coast, a small quarry produced tiny graphite crystals encased in marble. Commercial mining of graphite started in 2017 at Okanjande,

with the graphite flakes processed at a plant based at the old Okorusu fluorite mine (Schneider and Genis, 1992a). Another Namibian graphite deposit at Aukam is located approximately 40 km south-east of Aus.

The Lynx mine was the main producer of graphite in **Zimbabwe**, the graphite occurring as flakes in a biotite-sillimanite-garnet gneiss (Muchemwa, 1987). Other graphite deposits in the district include Graphite King, Juma and Silaka Kaswaya. Graphite schists are known from the Dete and Globe deposits in the Hwange district.

Graphite is found in **Botswana** in dolomites and graphitic schists at Moshaneng and in similar rocks close to the Bushman mine, Phudulooga area.

Graphite occurs in **Eswatini** in schists on the farm Sterkstroom in the Hlatikulu district.

Figure 465
Graphite crystals on calcite, 3.7 cm. Marble Delta, KwaZulu-Natal, South Africa. BRUCE CAIRNCROSS SPECIMEN AND PHOTO.

Grossular ◆ $Ca_3Al_2(SiO_4)_3$

Grossular, one of the garnet group of minerals, crystallizes in the cubic system, has a hardness of 6.5 to 7, specific gravity of 3.4 to 3.6, a white streak and vitreous to resinous lustre. This calcium- and aluminium-bearing variety of garnet can form beautifully coloured crystals that are often orange, red, yellow or green dodecahedrons. Green and red varieties are the most

Figure 466 A polished slice of green 'Transvaal jade' with black chromite seams, 14.2 cm. Rustenburg district, South Africa. BRUCE CAIRNCROSS SPECIMEN AND PHOTO.

desired, but grey grossular has little value. In southern Africa, grossular occurs in metamorphosed calc-silicates and marbles.

'Transvaal jade' is a misnomer for a variety of grossular found in the Rustenburg district, **South Africa** (Hall, 1924; Page, 1970). The name has undergone various iterations over time, including hydrogrossular garnet, that is, grossular with part of the orthosilicate ions replaced by hydroxide ions, which is essentially the same as hibschite $Ca_3Al_2(SiO_4)_{3-x}(OH)_{4x}$, a variety of grossular. Hibschite is no longer an IMA-approved species, but is considered an informal intermediate composition within the grossular-katoite series ($Ca_3Al_2(SiO_4)_3$ to $Ca_3Al_2[\square(OH)_4]_3$). The 'Transvaal jade' moniker came about due to the green varieties of the hibschite resembling that of oriental jade. Dark green and pink-red varieties are the most desirable, while the white-grey material has little commercial value. The deep pink to red coloration is caused by the presence of Mn^{2+} ions in the hydrogrossular structure. The bright green variety results from the presence of chromium Cr^{3+} ion.

Figure 467 The most desirable shade of 'Transvaal jade' is red-pink, such as in this 6.2-cm specimen. Brits district, South Africa. BRUCE CAIRNCROSS SPECIMEN AND PHOTO.

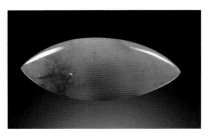

Figure 469 A translucent marquis cut, 8.15-carat 'Transvaal jade' showing colour zonation. BRUCE CAIRNCROSS SPECIMEN AND PHOTO.

Figure 468 Multicoloured layers of 'Transvaal jade' with scattered grains of black chromite, 13 cm. Rustenburg district, South Africa. BRUCE CAIRNCROSS SPECIMEN AND PHOTO.

Figure 470 A rough colour-zoned translucent specimen of 'Transvaal jade', 4.8 cm. Brits district, South Africa. BRUCE CAIRNCROSS SPECIMEN AND PHOTO.

The following coloured varieties of 'Transvaal jade'/grossular are found at the Rustenburg deposits on Buffelsfontein 465 JQ and Turffontein 462 JQ, around 70 km west of Pretoria: deep sea-green (translucent to spotted), vivid pure dark green (opaque), dull sea-green (translucent to opaque), bright apple-green, dull grey-green to bright yellowish-green (opaque), light to dark pink, mother-of-pearl to pale bluish and cream. There is a thick zone of hydrogrossular at Buffelsfontein, consisting of pink and green layers criss-crossed by lenses and seams of chromite. These deposits were exploited during the 1920s when 700 kg of hydrogrossular yielded £362 (Haughton, 1936). This gemstone material has subsequently been extracted, and its physical features make it ideal for carvings, cabochons, *objets d'art* and other jewellery.

A somewhat similar deposit in the Steelpoort district of Limpopo has been known for many years but has received

Figure 471 A Rough samples of various coloured 'Transvaal jade' backlit to show transparency prior to being shaped and polished. The largest piece is 4 cm. **B** Several polished cabochons of 'Transvaal jade' on a slab of material containing a chromite seam, illustrating the colour spectrum of this material. Brits district, South Africa. ERIC FARQUHARSON SPECIMENS B, BRUCE CAIRNCROSS SPECIMENS A AND PHOTOS.

scant attention. This site was discovered in 1907 on the farm Hendriksdal 216 JT, 17 km west of Fort Burger. Here, the 'jade' occurs in thin veins, several centimetres thick. It is pale green and is also closely associated with chromite layers. Similar material has been found as water-worn pebbles in alluvial deposits south of the Limpopo River at the old Seta diamond mine, 50 km west of Musina. These deposits contain pale green to grey grossular together with red alluvial corundum crystals.

Gemmy orange to red grossular has been recorded from Marble Delta, KwaZulu-Natal. Grossular has also been reported from the Soutpansberg, Mokopane and Letaba districts of Limpopo, and also from a few localities in the Northern Cape.

In **Namibia**, grossular comes from the Ais dome skarn deposit in the Omaruru district and from a few other skarn deposits in the Karibib district.

Grossular is relatively widespread in **Zimbabwe** as cinnamon-brown crystals in various pegmatites, calc-silicate rocks, granites and their respective metamorphic equivalents.

Figure 472 Grossular garnet overgrown on a black andradite core (not visible), 4 cm. North-west of Omaruru, Namibia. BRUCE CAIRNCROSS SPECIMEN AND PHOTO.

Grunerite ◆ $\square\{Fe^{2+}_2\}\{Fe^{2+}_5\}(Si_8O_{22})(OH)_2$

Grunerite crystallizes in the monoclinic system, has a hardness of 3.5 to 4, specific gravity of 3.77, and vitreous lustre, although asbestiform habits are silky.

Asbestiform grunerite is a fibrous variety of grunerite found in **South Africa** in banded iron formations at Penge in Limpopo and Rhenosterfontein, 20 km south-east of Zeerust. It is also known from the Lydenburg/Polokwane areas. Fibres of this mineral can be over 20 cm long, occurring in layered, continuous seams that may extend over long distances. The informal name 'amosite' (proposed by the geologist AL Hall in 1918) refers to Amosa Ltd, (**A**sbestos **M**ines **of S**outh **A**frica), the company that mined this asbestiform mineral (Beukes and Dreyer, 1986).

Figure 473 Grunerite 'amosite' asbestos, with extraordinarily long (37 cm) fibres. WARREN TAYLOR RAINBOW OF AFRICA COLLECTION, MARK MAUTHNER PHOTO.

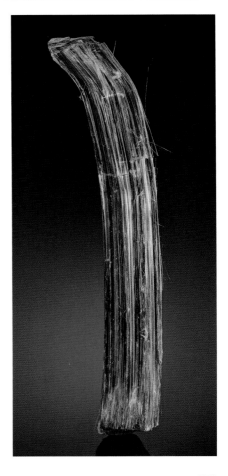

Figure 474 Grunerite 'amosite' asbestos, 15.8 cm. Penge mine, South Africa. BRUCE CAIRNCROSS SPECIMEN AND PHOTO.

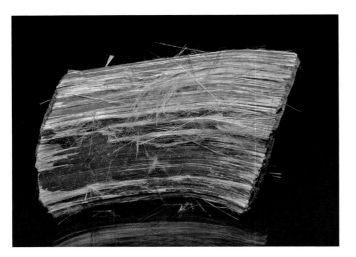

Gypsum ◆ CaSO$_4$·2H$_2$O

Gypsum crystallizes in the monoclinic system, has a hardness of 2, specific gravity of 2.32, a white streak and vitreous to pearly lustre. Gypsum is a common, soft, colourless to white mineral that cleaves very easily. It is easy to peel pieces off pure gypsum crystals. A variety of gypsum is called 'desert rose' because the mineral crystallizes in blades containing sand and clay, which cluster together, resembling rose petals. Another variety, selenite, has semi-transparent to transparent well-formed crystals. Gypsum has industrial uses in the manufacture of builders' plaster, gypsum boards and other construction materials. Gypsum is found in many arid regions in southern Africa, where evaporation in saline pans causes the mineral to crystallize. Some ore deposits, such as Tsumeb and the Kalahari manganese field, contain secondary gypsum.

The main **South African** gypsum deposits are in low-lying areas north-west of Kimberley and in saline pans west of Vanrhynsdorp, where the mineral is found as 'desert roses'. Beige 'desert roses', tens of centimetres in diameter, have been found at Windsorton. The Bushmanland region also has gypsum deposits. Attractive tan-coloured 'desert roses' come from the Tugela Valley beyond Kranskop in the Greytown district, KwaZulu-Natal. Some saline pans in the Free State contain gypsum.

Gypsum was sometimes found as large, sword-like, well-formed crystals over 1 m long, as at the Okiep copper mines (Cairncross, 2004b). Transparent, colourless massive gypsum is known from Hotazel, Langdon-Annex and the N'Chwaning

I and II and Wessels mines in the Kalahari manganese field (Cairncross *et al.*, 1997; Cairncross and Beukes, 2013). Here, euhedral crystals are rare; partially dissolved crystals and crystal aggregates are more common. Large crystals rarely survive blasting at the mines. Some large, well-developed tabular crystals measuring 29 cm come from N'Chwaning II.

(A)

Figure 475 Two gypsum specimens from the Kalahari manganese field, South Africa. **A** Terminated gypsum crystal, 23 cm, Wessels mine, South Africa; **B** gypsum with yellow ettringite, 7.3 cm, N'Chwaning II mine. DESMOND SACCO SPECIMEN (A); BRUCE CAIRNCROSS SPECIMEN (B) AND PHOTOS.

(B)

Figure 476 A spray of gypsum crystals from the Vergenoeg fluorite mine, South Africa. Field of view 3.1 mm. WOLF WINDISCH SPECIMEN AND PHOTO.

Figure 477 Large gypsum, 29.2 cm. Nababeep mine, South Africa. BRUCE CAIRNCROSS SPECIMEN AND PHOTO.

Figure 478 A 9-cm 'desert rose' from Gannavlakte, Free State, South Africa. BRUCE CAIRNCROSS SPECIMEN AND PHOTO.

The west coast of **Namibia**, from Lüderitz to north of Walvis Bay, is famous for its green or tan 'desert roses'. They are found close to the beach, where sand from unconsolidated sand dunes is trapped in the crystals during evaporation. There are several localities along the coastline further to the north, between Cape Cross and Swakopmund, where gypsum is found in layers just below the surface (Schneider and Genis, 1992b). Large, well-formed gypsum crystals have been found at the Tsumeb mine and at Rosh Pinah.

Figure 479
Rosette clusters of gypsum, 10.8 cm. Rosh Pinah, Namibia. ALLAN FRASER SPECIMEN, BRUCE CAIRNCROSS PHOTO.

Figure 481
Large plates of gypsum forming a robust 'desert rose', 22 cm. Collected in 1978 in the Lüderitz district, Namibia. BRUCE CAIRNCROSS SPECIMEN AND PHOTO.

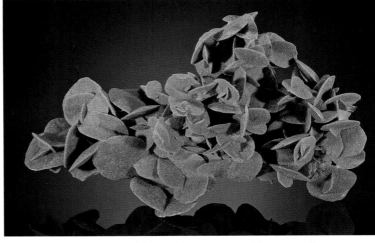

Figure 480 A large but delicate 39-cm gypsum 'desert rose'. Lüderitz district, Namibia. WARREN TAYLOR RAINBOW OF AFRICA COLLECTION, MARK MAUTHNER PHOTO.

Figure 482 Brown gypsum 'desert rose', 22.2 cm. Lüderitz district, Namibia. BRUCE CAIRNCROSS SPECIMEN AND PHOTO.

Economic deposits of gypsum are relatively scarce in **Zimbabwe**. The only deposit of any consequence is on Nottingham Farm in the Beitbridge district, where lumps, fragments, masses and selenite crystals of gypsum are found in Permian-age mudstones. The mineral is relatively widespread in other Karoo-age rocks in the Gokwe and Sebungwe districts. Uneconomic massive and crystalline gypsum is found as a gangue mineral at the Mangula copper mine and the Antelope asbestos mine.

The so-called Foley gypsum deposits are found 50 km west of Tonota and in the surrounding area in **Botswana**. Crystals of gypsum form in the soil as well as in mudstones. Layers of gypsum occur 40 km north of Ngware and at Sua Pan.

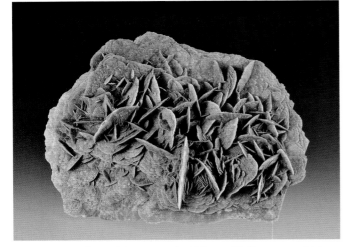

Figure 483 Green 'desert rose', 18.5 cm. Unlike most, which have sand incorporated with the gypsum crystals, this specimen has green, algal-stained clay. Collected from a clay-rich saline deposit near Lüderitz, Namibia. BRUCE CAIRNCROSS SPECIMEN AND PHOTO.

Halite ◆ NaCl

Halite crystallizes in the cubic system, has a hardness of 2, specific gravity of 2.17, white streak, and vitreous lustre. Halite, commonly known as rock salt, is colourless to white and invariably forms perfect cubic crystals. When algae is present in the water, the halite assumes an attractive pink colour. Halite crystallizes in arid areas, particularly on the west coast of Namibia. Conditions necessary for the precipitation of halite are: a shallow depression into which the salt solution can drain, an impermeable clay floor that prevents seepage and, where possible, an underground salt-bearing formation. Halite is most commonly used to season food, but also has chemical uses in the manufacture of chlorine, caustic soda, paper, glass, soap and detergents, and in food preservation.

Figure 484 Two halite specimens formed on the west coast of Namibia. The pale pink colour is caused by algae. **A** 16.5 cm; **B** 11.5 cm. BRUCE CAIRNCROSS SPECIMENS AND PHOTOS.

Figure 485 Soutpan, located north of Upington, South Africa. This evaporitic pan is seen in winter when halite and other salts have precipitated on the pan surface. Following summer rainfall, the salts dissolve in the pan water, to reappear during the next dry period. BRUCE CAIRNCROSS PHOTO, 1996.

Figure 486 A hypersaline pond on the west coast of Namibia, between Swakopmund and Cape Cross. Halite crystallizes on the margins via evaporation and is then harvested. Heaps of halite are piled along the fringes of the pond. BRUCE CAIRNCROSS PHOTO, 2017.

In **South Africa**, halite is found in evaporitic saltpans in low-rainfall regions from Calvinia through to Kimberley and northwards to Vryburg. Some farmers harvest the halite for cattle and game licks. Coastal saltpans are found north and east of Cape Town, up to Mossel Bay, and in areas around Gqeberha (Port Elizabeth), as at Koega. Several large saltpans are located about 100 km north/north-west of Upington.

Like gypsum, halite tends to be concentrated along the coastline in **Namibia**, between Walvis Bay and Cape Cross (Schneider and Genis, 1992b). The salt is periodically harvested from evaporation pans that are artificially maintained by commercial salt mining operations close to Swakopmund. Saltpans like the Arandis pan, which are similar to those in the arid regions of South Africa, occur in Ovamboland north of the Etosha Pan. In Namibia, salt is used to surface some roads along the coast north of Swakopmund.

Halite is relatively scarce in **Zimbabwe**. It can be found as encrustations at mineral springs in the Wankie and Sebungwe districts and in the Sabie valley.

Sua Pan, east of Maun, is the largest of the evaporation pans found in **Botswana** and has been mined for several evaporitic minerals. Halite forms in crusts and layers here and in smaller pans.

Hausmannite ◆ Mn²⁺MnO₄

Hausmannite crystallizes in the tetragonal system, has a hardness of 5.5, specific gravity of 4.84, a brown streak, and metallic lustre. Hausmannite is one of the ore minerals exploited for manganese and is not commonly found in well-formed crystals. It is black to silver-black and has a characteristic brown streak. Crystals form in a variety of different habits, the pagoda-like habit of many specimens from the manganese mines in South Africa being among the most attractive and the most sought after. Manganese metal is used extensively in the manufacture of ferromanganese and other alloys and in dry-cell batteries, as well as in certain chemicals. Hausmannite occurs in economic deposits, usually sedimentary-hosted.

Apart from minerals of the braunite group, hausmannite is the most common manganese oxide in the Kalahari manganese field, **South Africa**. Aesthetic specimens of well-formed, jet-black hausmannite crystals are found at N'Chwaning I, II, III and Wessels mines (Von Bezing *et al.*, 1991; Gutzmer and Cairncross, 2002; Cairncross and Beukes, 2013; Cairncross, 2018a). Associated minerals include andradite, baryte, calcite, celestine, datolite, gaudefroyite, hematite, kutnahorite, and sturmanite. In January 1999, a discovery of some fine hausmannite was made at N'Chwaning II mine. The crystals are typically pseudo-octahedral pagoda shaped, extremely lustrous and up to three centimetres on edge. While hausmannite crystals periodically appear in the Kalahari manganese field, this find has been associated with finely crystalline andradite garnet, small platelets of white baryte and, importantly, with very good, gemmy, rose-pink datolite crystals. The last-mentioned combination was rare and the contrast of the highly lustrous, black hausmannite with the red garnet and pink datolite is vivid.

In late 2002, a small find of hausmannite occurred at N'Chwaning II mine. The habit is the now-familiar pagoda shape typical of the material from this region. But the size of the crystals was rather remarkable – some were up to 8 cm on edge. Most are highly lustrous and black, while others have a tenacious orange coating of iron oxides. The Kalahari manganese field hausmannite specimens are among the finest known in the world.

Hausmannite is found at the Kombat mine, **Namibia**, which has a considerable array of esoteric manganese-bearing minerals. The Otjosondu manganese field, located 150 km north-east of Okahandja, contains the largest concentration of hausmannite in the country.

Figure 487 Typical highly lustrous pagoda-like crystals of hausmannite: **A** is associated with pink datolite, 5.2 cm; **B** 2.8 cm. N'Chwaning II mine, South Africa
BRUCE CAIRNCROSS SPECIMENS AND PHOTOS.

Figure 488 The N'Chwaning mine headgear at sunset, Kalahari manganese field, South Africa. BRUCE CAIRNCROSS PHOTO, 2011.

Hematite ◆ α-Fe$_2$O$_3$

Hematite crystallizes in the hexagonal system, has a hardness of 5 to 6, specific gravity of 5.26, a red to brown-red streak, and metallic to sub-metallic lustre. Hematite is a relatively common mineral that can occur as very finely disseminated grains, forming banded iron formation rocks composed purely of hematite. It sometimes forms beautiful bright silver crystals, and may also occur in a platy or micaceous habit, which is referred to as specular hematite, specularite or 'blinkklip' (glittering stone), a black iron ore with a brilliant glitter.

Iron from hematite is used to make steels and alloys such as ferroalloys, ferromanganese and ferrosilicon. Historically, hematite in the form of specularite and red ochre was prized as a cosmetic by indigenous southern Africans. Specularite was rubbed into the hair to make it sparkle, and red ochre was commonly smeared on the body or sprinkled on corpses in the hope that they would revive. Where ochre – a powdery, greasy mixture of hydrated iron oxides – was not available, massive hematite was pounded into a powder and mixed with fat. Specularite was much harder to obtain and, therefore, more valuable.

In **South Africa**, the earliest mining activities exploiting hematite at Doornfontein in the Northern Cape have been dated at 2000 BC, or even earlier, and Khoisan hunters and gatherers, who periodically mined the deposit, are thought to have extracted specularite before AD 80. Hematite at Doornfontein was not the only source of pigment in the region. Ancient workings were found at Sishen, and a famous old Gatkoppies mine is located near the Groenwater River close to Postmasburg. The old mine was visited by Europeans as early as 1801 and is described by a number of early travellers. In 1804, the German physician Hinrich Lichtenstein visited Gatkoppies to gain information about the Koranna and Tswana. So, too, did the famous explorer and naturalist William Burchell, who journeyed through the region in 1812 and reported that the Tswana knew specularite as

'*sibilo*'. Mining methods at Gatkoppies were haphazard, and Burchell (1822) recorded deaths from cave-ins. The site was an important economic centre, yielding specularite that was distributed all over southern Africa.

Banded iron formation is one of the country rocks of the Kalahari manganese deposits, and secondary hematite crystals are frequently found in vugs and fractures at N'Chwaning I and II and Wessels mines. One of the most famous discoveries was at Wessels mine in May 1988, when hundreds of spectacular hematites were encountered (Cairncross and Beukes, 2013). These were associated with small, red andradite garnets, calcite and baryte. Tabular crystals from this pocket measured up to 30 cm in diameter, and 5–10-centimetre crystals were common. Some crystals had mirror-bright, silver-grey shiny faces, forming stunning specimens.

Hematite specimens from the Kalahari manganese field date back to the early 1960s, coinciding with the first rhodochrosite discoveries, and even prior to that with specimens originating from the old Black Rock mine. During October 1996, several dozen 'floaters' of tabular hematite crystals were collected at Wessels mine. These are highly lustrous and some display a threefold propeller-like twinning pattern on the main crystal faces. Others are composed of an overgrowth of shiny, euhedral hematite crystallized over an older generation of corroded crystals. Some botryoidal hematite associated with bright yellow scalenohedral calcite a few millimetres in size were collected at N'Chwaning II mine in 2000.

During September 2001, several hundred hematite specimens were collected at N'Chwaning II, yielding some outstanding specimens. The largest of these are tennis-ball sized and multifaceted. Some of the crystal faces are extremely lustrous, while other faces have a matte texture, providing a pleasing contrast. Other crystals have very lustrous, silver crystal faces all round. Associated minerals are white calcite, white baryte and blue celestine.

Figure 489 Two iron ore mines in the Northern Cape, South Africa: **A** Sishen; **B** Beeshoek. The red colour is caused by hematite, the main ore mineral. BRUCE CAIRNCROSS PHOTOS, 1996.

In 2005, very unusual pseudohexagonal hematite was discovered at Wessels mine in the Kalahari manganese field. These are columnar crystals several centimetres in length, packed tightly together in bundles or forming circular masses. Associated minerals are andradite garnet and strontianite. In February 2010, a pocket of hematite-calcite-andradite at N'Chwaning II was discovered. The hematite crystals are not large, mostly less than a centimetre in diameter, but they have brilliant lustre and complex forms, resembling modified dodecahedra. The hematite is perched on drusy, bright red lustrous andradite together with calcite that is invariably twinned and contains an unknown included white acicular mineral.

Figure 490 Two hematite crystals associated with black gaudefroyite, 4.1 cm. N'Chwaning II mine, South Africa. BRUCE CAIRNCROSS SPECIMEN AND PHOTO.

Figure 491 A complex hematite crystal, 3.2 cm. N'Chwaning II mine, South Africa. BRUCE CAIRNCROSS SPECIMEN AND PHOTO.

Figure 492 Unusual elongate hexagonal hematite crystals with red andradite, calcite and strontianite, 12 cm. Wessels mine, South Africa. DESMOND SACCO SPECIMEN, BRUCE CAIRNCROSS PHOTO.

Figure 493 Two hematite specimens from the famous discovery at the Wessels mine (see text), South Africa: **A** a 9.5-cm hematite; **B** a 'floater' specimen, complete all round, of bright silver hematite, red andradite and white baryte, 10.5 cm. BRUCE CAIRNCROSS SPECIMENS AND PHOTOS.

Ⓐ

Ⓑ

Figure 494 Barrel-shaped hematite partially coated by red andradite, with some white calcite, 8 cm. N'Chwaning II mine, South Africa. BRUCE CAIRNCROSS SPECIMEN AND PHOTO.

Figure 495 Several hematite rosettes attached to four quartz crystals, 6.8 cm. Messina mine, South Africa. BRUCE CAIRNCROSS SPECIMEN AND PHOTO.

Figure 496 Solid, columnar-like hematite, 6.7 cm. Postmasburg region, South Africa. BRUCE CAIRNCROSS SPECIMEN AND PHOTO.

Figure 497 Another form of hematite, with bright red inclusions inside quartz, 6.5 cm. Orange River region, South Africa. BRUCE CAIRNCROSS SPECIMEN AND PHOTO.

At the Musina copper mines, hematite occurred as specularite and small 'eisen rosen' (iron roses). These occur as individual rosette-like balls included in quartz crystals or as free-standing groups. Hematite in Messina ore lodes imparts a red coloration to many associated minerals, particularly as inclusions in quartz crystals. Specularite is commonly associated with quartz as masses of fine platelets either coating the crystal faces or as inclusions. The Harper and Messina No. 5 Shaft mines contained specularite. Iron oxide was scarce at the Campbell mine, although tabular specular hematite crystals 3 cm in diameter have been found in some lodes. Similar crystals, 3 cm in diameter and 8 mm thick, came from a vug in the 21–760 stope in the Messina mine,

where aggregates of flaky specularite locally filled vugs up to 2 m in length. Attractive reniform masses of hematite are found at the Vergenoeg fluorspar mine north-east of Pretoria.

Hematite-included quartz and hematite-coated quartz come from deposits along the Orange River in the Northern Cape. Beautiful vibrant blood-red specimens, some sceptered and others doubly terminated, were collected at Onseepkans in 2001 and 2002.

The Aties iron mine in the Western Cape (Cole *et al.*, 2014) has produced interesting stalactitic and geode-filled hematite specimens. The deposit is located 12 km east of Vredendal and was exploited from 1975 to 1996 for use as flux in the cement industry.

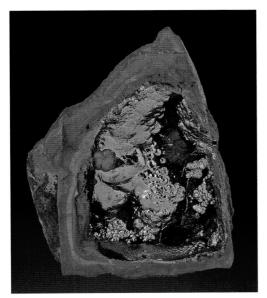

Figure 498 A cavity lined with black, lustrous hematite, 9.4 cm. Aties iron mine, Western Cape, South Africa. ROB SMITH AFRICAN GEMS AND MINERALS SPECIMEN, BRUCE CAIRNCROSS PHOTO.

Figure 499 A rare hematite pseudomorph after mimetite, 4 cm. Tsumeb mine, Namibia. BRUCE CAIRNCROSS SPECIMEN AND PHOTO.

Figure 500 Highly lustrous hematite stalactite, 12.3 cm. Aties iron mine, Western Cape, South Africa. ROB SMITH AFRICAN GEMS AND MINERALS SPECIMEN, BRUCE CAIRNCROSS PHOTO.

Hematite-rich iron ore has been mined from syenites and carbonatites in the Kalkveld Complex on the farm Eisenberg 78 in the Otjiwarongo district, **Namibia**, and both hematite and goethite are abundant in the Otjosondu manganese field. Basalt cavities in the Goboboseb Mountains west of the Brandberg have produced clear quartz crystals with inclusions of tiny bright red hematite flakes (see figure 710). In the Erongo Mountains, miarolitic cavities, tiny bright silver crystals, are sometimes found with quartz and feldspar. Large beds of banded iron formation that outcrop in south and south-east Namibia consist primarily of hematite and siderite. In southern Namibia, hematite is found included as microscopic red crystals inside quartz from pegmatites north of the Orange River. These crystals colour the quartz a vibrant red and orange, producing visually stunning specimens.

Iron ore has been mined for centuries in **Zimbabwe** and virtually every substantial outcrop that had potential for iron production was exploited. The Mashona produced a variety of iron tools, weapons and implements from smelting in clay furnaces. Hematite occurs in the major iron-ore deposits distributed in the Archaean greenstone belts. It has been mined in a number of districts, including the Kwekwe, Gweru, Chiredzi and Masvingo districts. Attractive specimens of reniform hematite were found in the Yank mine in the Kadoma district where banded iron formations outcrop. Specularite often occurs as minute inclusions in quartz crystals.

Most of **Botswana's** hematite is found in banded iron formations. These outcrop either in the Tati schist belt region in the north-east, or in the somewhat younger sedimentary rocks in the southern extremity of the country. The latter occur south of Kanye and south-east of Lobatse, where there are extensive deposits of banded iron formation. The former are represented by the deposits at Matsiloje.

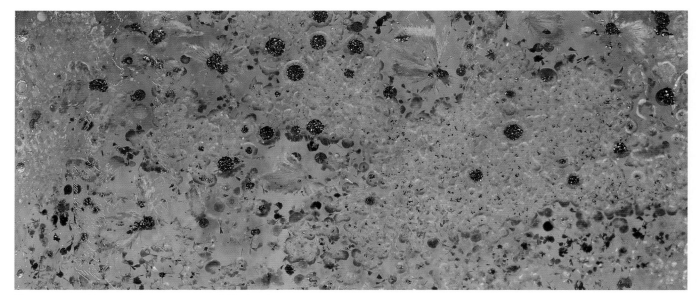

Figure 501 Close-up of a quartz crystal filled with silver hematite and orange goethite. Devils Reef, Eswatini. Field of view 1.6 cm. BRUCE CAIRNCROSS SPECIMEN AND PHOTO.

The largest deposit of hematite in **Eswatini** was mined at Ngwenya, 25 km north-west of Mbabane. Hematite was the main iron ore mineral, although magnetite, siderite and goethite were also present. Very attractive pseudomorphs of hematite after pyrite come from the Devil's Reef gold mine area. Cubes up to several centimetres on edge have been replaced by minute, steel-grey, sparkling hematite crystals, and specularite was abundant at the mine. Iron Hill, 2 km south of the Havelock mine in the Pigg's Peak district, contains significant iron ore. Hematite also exists in deposits at Gege and Maloma.

Figure 502 Silver hematite mixed with brown goethite replacing a cubic pyrite crystal, 3.2 cm. Devil's Reef, Eswatini. BRUCE CAIRNCROSS SPECIMEN AND PHOTO.

Heterosite ◆ $(Fe^{3+},Mn^{3+})PO_4$

Heterosite crystallizes in the orthorhombic system, has a hardness of 4 to 4.5, specific gravity of 3.4, a reddish-purple streak, and dull lustre. Although a relatively rare phosphate, heterosite is included here because of its very distinctive colour, one of the few purple mineral species. It does not form large distinctive crystals, occurring instead as solid lumps. Heterosite forms as a secondary iron-manganese phosphate in some granitic pegmatites, frequently replacing triphyllite.

Heterosite is found in **South Africa** from the Noumas and Straussheim pegmatites in the Northern Cape.

The Sandamap pegmatites in **Namibia** are famous for beautiful, richly coloured heterosite. Much of what has been called purpurite is in fact heterosite.

Heterosite is rare in **Zimbabwe**, but is recorded from the Sabi Star pegmatite in the Buhera district and the Wankie area.

Figure 503 Bright purple heterosite, 9.8 cm. This material was incorrectly identified as purpurite. Sandamap pegmatite, Namibia. BRUCE CAIRNCROSS SPECIMEN AND PHOTO.

Hydroxyapophyllite-(K) ◆ $KCa_4(Si_8O_{20})(OH,F)\cdot 8H_2O$

Hydroxyapophyllite-(K) crystallizes in the tetragonal system, has a hardness of 4.5 to 5, specific gravity of 2.37, a white streak, and pearly to vitreous lustre. Hydroxyapophyllite-(K) forms tabular crystals that have a characteristically pearly lustre. The mineral is typically colourless to white. Hydroxy refers to the OH (hydroxyl) content (see also fluorapophyllite). Hydroxyapophyllite-(K) occurs as a secondary mineral in some ore bodies.

A common mineral from the Kalahari manganese field, **South Africa**, where crystals up to 15 cm on edge are known to occur (Cairncross and Beukes, 2013). Hydroxyapophyllite-(K) is relatively common in N'Chwaning and Wessels mines. In mid-2002, several hundred specimens of white hydroxyapophyllite-(K) were collected from N'Chwaning II mine. Most of the specimens were in the miniature size range, but some large matrix cabinet specimens were collected, as well as a few large plates, up to 30 cm on edge. The individual crystals were remarkably uniform in size, averaging one centimetre. The only associated mineral was calcite. In the past, hydroxyapophyllite-(K) has been associated with rare occurrences of crystalline sugilite.

The Belfast granite quarry in Mpumalanga has produced extremely flat, colourless, transparent tabular crystals from a fault cutting through Bushveld Complex gabbro that is mined for dimension stone.

Figure 504 Two specimens from the same discovery. at N'Chwaning II mine, South Africa: **A** a large specimen of intergrown hydroxyapophyllite-(K) crystals, 22 cm; **B** a more free-standing crystals on matrix, 8 cm. BRUCE CAIRNCROSS SPECIMENS AND PHOTOS.

Figure 505 Highly lustrous hydroxyapophyllite-(K) crystals: **A** 1.8 cm; **B** 5.4 cm. N'Chwaning II mine, South Africa. BRUCE CAIRNCROSS SPECIMENS AND PHOTOS.

Figure 506 Tabular white hydroxyapophyllite-(K) crystals with pink schizolite and black aegirine, 5.7 cm. Wessels mine, South Africa. BRUCE CAIRNCROSS SPECIMEN AND PHOTO.

Figure 507 Tabular hydroxyapophyllite-(K) crystals coated by a brown layer, 5.2 cm. N'Chwaning II mine, South Africa. BRUCE CAIRNCROSS SPECIMEN AND PHOTO.

Figure 508 Two specimens from the Bushveld Complex: **A** highly tabular crystals with red, hematite-included quartz; many tiny chalcopyrite crystals are scattered on the crystals; field of view 9.4 cm; **B** hydroxyapophyllite-(K) with well-formed, brassy chalcopyrite crystals and quartz, some with red hematite, 6.7 cm. BRUCE CAIRNCROSS SPECIMENS AND PHOTOS.

Ilmenite ◆ Fe²⁺TiO₃

Ilmenite crystallizes in the hexagonal system, has a hardness of 5 to 6, specific gravity of 4.72, black streak, and metallic lustre. Ilmenite is a tough, hard ore mineral that is exploited for its titanium content. Titanium, in the form of titanium dioxide, is used as a paint pigment; no other compound has such a brilliant white colour. Titanium is also used in the manufacture of special titanium metals and in alloys in the aerospace industry, ceramics, oil refining and sunscreen lotions, and as a whitening agent in foodstuffs. Because titanium is chemically inert, it is widely used in the manufacture of prosthetics and artificial teeth implants. Since ilmenite does not easily disintegrate when transported as sediment, it survives the rigour of river transport and becomes concentrated in beach sands in sufficient quantity to be mined. It is an accessory mineral found in mafic and ultramafic rocks.

Ilmenite is commonly found, as at Richards Bay, **South Africa**, on coastal beaches and dune sands (Hugo and Cornell, 1991). It is exploited by Namakwa Sands on the Namaqualand coast and is common on the KwaZulu-Natal south coast at Umgababa and along the Eastern Cape coastline. Ilmenite is also common in metamorphic rocks, such as the marbles at Marble Delta in KwaZulu-Natal, and igneous rocks such as the Glenover, Goudini and Phalaborwa carbonatite pipes. In the Northern Cape, ilmenite is found in some quantity in a pegmatite between Kap and Riembreek. In the Bothaville district, ilmenite-rich sandstones were once prospected as a possible source of titanium, but the deposit was never developed.

Ilmenite is an accessory mineral in mafic and ultramafic rocks in **Namibia**, such as in the Kunene Complex in the Kaokoveld. Ilmenite is one of the main detrital minerals found in so-called heavy-mineral sands or placer deposits on the coastline in the Swakopmund district. Namib Desert aeolian dunes also contain substantial amounts of ilmenite. Bright orange, well-formed crystals up to 8 cm were found in

Figure 509 A section dug in the beach at Richards Bay, South Africa, exposing the sedimentary layering, with concentrations of dark heavy minerals visible. These include ilmenite, rutile, zircon and others, such as garnet and tourmaline. Camera lens cap for scale. BRUCE CAIRNCROSS PHOTO, 1988.

the Erongo Mountains (Cairncross and Bahmann, 2006a). These are weakly radioactive and have an orange coating of a mixture of rutile and iron oxides and iron hydroxides.

Ilmenite is found as detrital crystals in Kalahari sands and ancient Karoo-age sandstones in **Zimbabwe**, and in a wide variety of igneous and metamorphic rocks, pegmatites and quartz veins. Euhedral crystals come from the Lutope tin-bearing pegmatite, while a magnesium-rich ilmenite is found in the Clare, Wessels and Colossus kimberlites.

Ilmenite is often encountered in some alluvial gravels in **Eswatini**. The tin-bearing pegmatites south-west of Mbabane contain ilmenite.

Ilmenite is found along with other heavy minerals, such as rutile and garnet, in the coastal sands of **Mozambique** (Lächelt, 2004).

Figure 510 The heavy-minerals sand operation at Richard's Bay Minerals in KwaZulu-Natal, South Africa. Ilmenite, rutile and zircon are concentrated on the floating dredger. BRUCE CAIRNCROSS PHOTO, 1988.

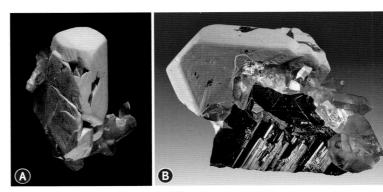

Figure 511 Two yellow-coated ilmenite crystals with quartz and black schorl: **A** 5.1 cm; **B** 5.6 cm. Erongo Mountains, Namibia. ULI BAHMANN SPECIMENS, BRUCE CAIRNCROSS PHOTOS.

Inesite ◆ $Ca_2MnSi_{10}O_{28}(OH)_2 \cdot 5H_2O$

Inesite crystallizes in the triclinic system, has a hardness of 5.5, specific gravity of 3.02, white streak, and vitreous to silky lustre. Inesite commonly forms pink to red, acicular, hard, fibrous crystals that occur in aggregates. Rare, transparent, platy crystals may also occur. Inesite occurs as a secondary mineral associated with some manganese deposits.

Inesite is a mineral that is not particularly common in the Kalahari manganese field, **South Africa**, but Wessels mine has produced some of the finest examples of this mineral found anywhere in the world (Cairncross and Beukes, 2013). During the mid-1970s and early 1980s, small, bladed crystals and groups of these crystals forming grape-like bunches of inesite were found. In May 1992, some superb specimens were discovered. These deep-red, transparent, thin, tabular crystals are arranged in radiating fan-like groups and are associated with several other minerals, including pink datolite, natrolite, tobermorite, orlymanite, rhodonite, xonotlite, and white thomsonite, which fluoresces yellow under short-wave ultraviolet light.

(A)

(B)

Figure 513
Acicular inesite crystals: **A** 8.6 cm; **B** 20 cm. N'Chwaning II mine, South Africa. BRUCE CAIRNCROSS SPECIMENS AND PHOTOS.

(A)

Figure 512 A selection of inesite specimens from the Wessels mine in the Kalahari manganese field showing the diversity of habits and associations: **A** pink inesite with white calcite and khaki orlymanite, 3.5 cm; **B** three spheres of inesite with brown orlymanite, 3.2 cm; **C** bladed transparent inesite with white natrolite, 3.2 cm. BRUCE CAIRNCROSS SPECIMENS AND PHOTOS.

(B)   (C)

From April to June 1996, several thousand specimens of small, acicular crystals of inesite were discovered at N'Chwaning II mine. These specimens have a very simple appearance, consisting of grass-like mats of inesite covering highly brecciated manganese ore. Most consisted of flat plates covered by a mat-like coating of pale brown to cinnamon-brown crystals. The largest preserved specimen from this find measures 49 × 35 cm. Inesite of a similar habit was found as far back as the early mid-1970s, where the acicular crystals were associated with calcite.

Dark red ferroan (iron-rich) inesite associated with calcite and xonotlite occurs with the rare orange prehnite specimens (Cairncross *et al.*, 2000). During mid-2000, massive aggregates of interlocking plates of red inesite up to 10 cm were collected at N'Chwaning II mine. This material superficially resembles the pink manganodiaspore from the Postmasburg region.

Jeremejevite ◆ $Al_6B_5O_{15}(F,OH)_3$

Jeremejevite crystallizes in the hexagonal system, has a hardness of 7.5, specific gravity of 3.29, white streak, and vitreous lustre. Jeremejevite is a rare gemstone. It can form very attractive, prismatic, blue crystals. Those discovered in Namibia are considered the finest in the world. Jeremejevite is found in pegmatites and alluvium weathered from these pegmatites.

The only southern African country currently known to possess jeremejevite is **Namibia**. It was discovered in Namibia in 1973, in granite at Mile 72 north of Swakopmund (Wilson *et al.*, 2002). Most of the crystals were pale yellow, but some were a beautiful cornflower-blue and of gem quality. These are considered the finest known. A few of this rare type have found their way into the collections of museums such as the Council for Geoscience collection, Ditsong Museum in Pretoria and the Smithsonian Institution in the USA.

Some fine, deep blue jeremejevite crystals – the most spectacular up to 10 cm long and 1 cm thick – were found in 2001 at the Erongo Mountains in Namibia, in miarolitic cavities that also yield excellent schorl tourmaline and aquamarine (Cairncross and Bahmann, 2006a; Von Bezing *et al.*, 2008; 2014; 2016). Gemstones have been faceted from these crystals. Most of the crystals, however, were colourless to pale yellow and less than 5 mm on edge, and many hundreds were collected by artisanal diggers. There have been more discoveries of jeremejevite from Erongo during the early 2000s. In 2007, some fine, large blue crystals up to 8 cm were found, and in 2010 more were discovered, some of these occurring in hollow, leached-out cavities, with the crystals themselves highly leached. The largest jeremejevite crystal found to date measures 16.5 cm (Moore, 2013).

Figure 514 Ninety-seven jeremejevite crystals, largest 1.4 cm. Erongo Mountains, Namibia. BRUCE CAIRNCROSS SPECIMENS AND PHOTO.

Figure 516 Two jeremejevite crystals: (left) 3.1 cm and (right) 2.7 cm. Erongo Mountains, Namibia. WARREN TAYLOR RAINBOW OF AFRICA COLLECTION, MARK MAUTHNER PHOTO.

Figure 515 A view from Erongo Mountains, looking south-south-west. The figure in the red trousers is standing on excavations that yielded many jeremejevite crystals in 2001. BRUCE CAIRNCROSS PHOTO, 2005.

Figure 517 A transparent jeremejevite crystal in matrix, 2.6 cm. Mile 72, Namibia. BRUCE CAIRNCROSS SPECIMEN AND PHOTO.

Figure 518 Two faceted jeremejevite gemstones: A 3.94 carats (9 mm); B 1.65 carats (1.2 cm). Erongo Mountains, Namibia. WARREN TAYLOR RAINBOW OF AFRICA COLLECTION, MARK MAUTHNER PHOTOS.

Figure 519 Jeremejevite crystals on feldspar matrix, largest 4 cm. Mile 72, Namibia.
COUNCIL FOR GEOSCIENCE, PRETORIA, SOUTH AFRICA SPECIMEN, BRUCE CAIRNCROSS PHOTO.

Kyanite ◆ Al_2SiO_5

Kyanite crystallizes in the triclinic system, has a hardness of 4 to 7.5, specific gravity of 3.53 to 3.67, a white streak and pearly to vitreous lustre. Kyanite forms a trilogy with sillimanite and andalusite, three metamorphic minerals with identical chemical composition. They are polymorphs of one another, crystallizing in different crystal systems. Kyanite forms elongate, bladed crystals of a characteristic blue, but can be green, yellow or orange (Cairncross and Dixon, 1995). Kyanite is a mineral of metamorphic rocks formed under high pressure and relatively low temperature. It is used as a refractory mineral because it converts to mullite and silica on calcination. Mullite is able to withstand temperatures of over 1,500°C.

Rare gem kyanite crystals are evident in shales interbedded in metamorphosed quartzite south of Nkandla, KwaZulu-Natal, **South Africa**. Intrusion of granites into the shale has resulted in metamorphism of the alumina-bearing shales into mica-quartz-kyanite schist. The rock also contains minor accessory minerals such as ilmenite and goethite. Historical reports suggest that some of the kyanite crystals are of gem quality, but such specimens are probably rare. Euhedral, gemmy, orange, green, blue and colourless crystals have been found near Port Nolloth in the Northern Cape. Blue crystals have been found in metamorphosed strata in the Witwatersrand gold mines, the Vredefort district in the Free State, the Marico district of the North West and on the farm Blaauvlei near Vanrhynsdorp in the Western Cape. Kyanite associated with bright green fuchsite (chromium-rich muscovite) comes from Phalaborwa.

Beautiful translucent kyanite crystals are found in quartz veins in schist between the old Gorob mine offices and the Kuiseb River, Namib Desert Park, **Namibia**. Kyanite was commercially mined in the Rehoboth district on farms Vogelpan 297 and

Figure 520 Blue kyanite with green fuchsite, a chrome-rich variety of muscovite, 7.8 cm. Loolekop hill, Phalaborwa, South Africa. BRUCE CAIRNCROSS SPECIMEN AND PHOTO.

Figure 521 Kyanite fashioned into a cabochon and a faceted gemstone: **A** 2 cm; **B** 3.64 carats (1.3 cm). Gamsberg district, Namibia. BRUCE CAIRNCROSS SPECIMEN AND PHOTO A, WARREN TAYLOR RAINBOW OF AFRICA COLLECTION, MARK MAUTHNER PHOTO B.

Figure 522 A polished slab of kyanite on quartz, 19.4 cm. Namib Desert, Namibia. BRUCE CAIRNCROSS SPECIMEN AND PHOTO.

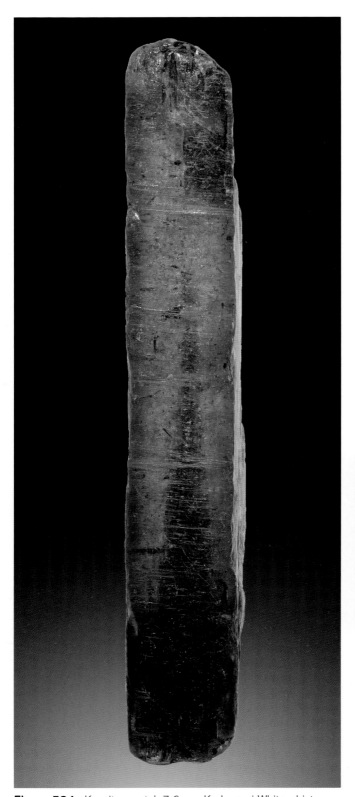

Uisib 298 (Schneider and Watson, 1992). At Vogelpan, about 80 km north of Rehoboth Station, lenses of kyanite occur in mafic dykes and gneisses. On farm Bethlehem 27 (Windhoek district), Kyanite Kop, a hill about 100 m in diameter and 30 m high, was commercially exploited for kyanite. The hill contains corundum, diaspore, minor rutile, tourmaline and quartz, surrounded by gneiss. Several kyanite localities occur in the Windhoek and Rehoboth districts.

Large kyanite deposits are known from metamorphic terrains in **Zimbabwe**, such as those in the north, notably in the Mwami area north-east of Karoi. Over 15,000 tonnes of kyanite was mined from eluvial deposit at Chipungwe, in the Karoi district, where kyanite pseudomorphs after chiastolite are found in graphite-bearing schist. Beautiful gem-quality blue kyanite has been exploited at Prylin ('Kyanite Hill'), in the Rushinga district, Mashonaland Central. Similarly, beautiful blue kyanite, some gem quality, come from the Kadunguri Whiteschists, Mashonaland West. Several kyanite deposits have been mined in the extreme eastern section of the country, east of Mutoko, close to the Mozambique border.

Kyanite was commercially mined at Halfway Kop, 15 km south-east of Francistown, **Botswana**. The hill is composed primarily of kyanite, which occurs in three different forms: individual crystals up to 8 cm, smaller crystals in kyanite-rich schist, and small, asbestiform-like, fibrous crystals.

Kyanite has been mined at Nhazonia, close to the eastern **Mozambique**-Zimbabwe border, approximately 200 km north-west of Beira.

Figure 523 A 4-cm kyanite with several smaller kyanites in quartz. Kadunguri Whiteschists, Mashonaland West, Zimbabwe. BRUCE CAIRNCROSS SPECIMEN AND PHOTO.

Figure 525 A mass of intergrown kyanite, 4.7 cm. Tsodilo Hills, Botswana. DAVID CARTER SPECIMEN AND PHOTO.

Figure 524 Kyanite crystal, 7.6 cm. Kadunguri Whiteschists, Mashonaland West, Zimbabwe. BRUCE CAIRNCROSS SPECIMEN AND PHOTO.

Lepidolite ◆ $K(Li,Al)_3(Si,Al)_4O_{10}(F,OH)_2$

Lepidolite crystallizes in the monoclinic system, has a hardness of 2.5 to 3, specific gravity of 2.8 to 3.3, a white streak, and pearly lustre. According to the IMA, lepidolite is no longer a valid species, rather a series with polylithionite and trilithionite the two end-members. For simplicity, the name lepidolite is used here. Lepidolite is a member of the mica group, one of the minerals mined for lithium. It is a lilac- to mauve-coloured mica, its attractive coloration caused by the presence of lithium. It forms platy, hexagonal crystals that can be up to several centimetres in diameter. It is important to note that lepidolite per se is often lithian muscovite, and that the purple colour is caused by traces of manganese and not lithium. Lepidolite is restricted to lithium-rich pegmatites, where it occurs in association with other lithium-bearing species such as spodumene.

Lithium-ion batteries are rechargeable and have multiple uses in mobile phones, cameras, portable electronics, electric vehicles and in military and aerospace applications. Lithium is used in air conditioners, welding fluxes, bleaches and sanitary agents. One of its main uses is as an additive in glass and ceramics to increase the strength of these materials and reduce their thermal expansion properties. The compound lithium carbonate ($LiCO_3$) is used as a medicine to treat depression.

Compact and fine crystals of lepidolite occur in a few pegmatites in the Northern Cape, **South Africa**, where it is found with other lithium species including spodumene, petalite and amblygonite. In Mpumalanga, lithium-bearing pegmatites outcrop on the farms Oshoek 212 IT and Houtbosch 189 IT, a few kilometres west of the Eswatini border.

In **Namibia**, lepidolite is found in many pegmatites in the Karibib-Usakos area (Von Bezing *et al.*, 2016). Lepidolite was abundant at the Helikon and Rubikon mines, where lithium, caesium, rubidium and beryllium-containing minerals were mined. At the Karlsbrunn pegmatite, 6.5 km south/south-west of the Albrechtshöe railway siding, high-grade lepidolite occurred with beryl and cleavelandite feldspar and rare petalite crystals that were up to 2.5 m long. A lepidolite-bearing pegmatite on the farms Umeis 110 and Kinderzitt 132 in the Warmbad district had red elbaite (rubellite) tourmaline, spodumene and tantalite-(Fe). Other noteworthy lepidolite-rich pegmatites are Neu Schwaben and Etiro. At Etiro, lepidolite was found with schorl, fluorapatite, monazite, topaz, native bismuth, brazialinite and other exotic pegmatite species. Lepidolite is also found in the tin-bearing pegmatites to the north and north-west of Karibib in the Cape Cross-Uis belt (for example, the Orawab pegmatite 500 m south of the Brandberg) and the Sandamap-Erongo belt (for example, the Onguati pegmatite) and in the Rössing area.

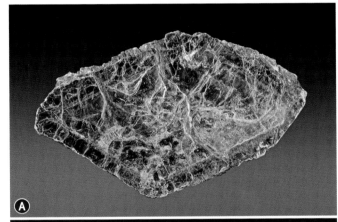

Figure 527 Two-colour zoned lepidolite crystals: **A** 6.5 cm; **B** 4.2 cm. These are most likely lithium-rich muscovite zoned with muscovite (see text for details). Karibib district, Namibia.
BRUCE CAIRNCROSS SPECIMENS AND PHOTOS.

Figure 526 The old Rubikon mine in the Karibib district. Lithium, beryllium and caesium minerals were exploited at this mine. BRUCE CAIRNCROSS PHOTO, 2016.

Figure 528 Hexagonal booklet of lepidolite in quartz, with pink elbaite tourmaline. Karibib district, Namibia. Field of view 2.6 cm. BRUCE CAIRNCROSS SPECIMEN AND PHOTO.

Figure 529 Fine-grained lepidolite with quartz and feldspar, 16.2 cm. Bikita pegmatite, Masvingo, Zimbabwe. PRIVATE COLLECTION, BRUCE CAIRNCROSS PHOTO.

Zimbabwe is a major producer of lithium, having produced 1,600 tonnes of lithium in 2018, much of it coming from one pegmatite, Bikita, in the eastern section of the Masvingo greenstone belt. Crystals are usually hexagonal, but here they formed rounded masses. Less abundant minerals mined for lithium at Bikita are amblygonite, petalite, spodumene and eucryptite. These minerals have been mined from other Zimbabwe pegmatites, for example at the Patronage mine (Goromonzi district) and the Vulcan mine (Mberengwa district). Lepidolite is often a by-product, as are topaz and quartz, at several other pegmatites that have been mined for tin, beryl or

tantalum. The Fungwe Gem beryl pegmatite, Rabbit Warren mine, Al Hayat claims and the Mistress mine were all sources of lepidolite. Lepidolite crystals 60 cm in diameter were found at the Benson pegmatite (Harare district), where beautiful purple crystals were found at the Mauve mine. Abundant lepidolite was associated with tantalite-(Fe) at Nel's Luck.

Lepidolite is found at the Prospect mine west of Francistown, **Botswana**, along with other pegmatite species such as spodumene, muscovite, quartz, potassium feldspar and albite.

In **Eswatini**, fine-grained lepidolite is found in a pegmatite near Kubuta with rubellite tourmaline crystals up to 12 cm long.

Magnesite ◆ $MgCO_3$

Magnesite crystallizes in the hexagonal system, has a hardness of 3.75 to 4, specific gravity of 3 to 3.1, white streak, and vitreous to dull lustre. Magnesite is a carbonate mineral that is mined for magnesium. It is white and exists as finely crystallized material or as encrustations. Large crystals of magnesite are rare in southern Africa, although Namibia has produced some noteworthy specimens. Magnesite is used as a refractory mineral. About 80% of the magnesia produced in South Africa is used in the iron and steel industries. Magnesite typically forms as a metamorphic alteration product of ultramafic rocks, with pyroxene and olivine providing the magnesium. Magnesite sometimes forms during the metamorphism that produces asbestos and serpentine.

Magnesite deposits are concentrated in Limpopo, **South Africa**, where the mineral has been recorded from Grasplaas 98 MT (Strydom, 1998). About 10 km east of Lilliput Siding, between Louis Trichardt and Musina, a vein of crystalline magnesite can be traced for 200 m (Wilke, 1965). Magnesite

also occurs in green serpentine about 6 km west of Musina station, and there are lens-shaped deposits of the mineral in serpentinite about 80 km north of the Murchison range near Leydsdorp in the Sutherland range. Other deposits occur in the Ellerton and Leonde mines in the Letaba district.

There is an important locality for massive magnesite in the eastern Barberton district, which includes part of the Crocodile River Valley between Kaapmuiden and Hectorspruit, especially near Magnesite Siding. In the Lydenburg district, magnesite occurs intermittently along a strip about 80 km from Welgevonden, 50 km north-west of Lydenburg, past Burgersfort and up the Olifants River near Malipsdrif. The chief deposits are at Aapiesboomen 295 KT and Aapiesdoorndraai 298 KT near Burgersfort. Magnesite is found in the North West on Roodekopjesfontein 15 JP in the Marico district. It is found in KwaZulu-Natal on Tugelarandt near Kranskop and at the Sitilo Mountains near Eshowe in small deposits in serpentinite.

Figure 530 Stained and partly weathered iron-rich magnesite, 2.6 cm. Vaalhoek, Bourke's Luck, Mpumalanga, South Africa.
PAUL MEULENBEELD SPECIMEN, BRUCE CAIRNCROSS PHOTO.

Figure 531 Massive magnesite with tiny, sparkling magnesite crystals, 6.7 cm. Magnesite Siding, Barberton district, South Africa. BRUCE CAIRNCROSS SPECIMEN AND PHOTO.

In **Namibia**, magnesite is found in the Windhoek and Rehoboth districts at Kransnek 269, close to the Aris Siding. Large, white, euhedral crystals up to 8 cm on edge occur here.

As in South Africa, most of the **Zimbabwe** deposits are distributed as veins in serpentinites, commonly associated with asbestos. However, the largest deposit, Barton Farm in the Kadoma district, is found in volcano-sedimentary rocks and lavas, and the magnesite is thought to have formed from the alteration of dolomites. Another large deposit, now worked out, occurred at Pande in the Beitbridge district.

Magnesite occurs in north-east **Botswana** in small, scattered deposits in serpentinites. It is also found at Khudumelapye in the western part of Moina Laagte.

Magnesite is occasionally found in **Eswatini** as an alteration product in ultramafic rocks on the western slopes of Ngwenya Mountain.

Figure 532 Two magnesite specimens from the Rehoboth district, Namibia: **A** 4.2 cm; **B** 7.2 cm. BRUCE CAIRNCROSS SPECIMENS AND PHOTOS.

Magnetite ◆ $Fe^{2+}Fe^{3+}O_4$

Magnetite crystallizes in the cubic system, has a hardness of 5.5 to 6.5, specific gravity of 5.18, black streak, and metallic to dull lustre. As the name indicates, magnetite is magnetic – one of its most easily identifiable characteristics. Magnetite often forms black to metallic octahedral crystals. It can be a source of iron and also titanium and vanadium, both of which are sometimes found in its crystal structure. It is an abundant and widespread mineral of igneous and metamorphic rocks, and some sedimentary rocks.

Magnetite is recorded from **South Africa** in crystals from banded iron formations in the Northern Cape, especially at the Prieska slate quarries, where it occurs as small octahedra. Entire koppies (hills) composed of pure, solid magnetite occur in the Bushveld Complex (Reynolds, 1986). Magnetite occurs as large, well-formed octahedral crystals in base metal deposits at Aggeneys. It is very common in the rocks of the Phalaborwa Complex, where it forms attractive octahedral crystals up to 5 cm on edge (Southwood and Cairncross, 2017).

Titaniferous magnetite is found in gneiss in the Kunene Complex, **Namibia**. Goethite/hematite pseudomorphs after magnetite occur in the Erongo pegmatites. These sharp-edged crystals are several centimetres in diameter. Magnetite quartzites are widespread in certain metamorphic terrains such as in the Matchless amphibolite belt in central Namibia.

Magnetite is common in many different rock types in **Zimbabwe**, particularly carbonatites such as at Shawa and Dorowa, where it very often forms magnetic, shiny black, octahedral crystals. Shabani chrysotile asbestos mine has good magnetite crystals. Magnetite is common at the Mphoengs pyrite deposit in the Bulilimamangwe district together with pyrite, chalcopyrite and marcasite.

Magnetite occurs with hematite, goethite and siderite at Ngwenya iron mine, **Eswatini**. It is found in quartzites, with goethite, at Gege and in an iron ore deposit at Maloma on the Mhlatuze River.

Figure 533
Octahedral magnetite crystal partly embedded in galena-chalcopyrite ore. Broken Hill mine, Aggeneys, South Africa. Field of view 2 cm. BRUCE CAIRNCROSS SPECIMEN AND PHOTO.

Figure 534 Partly weathered magnetite, 3.8 cm. Thabazimbi district, South Africa. BRUCE CAIRNCROSS SPECIMEN AND PHOTO.

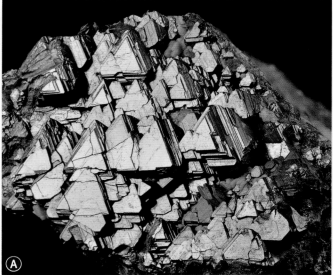

Figure 535 Magnetite crystals from the Palabora mine, South Africa: **A** series of stacked semi-parallel, flattened octahedral crystals, 2.5 cm; **B** large modified octahedral crystal, 4 cm. ALLAN FRASER SPECIMEN B, BRUCE CAIRNCROSS A AND PHOTOS.

Malachite ◆ $Cu(CO_3)(OH)_2$

Malachite crystallizes in the monoclinic system, has a hardness of 3.5 to 4, specific gravity of 4.05, pale green streak, and vitreous lustre. The beautiful dark green colour of malachite is the reason for its popularity as specimens, gemstones and lapidary items. Although malachite forms crystals, it is the massive form that is used by stone carvers and jewellers to produce ornaments, carvings, jewellery and *objets d'art*. Crystals of malachite are often acicular. A widespread secondary mineral of the oxidation zones of copper-bearing ore deposits, malachite occurs in many copper deposits and is often associated with azurite.

Malachite is a very common oxidation mineral of the copper sulphides at Musina, **South Africa**, where clusters of malachite needles up to 2 cm long have been found in crevices of chalcocite ore. Groups of prismatic malachite crystals are also found in parallel arrangement attached to quartz crystals. In these instances, the malachite is olive-green rather than the more common bright green. Malachite is common at the Stavoren tin mines, forming green stains and coatings on fluorite (Atanasova *et al.*, 2016). Cavities in the fluorite contain small, 1-mm acicular tufts, but larger sprays of coarser crystals are also found. Malachite may partially replace some stubby azurite crystals. The finest Stavoren malachite specimens were found in the Hillside quarry dump. Good specimens have been found at the Broken Hill, Willows, Kwaggafontein, Leeuwenkloof, Palabora, Vergenoeg, Argent and Albert mines, to name but a few.

Malachite is associated with many of the hundreds of copper deposits scattered throughout **Namibia**. Its characteristic green staining of rocks provides a valuable

▲ **Figure 536** Typical malachite staining on outcrop indicating the potential of copper mineralization in the area. KHAN MINE, NAMIBIA. BRUCE CAIRNCROSS PHOTO, 2017.

◀ ▼ **Figure 537** Two examples of malachite replacing azurite: **A** 3.6 cm; **B** field of view 4 cm. Tsumeb mine, Namibia. BRUCE CAIRNCROSS SPECIMENS AND PHOTOS.

Ⓐ

Ⓑ

clue to the presence of copper. Beautiful spheres of malachite up to 8 cm in diameter were fairly common at the Tsumeb mine, where malachite formed velvet-like mats of tiny fibrous crystals covering other minerals and rock (Von Bezing *et al.*, 2014). Malachite pseudomorphs after azurite are perhaps the most famous forms of malachite from the Tsumeb mine, with swirls of secondary malachite in layers partially or wholly replacing blocky crystals of dark blue azurite. Malachite was found as crystals and groups of crystals up to several centimetres on edge at the Onganja copper mine. In the northern Kaokoveld, the copper deposit near Omaue that produces dioptase and chrysocolla also contains fibrous and needle-like crystals of malachite. Apart from this locality, specimens can also be found at Kandesi, Okapanda and Okandawai (Schnaitmann and Jahn, 2010; Bowell *et al.*, 2013). Malachite may be found in certain pegmatites, as at Uis.

Figure 538 A near-perfect 8-cm sphere of soft malachite with white dolomite. Tsumeb mine, Namibia. DESMOND SACCO SPECIMEN, BRUCE CAIRNCROSS PHOTO.

Figure 539 Malachite crystals collected from an outcrop at the old Gorob mine, Namibia. Field of view 2.3 cm. BRUCE CAIRNCROSS SPECIMEN AND PHOTO.

Figure 540 Malachite crystals in a cavity of blue shattuckite, 5.2 cm. Kaokoveld Plateau, Namibia. BRUCE CAIRNCROSS SPECIMEN AND PHOTO.

Figure 541 Three habits of malachite from the old Onganja mine, Namibia: **A** cuprite crystals partly covered by globules of malachite, field of view 1.9 cm; **B** primary malachite crystals on calcite, 3.5 cm; **C** cluster of bladed azurite crystals pseudomorphically replaced by malachite, 8.2 cm. BRUCE CAIRNCROSS SPECIMENS AND PHOTOS.

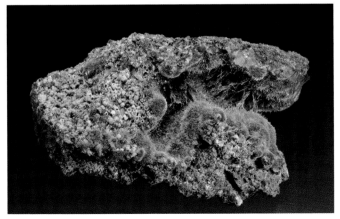

Malachite is found associated with many copper deposits in **Zimbabwe**, most of which are stratabound, i.e. interlayered in sedimentary rocks. Ancient African copper workings are known from some of the deposits, for example at the Alaska mine in the Makonde district. Malachite is reported from the following operations, most now defunct: Cedric, where malachite was the main copper ore, Anna (Makonde district), Mangula, Copper Queen (Gokwe district), Bowcop (Chegutu district), Copper Duke (Kadoma district), Edward (Chiredzi district), Elephant (Bikita district), Luca (Hwange district) and Zimbabwe Copper (Masvingo district), and from copper mines in the Chinhoyi district.

Malachite, fluorite and galena have been reported from quartz veins 6 km west of the Taupse River, **Botswana**. Scattered copper showings are reported west of Maun.

In **Eswatini**, malachite has been found at Hluti, south-west of Mabasa hill in a quartz vein. At Kubuta, an outcropping quartz vein on a ridge between the Sibowe and Lubuya streams contains malachite, bornite, galena and cerussite.

Figure 542 Fibrous malachite crystals lining a cavity in copper ore, 13 cm. Tsumeb mine, Namibia. BRUCE CAIRNCROSS SPECIMEN AND PHOTO.

Manganite ◆ $Mn^{3+}O(OH)$

Manganite crystallizes in the monoclinic system, has a hardness of 4, specific gravity of 4.33, red-brown to black streak and metallic to submetallic lustre. Not to be confused with magnetite, manganite can be distinguished by its jet-black colour and highly lustrous sheen. Southern African crystals are usually under 1 cm on edge and are well formed, although free-standing crystals are relatively scarce. Manganite occurs in low-temperature veins and in manganese deposits, usually as one of the main ore minerals.

In **South Africa**, manganite occurs at most of the mines and in all orebodies of the Kalahari manganese field (Nel *et al.*, 1986; Von Bezing *et al.*, 1991; Cairncross *et al.*, 2017). Crystals have been found in pockets at the Adams, Mamatwan, Hotazel, Wessels and N'Chwaning I and II mines. Associated minerals are quartz, chalcedony, calcite and rhodochrosite. The crystals from N'Chwaning II are associated with quartz and rhodochrosite crystals.

Manganite is found in abundance at the Otjosondu manganese fields, **Namibia** (Cabral *et al.*, 2011). It is a component of the ore at the polymetallic Kombat mine in the Otavi mountainland.

Manganite has been found 32 km north of Mount Darwin, and 32 km west/north-west of Sinoia, **Zimbabwe**. It occurs at the Patronage mine near Harare and at the Armadillo claims, Bubi district.

Manganite, staurolite, magnetite and garnet are reported to occur in **Eswatini** from a quartz vein in quartzite, 11 km east of Goedgegun, Nhlangano.

Figure 543 An aerial view of the workings on Black Rock hill in the Kalahari manganese field. BRUCE CAIRNCROSS PHOTO, 2011.

Figure 544 Manganite, 6.7 cm. N'Chwaning II mine, South Africa. BRUCE CAIRNCROSS SPECIMEN AND PHOTO.

Figure 545 Sub-parallel groups of manganite crystals with blue chalcedony, 11.2 cm. N'Chwaning I mine, South Africa. BRUCE CAIRNCROSS SPECIMEN AND PHOTO.

Figure 546
Somewhat unusual manganite in needle-like crystals, 5.2 cm. N'Chwaning II mine, South Africa.
DESMOND SACCO SPECIMEN, BRUCE CAIRNCROSS PHOTO.

Marcasite ◆ FeS$_2$

Marcasite crystallizes in the orthorhombic system, has a hardness of 6 to 6.5, specific gravity of 4.92, a greenish-black streak, and metallic lustre. Marcasite and pyrite are polymorphs, i.e. they have the same chemical composition, both are iron disulphides, but crystallize in different systems: pyrite is cubic, while marcasite is orthorhombic. Marcasite has a characteristic metallic brass colour. Chemically, it is relatively unstable and may oxidize over time into a white efflorescent powdery sulphate. Marcasite used to be a popular material in jewellery making. It is most common in limestone, dolomites and some shales.

Small, euhedral, pyramidal marcasite crystals sometimes occur in groups several centimetres in diameter in carbonaceous shales closely associated with coal seams in the Witbank Coalfield, Emalahleni, **South Africa**. Marcasite has also been reported from several orebodies and deposits, such as the Witwatersrand and Barberton goldfields, the Phalaborwa Complex, the Kalahari manganese field and the Argent mine between Johannesburg and Emalahleni. During dredging of the Richards Bay harbour, attractive clusters of marcasite and pyrite concretions up to 10 cm in diameter were removed from offshore sediments in the harbour area.

Beautiful brass-coloured marcasite crystals arranged in cockscomb shapes were found at the Rosh Pinah lead-zinc mine in southern **Namibia** (Cairncross and Fraser, 2012). These may be the finest marcasite specimens in southern Africa. The crystals occur in plates up to 30 cm or can consist of smaller clusters where the marcasite is often associated with sphalerite and baryte. Marcasite is reported from the Brandberg West mine and the Crystal tin mine 25 km north-west of Omaruru.

Marcasite is found in **Zimbabwe** at the Red Wing claims (Karoi district) and the Mphoengs pyrite deposit (Bulilimamangwe district), where it occurs with chalcopyrite, pyrite and pyrrhotite.

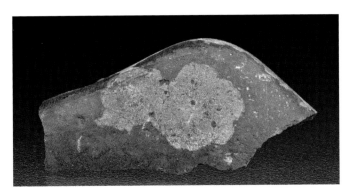

Figure 547 A thin layer of marcasite on the smooth surface of a broken pebble, 5.4 cm. Kleinkopje colliery, South Africa.
BRUCE CAIRNCROSS SPECIMEN AND PHOTO.

Figure 548
Brassy marcasite with yellow baryte, 6.3 cm. Rosh Pinah mine, Namibia.
BRUCE CAIRNCROSS SPECIMEN AND PHOTO.

▲ **Figure 549** Bright brassy marcasite crystals. Rosh Pinah mine, Namibia. Field of view 3 cm.
BRUCE CAIRNCROSS SPECIMEN AND PHOTO.

◄ **Figure 550** A large 25-cm plate of marcasite crystals. Rosh Pinah mine, Namibia. BRUCE CAIRNCROSS SPECIMEN AND PHOTO.

Mercury ◆ Hg

Mercury exists as a liquid at room temperature and therefore does not possess a crystal system, hardness or streak. It does have a high specific gravity of 14.48 and a metallic lustre. Although rarely found in nature, mercury is the only metal that exists as a liquid at normal atmospheric temperature and pressure. Mercury is a significant environmental pollutant, and its use has been severely curtailed in recent times. It is used in batteries and electronic apparatus and, before its toxicity was fully realized, was used in dentistry as an amalgam in tooth fillings. Mercury generally occurs as a component of mercury-bearing minerals such as cinnabar and orpiment. It occurs in low-temperature veins and some low-grade metamorphic rocks.

Native mercury is found as minute spherules and globules in schist from the Monarch cinnabar mine in **South Africa** and in other smaller deposits in the Murchison range (Pearton, 1986).

The only reported occurrence of mercury in **Namibia** comes from the Tsumeb mine, where it forms an amalgam with silver.

The Pilgrim claims in **Zimbabwe's** Bubi district contain mercury, associated with cinnabar. Artisanal gold miners in the country use mercury to extract gold, causing serious environmental problems.

Figure 551 Small droplets of mercury with red cinnabar. Monarch mine, South Africa. Field of view 2.4 cm.
BRUCE CAIRNCROSS SPECIMEN AND PHOTO.

Mesolite ◆ $Na_2Ca_2Al_6Si_9O_{30} \cdot 8H_2O$

Mesolite forms in the orthorhombic system, has a hardness of 5, specific gravity of 2.26, white streak, and vitreous to silky lustre. It is characteristically white to colourless and forms bladed to tabular crystals. Mesolite is one of the zeolite group of minerals and is often found in association with other zeolite minerals, namely analcime, natrolite, scolecite and stilbite. Mesolite is common in volcanic rocks, particularly basalt.

Beautiful specimens, ranging from delicate 1-cm acicular sprays to silky, snow-white composite crystals up to 2 cm in length, have been found at the Palabora mine, Limpopo, **South Africa** (Gliddon and Braithwaite, 1991; Southwood and Cairncross, 2017). The mineral completely lines cavities or projects through calcite or clear fluorapophyllite crystals that have a a greenish hue caused by underlying prehnite. These composite, acicular crystals, growing together in compact parallel growth, often have well-developed forms with distinct pyramidal or, sometimes, indistinct fibrous terminations. Small, clear, pseudocubic fluorapophyllite crystals occasionally decorate these mesolite crystals. Mesolite has also been recorded from amygdales in the Drakensberg basalts.

Figure 552 Elongate silky white mesolite crystals on pale green prehnite, 3.8 cm. Palabora mine, South Africa. BRUCE CAIRNCROSS SPECIMEN AND PHOTO.

Microcline ◆ KAlSi$_3$O$_8$

Microcline crystallizes in the triclinic system, has a hardness of 6 to 6.5, specific gravity of 2.55 to 2.63, a white streak, and vitreous to pearly lustre. Microcline is one of the potassium-bearing feldspars and occurs as white to cream-coloured crystals. It can form as very large crystals and masses in pegmatites where it is often mined for use in the ceramic and paint industries. A variety of microcline known as amazonite has a beautiful green coloration. Microcline is a widespread and common mineral of acidic plutonic rocks, such as pegmatites and granite, as well as syenite.

In **South Africa's** Limpopo province, amazonite has been found as good-quality crystals and masses on the farms Honeydew 86MR, Koedoesrand 199LR and Biesjesfontein 83MR in the Mokopane district, and black microcline crystals have been extracted from a syenite dyke on the farm Konstanz in the Letaba district. Amazonite is also found in pegmatites in the Pella and Pofadder areas in the Northern Cape and in the Baviaanskrans pegmatite (Hugo, 1970; 1986).

White and cream-coloured microcline is found in many pegmatites in the Karibib district, **Namibia**. It has been mined at the Rubikon and Helikon pegmatites and at the Etiro deposit. The distribution of microcline is similar to that of cassiterite (tin) and ferberite-scheelite (tungsten). Amazonite occurs in pegmatites south of Otjiwarongo on the farm Okanjande 145, and in the Maltahöhe district on the farm Neuhof 100. Good crystals have been collected from the pegmatites and miarolitic cavities at Klein Spitzkoppe and, more recently, from the Erongo Mountains.

Zimbabwe has several occurrences of microcline. At the Benson pegmatite, cream-coloured microcline crystals up to 20 cm on edge were fairly common, and amazonite feldspar was sometimes found of a quality suitable for lapidary purposes. Blue-green semiprecious amazonite comes from pegmatites in the north-east Zambezi Belt, notably in the Rushinga district, but also from the Mwami pegmatites (Karoi district), where the Dumbgwe, Kataha, Mago, Zunji and Ganyanhewe deposits were mined. At Ganyanhewe, feldspar was associated with gem-quality beryl, epidote and optical-quality quartz. Green amazonite was found at the Al Hayat claims (Bikita district) and the Mistress tantalite-(Fe) deposit (Harare district). Most of this amazonite was mined in the 1960s.

In **Botswana**, crystals of microcline, up to 15 cm, have been found in a pegmatite at Bodiakhudu, and the mineral occurs in other pegmatites west of Francistown. A pegmatite in the Timbale granite has crystals up to 20 cm on edge.

Microcline is common in ancient granites in **Eswatini**.

Amazonite comes from the border area between southern **Mozambique** and Zimbabwe.

Figure 553 Well-formed microcline crystals, 2.5 cm. Klein Spitzkoppe, Namibia.
BRUCE CAIRNCROSS SPECIMEN AND PHOTO.

Figure 554 A cleavage sample of microcline variety amazonite, 6.3 cm. Kenhardt district, Northern Cape, South Africa.
BRUCE CAIRNCROSS SPECIMEN AND PHOTO.

Figure 555 Pale green microcline variety amazonite crystal, 3.1 cm. Klein Spitzkoppe, Namibia. BRUCE CAIRNCROSS SPECIMEN AND PHOTO.

Millerite ◆ β-NiS

Millerite crystallizes in the trigonal system, has a hardness of 3 to 3.5, specific gravity of 5.41, a green-black streak, and metallic lustre. Not to be confused with milarite, millerite is a nickel sulphide which, when it occurs in sufficient quantity, can be mined for its nickel content. Crystals of millerite have a very characteristic and distinctive acicular habit, hair-like crystals often occurring in fan-shaped radiating groups. They are always brass-coloured. Millerite occurs in limestones such as calc-silicates/marbles and ultramafic rocks.

The nickel deposit at Pafuri, Limpopo, is the only locality in **South Africa** where large specimen-quality millerite occurs. The mineral forms attractive fan-like sprays that occur in extremely hard marble. This deposit has never been commercially exploited for nickel, but was the focus of several exploration programmes (Southwood, 1984), and millerite specimens from here are found in many southern African mineral collections. They were collected from outcrops and old exploration trenches. Tiny, often microscopic, millerite crystals are recorded from the Witwatersrand and Barberton goldfields and the Phalaborwa Complex. Major nickel-producing mines exploiting mineralization in the Bushveld Complex include Nkomati in Mpumalanga, Zebediela and Burgersfort nickel.

Traces of millerite were found at the Tsumeb mine and the Kaokoveld sodalite deposits, **Namibia**.

All of the known occurrences of millerite in **Zimbabwe** are in ultramafic rocks (Viljoen *et al.*, 1976). At the Damba mine in the Bubi district, millerite is found in komatiites, associated with pentlandite. The Epoch mine in the Insiza district and the Trojan mine in the Bindwa district are both said to contain millerite, as does the Railway Block chrome mine in the Shurugwi district.

Millerite is a relatively rare constituent of ores at Selebi-Phikwe, **Botswana**.

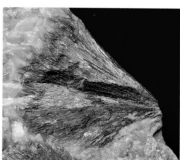

Figure 556 Close-up of a spray of millerite, 4.5 cm. Pafuri nickel deposit, Vhembe district, Limpopo, South Africa. BRUCE CAIRNCROSS SPECIMEN AND PHOTO.

Figure 557 A large matrix specimen of millerite collected from an outcrop, 21.6 cm. Pafuri nickel deposit, Vhembe district, Limpopo, South Africa. BRUCE CAIRNCROSS SPECIMEN AND PHOTO.

Mimetite ◆ Pb₅(AsO₄)₃Cl

Mimetite crystallizes in the hexagonal system, has a hardness of 3.5 to 4, specific gravity of 7.28, a white streak, and vitreous, resinous to subadamantine lustre. Mimetite can occur as perfect six-sided crystals or as divergent sprays of finely crystalline aggregates. The colour varies from orange, bright yellow or pale yellow to almost colourless. Mimetite is a secondary mineral that is found in some lead deposits, where it forms from the oxidation of lead.

Mimetite has been found in **South Africa** as small, clear yellow to colourless prismatic crystals at the Edendale lead mine, Stavoren tin mines and the Argent silver mine near Delmas in Gauteng (Atanasova *et al.*, 2016). It is never common, but good micromount to thumbnail specimens are known.

In **Namibia**, the famous Tsumeb mine in the Otavi mountainland has produced a plethora of mimetite specimens (Gebhard, 1999). Some of the world's finest known crystals of this rare mineral come from Tsumeb, where chablis-yellow, transparent crystals up to several centimetres in length, were discovered in the 1970s (Wilson, 1977). Gem-quality prismatic crystals over 3 cm long and groups of crystals were found, including some large cabinet- and museum-sized specimens that were coated with small, spiky mimetite crystals. Some of these weigh several kilograms and consist of clusters of yellow and yellow-orange acicular crystals. Others have stellate groups that are often associated with malachite.

Mimetite is rarely found in **Zimbabwe** and does not rival the material from Namibia. However, it is found in the Mutare area and from the Copper King gossan, where it is associated with pyromorphite and anglesite.

Figure 558 Two scanning electron microscope images of mimetite crystals: **A** Field of view 0.9 mm. Argent mine, South Africa; **B** field of view 0.76 mm. Slipfontein, Bushveld Complex, South Africa. WOLF WINDISCH SPECIMENS, MARIA ATANASOVA IMAGES.

Figure 559 Mimetite crystals associated with **A** copper, 3.5 cm, and **B** red cuprite. Tsumeb mine, Namibia. Field of view 1.8 cm. BRUCE CAIRNCROSS SPECIMENS AND PHOTOS.

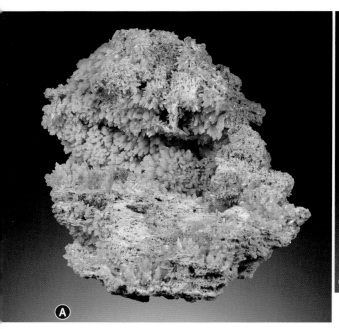

Figure 560 Two large mimetite specimens: **A** 20 cm; **B** 13 cm. Tsumeb mine, Namibia. DESMOND SACCO SPECIMENS, BRUCE CAIRNCROSS PHOTOS.

Figure 561 Well-formed, prismatic mimetite crystals. Tsumeb mine, Namibia. Field of view 1.6 cm. BRUCE CAIRNCROSS SPECIMEN AND PHOTO.

Molybdenite ◆ MoS$_2$

Molybdenite crystallizes in the hexagonal system, has a hardness of 1 to 1.5, specific gravity of 4.62 to 5, greenish streak, and metallic lustre. It is the only molybdenum-bearing mineral of economic importance. Crystals of molybdenite are easy to identify as they form flat, silver-grey, hexagonal plates and are very soft and sectile (can bend without breaking). They also tend to soil the fingers when handled. Graphite differs from molybdenite in that it has a black streak. Molybdenum is used as an alloying agent and a refractory material.

Molybdenite is mostly recovered from low-temperature, low-grade hydrothermal deposits such as quartz veins and pegmatites. It is also present in porphyry copper deposits and some granites. Molybdenite is found in southern African granites, pegmatites and porphyry copper deposits.

Molybdenite occurs in many ore deposits in **South Africa**, usually as a trace mineral (Cairncross and Dixon, 1995). The largest deposit occurred on the farm Groenvley 224 KR and Appingedam in the Zaaiplaats tin field north-west of Mokopane (Lenthall, 1974), associated with arsenopyrite, sphalerite, cassiterite, quartz, tourmaline and fluorite. Well-formed euhedral platelets up to 2 cm occur in the Bushveld granites at Houtenbek 194 JR, north-east of Pretoria, and several tonnes of molybdenite were mined before 1921 at the Stavoren tin mines (Atanasova *et al.*, 2016). At Stavoren 676 KS, molybdenite occurs with molybdite and cassiterite. Sandstones located to the south and south-east of Beaufort West in the Western Cape contain molybdenite mineralization. The Riviera deposit, 23 km north of Piketberg, has quartz-calcite molybdenite veins with scheelite and pyrrhotite. Molybdenite is also found in the Richtersveld, Northern Cape, west of Vioolsdrif.

Figure 562 Molybdenite crystals collected at Houtenbeck, South Africa. BRUCE CAIRNCROSS SPECIMEN AND PHOTO, 2016.

Molybdenite is reported from several geological deposits in **Namibia**, including pegmatites and porphyry copper deposits (Roesener and Schreuder, 1992). The finest hexagonal molybdenite crystals were found in the Onganja copper mine 65 km north-east of Windhoek. 'Rosettes' of molybdenite from this locality measure up to 8 cm in diameter, and good specimens were collected off the old mine dumps (Cairncross and Moir, 1996). Other noteworthy deposits that contain molybdenite are at the Natas mine, Haib copper deposit, Lorelei, south-east of Rosh Pinah, Davib Ost 61, Tsumeb, Klein Spitzkoppe, Navachab and the Krantzberg mine in the Erongo region.

In **Zimbabwe**, at the Lazeno deposit in the Chipinge district, a brecciated syenite contains sub-economic concentrations of molybdenite, associated with pyrrhotite, fluorite, sphalerite, chalcopyrite and scheelite. At the Molly molybdenite deposit in the same district, molybdenite crystals 1–2 cm in diameter occur in quartz veins in granite, together with the rare mineral powellite. At Gumboot, in the Makonde district, attractive specimens of transparent, glassy quartz are associated with molybdenite and scheelite. Molybdenite is also found at the Ball mine in the Rushinga district, where it is associated with scheelite.

Molybdenite is found disseminated in some of **Eswatini's** granites and pegmatites. It is found in quartz veins along the Komati River near the old Pigg's Peak-Mbabane road.

Figure 563 Molybdenite crystal in matrix from Appingedam, Zaaiplaats tin field, South Africa, 4.2 cm. BRUCE CAIRNCROSS SPECIMEN AND PHOTO.

Figure 564 Hexagonal molybdenite on calcite. Onganja mine, Namibia. Field of view 1.8 cm. BRUCE CAIRNCROSS SPECIMEN AND PHOTO.

Monazite-(Ce) ◆ (Ce,La,Nd,Th)PO$_4$

Monazite-(Ce) crystallizes in the monoclinic system, has a hardness of 5 to 5.5, specific gravity of 4.6 to 5.4, a white to pale yellow streak, and waxy, resinous lustre. Monazite-(Ce) is the most common species in this group of rare-earth element-containing minerals, the others being monazite-(La), monazite-(Nd), monazite-(Sm), and an unnamed Gd-dominant monazite. Crystals are characteristically yellow, orange or brown and have a distinctive waxy to resinous appearance. REEs are used in various ways. Thorium is used in nuclear reactors and in welding electrodes. Thorium-magnesium alloys are used in the aerospace industry. Other REEs add strength to steel and are used as catalysts and as polishing agents for glass, and in fluorescent lights and magnetic and electronic components. Monazite-(Ce) occurs as an accessory mineral in granites and gneiss, pegmatites, eluvial deposits and some carbonatites.

Monazite-(Ce) occurs as euhedral sharp-edged crystals up to 10 cm long in a number of pegmatites in the Northern Cape, **South Africa**. In the Witkop pegmatite, 3 km south of the farmhouse on Styr-Kraal, Kenhardt district, it is found as single, clove-brown crystals up to 25 mm in diameter or as smaller clusters, and is usually associated only with quartz and muscovite. Monazite-(Ce) occurs in a similar fashion at the Blesberg mine in the Noumas pegmatite, where it is associated with apatite crystals. In the Warmbad Noord pegmatites, the crystals are brownish-green. Several monazite-(Ce) pegmatites, such as Japie, Riemvasmaak and Blomerus, are found along the Bak River (Minnaar and Theart, 2006). The Glenover Complex, 80 km north/north-west of Thabazimbi, is a monazite-(Ce)-bearing pyroxenitic carbonatite. Crystals were found at the Mutue Fides tin fields in the Bushveld Complex.

Monazite-(Ce) occurs in some quantity on the farm Houtenbek 194 JR in the Groblersdal district, Mpumalanga, in large, brown prismatic crystals up to 7 cm long and 1 cm wide. Micro-crystals are known from Stavoren (Atanasova *et al.*, 2016). Steenkampskraal is one of the larger monazite-(Ce) deposits in the country (Knoper, 2010). This deposit, in southern Namaqualand, 80 km north-east of Vanrhynsdorp, consists of high-grade monazite-(Ce) emplaced along a subvertical shear zone in granite-gneiss. Granular monazite-(Ce) occurs here associated with apatite, chalcopyrite, malachite and other sulphides. This deposit, which was mined in the past for REEs, in particular thorium, probably owes its origin to high-temperature hydrothermal activity. On the eastern seaboard of KwaZulu-Natal, monazite-(Ce) is one of the detrital heavy minerals found in beach and dune sands.

The most well-known locality in **Namibia** for producing monazite-(Ce) specimens is a dolomitic carbonatite pegmatite that outcrops on the farm Eureka 99, about 38 km west of Usakos (Von Bezing *et al.*, 2014). Here, some very large, attractive, red-brown crystals, over 10 cm on edge, have been found. Other carbonatite and related rocks that contain monazite-(Ce) are the Marinkaskwela carbonatite in southern Namibia (Diehl, 1992c), the Ondumakorune Complex north-east of Kalkveld, and the Okorusu Complex (which contains green monazite-(Ce) crystals). Monazite-(Ce) is found at the Rössing uranium deposit and the De Rust pegmatite north of the Brandberg. Small orange micromount crystals have been found in the Erongo pegmatites. Monazite-(Ce) is also found in heavy-mineral sands on the coast.

Figure 565 Monazite-(Ce) crystal, 4.8 cm. Amen pegmatite, Goodhouse, South Africa. BRUCE CAIRNCROSS SPECIMEN AND PHOTO.

Figure 566 Resinous monazite-(Ce) crystals. Eureka mine, Namibia. Field of view 1.4 cm. BRUCE CAIRNCROSS SPECIMEN AND PHOTO.

Granite outcrops in the Zimunya area 25 km south of Mutare, **Zimbabwe**, contain abundant monazite-(Ce) crystals. In the Beitbridge district, metamorphic migmatites have monazite-(Ce) together with two other radioactive species, euxenite and thorite. Monazite-(Ce) is present as pale yellow to dark brown, flattened, tabular crystals or pseudo-octahedral crystals in gneisses and pegmatites at the Byerley uranium prospect, which is located next to the Beitbridge-Masvingo road. Monazite-(Ce) crystals up to 1 cm on edge are found in the pegmatite at the Ebonite claims 6 km south-west of Bikita. The Fungwe Gem beryl pegmatite also has monazite-(Ce), as does the Benson deposit in the Mutoko greenstone belt. Here, monazite-(Ce) is associated with zircon, uraninite, euxenite and other esoteric radioactive minerals, such as fergusonite and samarskite. The Bepe

pegmatite claims in the Byhera district have veins of monazite-(Ce) in quartz. The monazite-(Ce) is red-brown and makes up about 40% of the rock. Some alluvial monazite-(Ce) is found in the Dewure River sands, as well as in heavy-mineral deposits of ancient sedimentary sequences of the Karoo Supergroup sandstones. Both the Katete carbonatite north-west of Harare in the Binga district and the Dorowa carbonatite in the Buhera district contain monazite-(Ce).

In **Eswatini**, monazite-(Ce) is found in schists at the Mhlabane Mountain in the Gege district, associated with zircon, magnetite and rutile. Monazite-(Ce) is found with cassiterite in alluvial and eluvial gravels near Mbabane at the old McReedy's and Star tin mines.

Monazite-(Ce) occurs in heavy-mineral sands along the southern **Mozambique** coastline.

Mottramite ◆ $PbCu^{2+}(VO_4)(OH)$

Mottramite crystallizes in the orthorhombic system, has a hardness of 3 to 3.5, specific gravity of 5.9, a yellow streak, and vitreous lustre. Mottramite is a lead-copper vanadate. Crystals are small and tend to aggregate to form botryoidal masses and drusy crusts. They are usually black, dark grey or dark olive-green. Mottramite occurs in lead-zinc deposits that also contain vanadium.

Very small, greenish-black to black, scaly aggregates of intergrown mottramite crystals with a greasy lustre occur at the old Argent mine, east of Johannesburg, **South Africa**.

These are associated with more abundant pyromorphite crystals in a matrix of quartz. Mottramite has also been reported from the Edendale lead mine east of Pretoria and Olifantspoort mine in the Rustenburg district (Atanasova *et al.*, 2016).

Mottramite was one of the first vanadium-bearing species found in the Otavi mountainland, **Namibia**, in the early part of the twentieth century (Cairncross, 1997). It was originally discovered in the area in 1919, at Karavatu. Large masses of drusy mottramite came from the Tsumeb mine (Gebhard, 1999).

Figure 567 **A** Front and **B** rear view of the same 9.8-cm specimen of mottramite. This shows the colour variation of the species, from grey-black to olive-green. Tsumeb mine, Namibia. BRUCE CAIRNCROSS SPECIMEN AND PHOTOS.

Crystals can be very dark brown to black or light to dark green. The most common habit is arborescent clusters of crystals. Attractive masses of dark green mottramite were found at the Gross Otavi mine and Uitsab North in the Otavi mountainland. Mottramite occurs as dark green drusy coatings on other minerals in several of the copper prospects in the Kaokoveld, including the Christoff, Onderra and Okandawasi workings (Bowell *et al.*, 2013).

Figure 568 Drusy mottramite partially coating a boxwork of quartz, 12.5 cm. Kaokoveld plateau, Namibia. BRUCE CAIRNCROSS SPECIMEN AND PHOTO.

Figure 569 Mottramite displaying two different forms: **A** dendritic branching clusters of green mottramite, 7.2 cm; **B** arborescent clusters of grey-black crystals together with colloform masses, and tan calcite; field of view 4.5 cm. Tsumeb mine, Namibia. BRUCE CAIRNCROSS SPECIMENS AND PHOTOS.

Muscovite ◆ KAl$_2$(Si$_3$Al)O$_{10}$(OH,F)$_2$

Muscovite crystallizes in the monoclinic system, has a hardness of 2.5 to 4, specific gravity of 2.77 to 2.88, a white streak, and vitreous to pearly lustre. Muscovite is one of the mica group of minerals and is a phyllosilicate, i.e., it forms in sheets that can easily be peeled away from one another. It is a common rock-forming mineral. Muscovite is used in electrical apparatus and in the electronics industry. It is crushed and used as an additive in paint, plaster and wall coatings.

Muscovite is common in silica-rich igneous rocks and in some sandstones and shales. It is a very common constituent in granite, pegmatites and some schists. Fuchsite is a bright green chrome-rich variety of muscovite, and is one of the main minerals that cause the green colour of many ancient greenstone-belt rocks.

Sheets of muscovite over a metre in diameter occur in granitic pegmatites in Mpumalanga, **South Africa**, in the Phalaborwa area (Robb and Robb, 1986). The town of Mica, near Phalaborwa, takes its name from muscovite mica. The Union, Protea, Bantam and Impala Lily mines are among those that have exploited muscovite in this region. The mineral occurs in large crystals in coarsely crystalline quartz and feldspar, often associated with

Figure 570 A view of the feldspar-quartz-muscovite mine dumps at the Union mine, Mica district, South Africa. BRUCE CAIRNCROSS PHOTO, 1990.

Figure 571
A 40-cm sheet of muscovite collected off the dumps shown in figure 570.
BRUCE CAIRNCROSS SPECIMEN AND PHOTO

garnet and fluorapatite. Crystals are sometimes invaded by fine rutile needles that follow the muscovite crystal lattice in a beautiful, regular crystallographic pattern.

At the Musina copper mines, fine-grained sericite is a common coating on quartz and epidote crystals. Fuchsite occurs in many metamorphosed Archaean rocks in Mpumalanga and as crystals up to 10 mm in length at the Phalaborwa Complex.

Muscovite is ubiquitous in pegmatites in **Namibia**, either as individual, flaky, silver crystals or, occasionally, as classic hexagonal mica 'books'. It is also common in acidic igneous rocks, such as granites, and metamorphic schists. Most of the pegmatites in the Brandberg-Erongo-Uis-Karibib region contain abundant muscovite, as do the pegmatites in Tantalite Valley. Even though a common species, some remarkable and fine hexagonal books of muscovite have been collected from the miarolitic cavities of the Erongo Mountain granite. These vary in colour from silver-grey to brilliant yellow.

Muscovite was mined extensively for over 40 years from pegmatites in Proterozoic rocks in the Mwami mica field, north-east of Karoi, **Zimbabwe**. Only waste muscovite and discard material are now found on the dumps. Mining started here in 1919, peaking in the early 1950s. Most pegmatites that were exploited were in the Karoi district at mines such as Beckett, Catkin, Grand Parade, Hendren, Miami, Owl and Zonkosi. At the Hall's Own pegmatite, approximately 50 tonnes of muscovite were mined in the late 1940s, together with gem-quality quartz, aquamarine and tantalite.

Large crystals of muscovite are found in pegmatites north and south of Francistown, **Botswana**, and in the Vukwe area at Tantebane Kopjes. Crystals 8 cm thick and up to 13 cm in diameter were mined from a pegmatite in granite-gneiss 32 km south/south-east of Mochudi.

Muscovite is common in pegmatites, particularly one pegmatite near Gege, south of Nahlozane gorge, **Eswatini**, where crystals up to 9 cm thick occur.

Figure 572 Two different-coloured muscovite specimens: **A** with green fluorite and goshenite, 7.8 cm; **B** bright yellow crystals with green fluorite and black schorl, 4 cm. Erongo Mountains, Namibia. BRUCE CAIRNCROSS SPECIMENS AND PHOTOS.

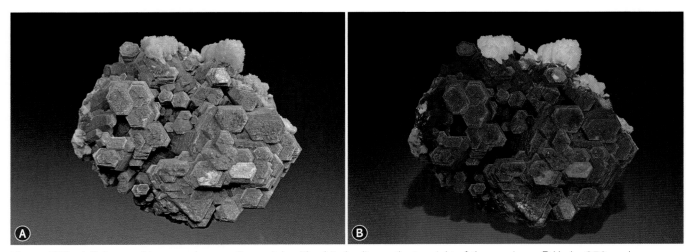

Figure 573 **A** Columns of hexagonal muscovite crystals with white hyalite on the top right of the specimen. **B** Under 350 nm long-wave ultraviolet light, the hyalite fluoresces bright yellow-green. Erongo Mountains, Namibia. BRUCE CAIRNCROSS SPECIMEN AND PHOTOS.

Nambulite ◆ $LiMn^{2+}_4Si_5O_{14}(OH)$

Nambulite crystallizes in the triclinic system, has a hardness of 6.5, specific gravity of 3.53, pale yellow streak, and subvitreous lustre. Nambulite crystals are typically reddish-brown to orange, or orange-yellow. It is a rare mineral, worldwide, particularly as high-quality collector specimens. It is included here, as the finest crystals in the world come from southern Africa.

In 1975, transparent, red gem-quality crystals were discovered at the Kombat mine in **Namibia** (Cairncross, 2020b). It was first thought to be rhodonite, another manganese-bearing mineral, but analyses showed the crystals to be nambulite, a species discovered in Japan only three years earlier. The Kombat crystals are superior to any others. The crystals can be bright vermillion-red and transparent.

◄ **Figure 574** Faceted nambulite crystal, 4.44 carats (1.1 cm). Kombat mine, Namibia. WARREN TAYLOR RAINBOW OF AFRICA COLLECTION, MARK MAUTHNER PHOTO.

Figure 575 ➤
A terminated nambulite crystal, 1.4 cm. Kombat mine, Namibia. BRUCE CAIRNCROSS SPECIMEN AND PHOTO.

Figure 576 A gem-quality nambulite crystal with several other small crystals, on gypsum with some brown brushite, 6.5 cm.
DESMOND SACCO SPECIMEN, BRUCE CAIRNCROSS PHOTO.

Natrolite ◆ $Na_2Al_2Si_3O_{10} \cdot 2H_2O$

Natrolite crystallizes in the orthorhombic system, has a hardness of 5 to 5.5, specific gravity of 2.26, white streak, and vitreous to pearly lustre. Natrolite is one of the zeolite group of minerals. It is relatively easy to identify, as it is always white to colourless and commonly forms radiating sprays of acicular or thin prismatic crystals. However, it can also be found as single, matchstick-like crystals (see the example from the Witwatersrand goldfield shown in figure 577). The mineral occurs in cavities (vesicles) in mafic lavas, notably basalt. Also as a secondary vein and vug-filling mineral in some orebodies.

In **South Africa**, the Chrome mine in Mokopane district, Limpopo, yielded a single pocket of aesthetically pleasing natrolite crystals. This was quite exceptional, as there is a general absence of collector-quality mineral specimens from the chrome and platinum mines of the Bushveld Complex. These dramatic snow-white, radiating crystals, contrasting starkly with a black matrix, were formed in a fault surface that cut through the chromite orebodies. In the Drakensberg basalts, natrolite forms radiating sprays in amygdales.

Beautiful sprays of natrolite are found as secondary minerals in some diamond mine kimberlites, as at Bultfontein and Wesselton. In the Palabora mine, natrolite occurs, rarely, as colourless, transparent or opaque prismatic crystals, up to 2 cm long, embedded in opaque, silky white, matted, radiating pectolite on fluorapophyllite (Southwood and Cairncross, 2017). The Wessels mine in the Kalahari manganese field has produced spectacular, large natrolite crystals associated with hydroxyapophyllite-(K), thomsonite

Figure 577 Natrolite on conglomerate matrix, 9.8 cm. Witwatersrand goldfield, South Africa. BRUCE CAIRNCROSS SPECIMEN AND PHOTO.

Figure 579 Natrolite, red inesite, white tobermorite and mauve datolite, 2.6 cm. Wessels mine, South Africa. BRUCE CAIRNCROSS SPECIMEN AND PHOTO.

Figure 578 Natrolite crystals from the Witwatersrand goldfield, South Africa. Field of view 4 cm. BRUCE CAIRNCROSS SPECIMEN AND PHOTO.

Figure 580 A mass of natrolite crystals. Witwatersrand goldfield, South Africa. Field of view 7 cm. BRUCE CAIRNCROSS SPECIMEN AND PHOTO.

and inesite (Cairncross and Beukes, 2013). Veins containing interlocking white natrolite crystals occur at Salpeterkop in the Karoo.

During the construction of the Hardap Dam, Keetmanshoop district, **Namibia**, beautiful large natrolite crystals, up to 10 cm long, were discovered in basalts, associated with other zeolite species such as mesolite and analcime (Von Bezing *et al.*, 2014). Natrolite crystals are found in cavities in the Aris phonolite south of Windhoek.

Natrolite occurs in cavities in Karoo basalts in western **Zimbabwe** and similar rocks in **Lesotho** and the south-western region of southern **Mozambique**.

Figure 581 A faceted 2.5-carat (1.3-cm) natrolite. Hardap Dam, Keetmanshoop district, Namibia. WARREN TAYLOR RAINBOW OF AFRICA COLLECTION, MARK MAUTHNER PHOTO.

Figure 582 Sprays and interlocking crystals of natrolite, 8.7 cm. Hardap Dam, Keetmanshoop district, Namibia. BRUCE CAIRNCROSS SPECIMEN AND PHOTO.

Olmiite ◆ CaMn[SiO₃(OH)](OH)

Olmiite crystallizes in the orthorhombic system, has a hardness of 5 to 5.5, specific gravity of 3.05, a pale pink to reddish-pink streak, and vitreous lustre. Olmiite is a relatively new mineral species discovered in South Africa during the early 2000s and formally described by Bonazzi *et al.* (2007). The crystals can take a variety of shapes, from single prismatic crystals to perfectly rounded aggregates. The colour is also variable, ranging from off-white, cream or brown to salmon pink or deep pink-red.

The mineral was discovered in the N'Chwaning II mine in the Kalahari manganese field, **South Africa**. This was one of the most important mineral discoveries (scientifically and volumetrically) in South Africa during the past 20 years (Cairncross and Beukes, 2017). The initial discovery yielded predominantly cream-coloured crystals and crystal aggregates, which were incorrectly identified twice – initially as baryte by the mine geologist, then later as poldervaartite. The specimens were subsequently described as a new mineral species, after it was discovered that the position of the Mn-cations occupy different divalent sites to that of poldervaartite. However, later analyses have shown that often olmiite and poldervaartite may be intimately intergrown and therefore both species may co-exist in single specimens. The volume of specimens that has entered the collector market has been prodigious, as has the habit and colour of the specimens and associated minerals.

Specimen sizes range from thumbnail- to museum-sized banded iron-formation slabs coated in olmiite. Two distinct varieties of olmiite were collected. The first and most common type has individual crystals that are opaque and cream, pale flesh pink to off-white. Many specimens have crystals bundled together into radiating colloform aggregates or are bowtie-shaped, with the crystal aggregates tapering into a central point. Many were collected on matrix, with the cream-coloured olmiite providing an attractive colour contrast on the black to ochre-brown matrix. Freestanding individual crystals with a high lustre are common.

The cream-coloured olmiite fluoresces deep red under short-wave ultraviolet light. Associated species are few. Acicular pale blue celestine crystals up to 5 mm occur on some specimens, but are rare (see figures 229 & 230). Micro-crystalline hematite, minor andradite, bultfonteinite and rare oyelite constitute some of the minerals in and on the oxidized matrix to the olmiite. More recently collected specimens are associated with pink bultfonteinite, celestine, baryte, thaumasite, datolite, caryopilite, gageite and minor andradite garnet.

Since the initial discovery in late 2001, thousands of olmiite specimens have been collected, many of outstanding quality. Even so, it remains unique to the Kalahari manganese field.

Figure 583 One of the best examples of spherical olmiite, 4.5 cm. N'Chwaning II mine, South Africa. DESMOND SACCO COLLECTION, BRUCE CAIRNCROSS PHOTO.

Figure 584
A large, single composite olmiite crystal on calcite and white bultfonteinite, 4 cm. N'Chwaning II mine, South Africa. DESMOND SACCO SPECIMEN, BRUCE CAIRNCROSS PHOTO.

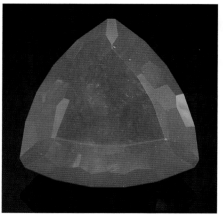

Figure 585 A rare and unusual faceted olmiite, 31.75 carats (2.1 cm). N'Chwaning II mine, South Africa. WARREN TAYLOR RAINBOW OF AFRICA COLLECTION, MARK MAUTHNER PHOTO.

Figure 586 Examples of olmiite, showing the diversity of habits, colour and associated minerals. **A** Barrel-shaped olmiite on calcite, 5.6 cm; **B** olmiite crystals with white calcite and black goethite, field of view 3.8 cm; **C** and **D** two different-coloured spheres of olmiite, 7.6 cm and 9.5 cm; **E** bowtie olmiite, 3.7 cm; **G** red-brown olmiite on brown gageite with small, transparent calcite crystals, 3.4 cm; **F** and **H** discrete olmiite crystals, some dark brown to black, others caramel-coloured on calcite, 11.4 cm and 13.4 cm. N'Chwaning II mine, South Africa. BRUCE CAIRNCROSS SPECIMENS AND PHOTOS.

Opal ♦ $SiO_2 \cdot nH_2O$

Opal is amorphous, has a hardness of 5.5 to 6.5, specific gravity of 2 to 2.5, a white streak, and vitreous, pearly, waxy or resinous lustre. Opal is not a true mineral, as such. It is either composed of cristobalite and/or tridymite or amorphous silica. Four types of opal are known: opal-CT containing cristobalite-tridymite, opal-C composed of cristobalite, opal-AG, the amorphous variety (amorphous-gel) and opal-AN, also amorphous (amorphous-network), found as hyalite. It occurs in many colours, the most famous varieties being 'precious opal' from Australia and 'fire opal' from Mexico. These gemstone varieties have not yet been found in southern Africa. 'Common opal', which occurs in southern Africa, is typically grey to white but can also be cream or light brown.

The mineral can be cut and polished. Botryoidal opaline silica is sometimes referred to as hyalite. Opal exists as a secondary vein-filling in greenstone belt serpentinites. It is common around the vents of hot springs and in gossans.

In **South Africa**, 'common opal' occurs in a variety of neutral colours and patterns in the Northern Cape near Postmasburg and Pella, the Pilanesberg, North West, and near Vivo, Limpopo, in the western part of the Soutpansberg. It is widespread in small deposits in other areas and is often found in gossans associated with ore deposits. Veins and pods of white, porcelain-like opal occur in some of the ultramafic rocks in the Barberton greenstone belt. Beautiful, brown- and white-streaked, opalized wood is found in sedimentary deposits close to Mtubatuba in north-east KwaZulu-Natal. Silicified tree trunks several metres long and over a metre wide are found in sedimentary rocks in the foothills of the Drakensberg in northern KwaZulu-Natal and the north/north-east Free State.

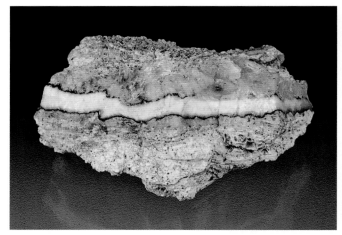

Figure 587 A vein of white opal in weathered serpentinite, 5.2 cm. Stolzburg Complex, Mpumalanga, South Africa. BRUCE CAIRNCROSS SPECIMEN AND PHOTO.

Figure 588
Cream opal, displaying typical conchoidal fracture pattern, 5 cm. Barberton district, South Africa. BRUCE CAIRNCROSS SPECIMEN AND PHOTO.

Figure 589 Two examples of the mottled brown-cream opalized wood from the Mtubatuba district, South Africa: **A** (left) 11.5 cm; **B** a cut and polished slice, 16.5 cm. BRUCE CAIRNCROSS SPECIMENS AND PHOTOS.

Beautiful turquoise blue botryoidal opal-CT was collected at Mesopotamia, west of Khorixas, **Namibia**. Yellow and yellow-green botryoidal opal-AN hyalite is found coating feldspar, schorl tourmaline, quartz and aquamarine in pegmatites in the Erongo Mountains. This Erongo material fluoresces an intense yellow-green under ultraviolet light (Cairncross and Bahmann, 2006a). Opalized and silicified wood is found in the Khorixas district at the Petrified Forest. Large prehistoric tree trunks have been totally replaced, molecule for molecule, by the silica.

In **Zimbabwe**, pale green to white veins of opal (unspecified) are relatively common in many greenstone-belt serpentinites. Opal is also found in veins in the ultramafic rocks of the Great Dyke. So-called wood opal (petrified wood) is found in the Zambezi Valley where it has weathered out of the enclosing Karoo sedimentary strata. Wood opal is also found on the Mafungabusi plateau.

As in neighbouring South Africa, localized pockets of porcelainous white opal is found in ultramafic portions of the Barberton greenstone belt in the north-west area of **Eswatini**.

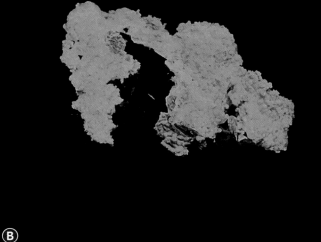

Figure 591 Yellow hyalite on black schorl, 6.4 cm: **A** viewed under normal light, and **B** highly fluorescent under 365 nm long-wave ultraviolet light. Erongo Mountains, Namibia.
BRUCE CAIRNCROSS SPECIMEN AND PHOTOS.

Figure 590 Opal (Opal-CT), 4.6 cm. Mesopotamia copper valley, Namibia. GERHARD LOUW SPECIMEN AND PHOTO.

Figure 592 Porcelainous white opal (polished front surfaces) from Mashamba, Zimbabwe. The piece on the right is 8 cm.
BRUCE CAIRNCROSS SPECIMENS AND PHOTO.

Orthoclase ◆ KAlSi$_3$O$_8$

Orthoclase is a member of the feldspar group of minerals. It crystallizes in the monoclinic system, has a hardness of 6 to 6.5, specific gravity of 2.55 to 2.63, white streak and vitreous to pearly lustre. Orthoclase is a potassium-bearing feldspar. It has a similar chemical composition to microcline, but crystallizes in the monoclinic system. It commonly forms large, well-shaped crystals in pegmatites, where it is associated with quartz, mica and other feldspars such as microcline and albite. Orthoclase is a rock-forming mineral commonly found in felsic igneous rocks, syenites and some sandstones, notably arkose, which has a high feldspar content. The distinctive red-orange colour of arkose is caused by orthoclase.

This feldspar occurs as well-formed crystals that are commonly associated with quartz in pegmatites at the Bushveld tin mines, **South Africa**. Many granites, syenites and gneisses in the Limpopo, Mpumalanga, KwaZulu-Natal, Western Cape and Northern Cape provinces contain orthoclase. Aesthetic euhedral crystals are found together with chalcopyrite crystals in some mines in the Okiep copper district. A large deposit of unakite (a rock composed of pink orthoclase and green epidote, which is used for lapidary purposes) occurs near Neilersdrif, between Kakamas and Keimoes, in the Northern Cape. Unakite can also be found close to the northern border with Zimbabwe.

Orthoclase is common in many of **Namibia's** granites and gneisses. It is often found in pegmatites as attractive, well-formed crystals associated with the usual suite of pegmatite minerals, such as schorl, quartz, microcline, topaz and tourmaline.

Orthoclase feldspar occurs in pegmatites in the Karoi district, **Zimbabwe**. In the late 1980s, the St Ann's mine produced some very aesthetic off-white to pale pink interlocking orthoclase crystals. Many were associated with acicular schorl tourmaline and some specimens were studded with small, transparent, blue topaz crystals. Orthoclase is common in much of the granite terrain that outcrops across large tracts of central and eastern Zimbabwe. A gem variety of orthoclase is said to come from the Fungwe area. Orthoclase is present in the rocks of the Shawa and Dorowa carbonatites.

A conspicuous 6-km ridge of syenite west of Francistown, **Botswana**, is composed of over 95% orthoclase. Orthoclase has a similar distribution pattern to that of microcline, and can be found in the pegmatites in that area.

Figure 593 Well-formed orthoclase crystals, 5.4 cm. Klein Spitzkoppe, Namibia. BRUCE CAIRNCROSS SPECIMEN AND PHOTO.

Figure 594 Orthoclase from Klein Spitzkoppe, Namibia. Naturally corroded cores of the feldspar are overgrown by a later, second generation, 5.6 cm. BRUCE CAIRNCROSS SPECIMEN AND PHOTO.

Papagoite ◆ CaCu[H₃AlSi₂O₉]

Papagoite crystallizes in the monoclinic system, has a hardness of 5 to 5.5, specific gravity of 3.25, light blue streak, and vitreous to dull lustre. Papagoite is a rare copper-bearing silicate hydroxide that has a vibrant blue colour. It is included here because famous collector-quality specimens come from South Africa.

The Musina copper mines in **South Africa** have produced some extraordinary papagoite specimens that are most commonly, but not exclusively, found as inclusions in quartz crystals (Cairncross, 1991, 2022). These inclusions are typically located in the terminations of the quartz, but can also be dispersed throughout the crystal. The papagoite occurs as minute elongate, fibrous crystals that often form radiating haloes. Associated species include other blue copper minerals such as ajoite and shattuckite, as well as copper, epidote, hematite, kaolinite and piemontite. Rare, solid masses of tightly packed platy crystals are also known. Some quantitative analyses have been undertaken on the dark blue minerals included in the quartz and most are papagoite (Matsimbe, 2019). It is important to note that most papagoite and shattuckite included in quartz crystals has not been quantitatively identified, but rather named based on the colour and habit of the crystals.

Figure 595 Rare crystals of papagoite confirmed by X-ray diffraction analysis, 2 cm. Messina mine, South Africa.
BRUCE CAIRNCROSS SPECIMEN AND PHOTO.

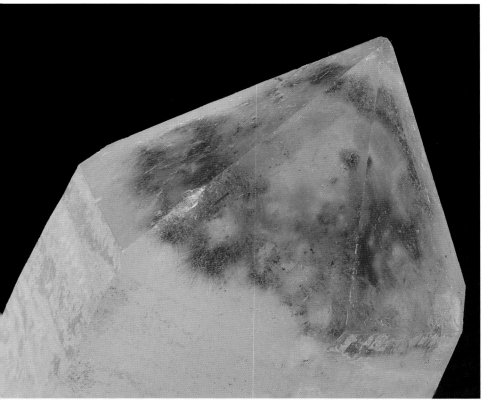

Figure 596 A cut and polished section of hexagonal quartz with included minerals, 4 cm. The dark blue fibrous crystals were analysed and are papagoite. The lighter blue material is ajoite. Messina mine, South Africa.
BRUCE CAIRNCROSS SPECIMEN AND PHOTO.

Figure 597 The termination of a quartz crystal included with blue papagoite. Messina mine, South Africa. Field of view 6.3 cm. BRUCE CAIRNCROSS SPECIMEN AND PHOTO.

Pectolite ◆ $NaCa_2Si_3O_8(OH)$

Pectolite crystallizes in the triclinic system, has a hardness of 4.5 to 5, specific gravity of 2.74 to 2.88, a white streak, and silky to vitreous lustre. Pectolite is white and occurs as distinctive fibrous crystals that often aggregate into compact spherical groups. Some pectolite crystals are very soft and sectile. Pectolite occurs in amygdales in basalts, and rarely as a secondary mineral in veins and cavities in economic orebodies and kimberlites.

At the Palabora mine, **South Africa**, pectolite is found in open cavities as delicate needles up to 3 cm in length, coated with minute, clear, fluorapophyllite crystals (Gliddon and Braithwaite, 1991; Southwood and Cairncross, 2017). Pectolite was found in the Ramp Dyke in small cavities, most of which are completely filled with matted needles of

pectolite. Very soft, radiating balls up to 5 mm in length have been found in the Main Dyke. In the Kalahari manganese field, pectolite is often found associated with other minerals at the N'Chwaning II and Wessels mines. Sprays of pectolite have also been recorded from Bloemhof in the North West. Attractive snow-white balls of pectolite were found in kimberlite pipes, notably at the Dutoitspan mine and at Jagersfontein. Pectolite occurred with an unusual black variety of apophyllite at Jagersfontein.

In **Zimbabwe**, pectolite is found, rarely, as small, needle-like crystals in cavities in Karoo basalts.

Figure 599 White pectolite with unusual black fluorapophyllite crystals, largest 2 cm. Jagersfontein diamond mine, South Africa. MCGREGOR MUSEUM SPECIMEN, BRUCE CAIRNCROSS PHOTO.

Figure 598 Interlocking, acicular natrolite crystals surrounded by radiating aggregates of pectolite. Palabora mine, South Africa. Field of view 3.2 cm. BRUCE CAIRNCROSS SPECIMEN AND PHOTO.

Figure 600 White pectolite, orange prehnite, hydroxyapophyllite-(K) and pink ferroan inesite, 10.2 cm. N'Chwaning II mine, South Africa. PAUL BOTHA SPECIMEN, BRUCE CAIRNCROSS PHOTO.

Phlogopite ◆ $KMg_3Si_3AlO_{10}(F,OH)_2$

Phlogopite crystallizes in the monoclinic system, has a hardness of 2 to 2.5, specific gravity of 2.76 to 2.9, white streak, and pearly lustre. Phlogopite is a member of the mica group of minerals, a sheet silicate that forms flat, platy crystals typical of micas. Crystals have a characteristic golden-orange to red colour, unlike silver-white mica muscovite and dark brown biotite. Phlogopite is one of the rock-forming species in certain igneous and metamorphic rocks, for example carbonatites and marbles.

Phlogopite occurs in Limpopo, **South Africa**, at the Palabora mine, where it is particularly abundant. It occurs in large brown, hexagonal crystals up to 20 cm in diameter. It is also found in many kimberlites.

Phlogopite occurs in some of the **Namibian** carbonatites. In the Rössing district, crystals up to 8 cm in diameter are found in amphibolite schist. Clusters of phlogopite, with crystals up to 6 cm in diameter, are found in a pegmatite hosted in marble on the farm Vergenoeg 92 in the Karibib district.

As at other southern African localities, phlogopite is common in the carbonatites of **Zimbabwe**, such as Shawa, as well as at the Colossus kimberlite. Phlogopite is found with alexandrite at the Novello claims in south-east Zimbabwe.

Figure 601 Phlogopite, 13.5 cm. Palabora mine, South Africa.
BRUCE CAIRNCROSS SPECIMEN AND PHOTO.

Figure 602 Hexagonal phlogopite on salmon-coloured calcite: **A** 6.2 cm; **B** 4.5 cm. Found in 2008 east of Swakopmund, Namibia.
BRUCE CAIRNCROSS SPECIMENS AND PHOTOS.

Platinum ◆ Pt

Platinum crystallizes in the cubic system, has a hardness of 4 to 4.5, specific gravity of 21.44, a steel-grey streak, and metallic lustre. Native platinum (Pt) is rare in nature, and it usually combines with other elements such as sulphur to form other mineral species. Platinum is even heavier than gold, its weight being a distinguishing characteristic. Superficially, platinum crystals may resemble galena, but the specific gravity of platinum is three times greater. Platinum has a very high melting point and is very malleable and ductile. Commercially, it is known as 'white gold', and is used extensively in the jewellery trade and for certain coinage, and plays an important role in catalytic converters. It is used in the manufacture of glass and fibreglass, in certain electrical components, and to convert hydrogen to heat energy in fuel cells. Platinum occurs in ultramafic rocks, such as those in the Bushveld Complex and the Great Dyke. Crystals of platinum can weather from these rocks and be deposited with alluvium.

In **South Africa**, tiny crystals of platinum up to 1 mm on edge are found in the rocks of the Onverwacht dunite pipe. The mineral is panned from alluvium near the pipe. Platinum also comes from the Mooihoek, Twyfelaar and Driekop pipes in the Bushveld Complex. The small crystals are well formed and occur as cubes, octahedra or combinations thereof. Microscopic grains are fairly common in the platinum reefs of the Bushveld Complex, the most famous being the Merensky Reef in the eastern and western Bushveld Complex.

In **Zimbabwe**, native platinum has been found at the Mimosa platinum mine in the Shurugwi district, the Umtebekwa valley (as alluvial platinum), the 'potato' reef in the Great Dyke and as alluvial nuggets in the Somabula diamondiferous deposits.

Figure 603 Tiny (up to 1 mm) grains of bright silver platinum mixed with black chromite and magnetite. Driekop mine, Limpopo, South Africa. Field of view 1.5 cm. BRUCE CAIRNCROSS PHOTO.

Poldervaartite ◆ $Ca(Ca_{0.5}Mn)(SiO_3OH)(OH)$

Poldervaartite crystallizes in the orthorhombic system, has a hardness of 4, specific gravity of 2.91, a white streak, and vitreous to pearly lustre. Crystals are pale tan and typically cluster together. Poldervaartite is a type-locality mineral from Wessels mine in South Africa and was reported as having been found only once, but there have been several occasions when this rare mineral was encountered at the mine. Associated minerals include andradite, calcite, bultfonteinite, hematite and henritermierite. Olmiite specimens discovered in 2001 were thought to be poldervaartite, but further analyses carried out showed this not to be the case. It should be noted, however, that olmiite crystals very commonly display light cream to dark orange colour-zoned layers at a macroscopic and microscopic level. These appear to indicate chemical variants and repetitive interlayering of poldervaartite and olmiite in the same crystals; therefore, both species are most likely present in single specimens, a phenomenon similar to the interlayering of ettringite and sturmanite (see olmiite).

Figure 604 Poldervaartite crystals, possibly intermixed with olmiite, and minor red andradite, 2.6 cm. Wessels mine, South Africa. BRUCE CAIRNCROSS SPECIMEN AND PHOTO.

Prehnite ◆ $Ca_2Al_2Si_3O_{10}(OH)_2$

Prehnite crystallizes in the orthorhombic system, has a hardness of 6 to 6.5, specific gravity of 2.9 to 2.95, a white streak, and vitreous to pearly lustre. It occurs as beautiful, spherical, crystal aggregates, commonly a vibrant light green to apple-green. Prehnite was the first mineral to be named and described from South Africa, and it was also the first mineral to bear the name of a person, Hendrik von Prehn (1733–1785). Prehnite is a low-temperature mineral that occurs in certain lavas such as basalts. It is commonly associated with dolerite intrusions.

Prehnite is believed to have been discovered in the Karoo dolerites of the Cradock district, Eastern Cape, **South Africa**, although it is known from other dolerites in the Karoo. At the Palabora mine in the main dolerite dyke fracture zone, prehnite forms characteristic pale green crusts of compact radiating blades up to 1 cm thick with terminations on the surface of the crust (Gliddon and Braithwaite, 1991; Southwood and Cairncross, 2017). Small, individual, platy crystals up to 3 mm across were occasionally found on dolerite. There are similar occurrences of prehnite in some of the dolerite quarries in KwaZulu-Natal, as at the Coedmore quarry in Durban.

At Soetwater, near Calvinia in the Northern Cape, a farmer looking for Iceland-spar calcite in caves in the dolerite sills in the 1970s found an unusual occurrence of prehnite in the shape of

Figure 605 Prehnite from the Premier mine, South Africa. Field of view 1.8 cm. BRUCE CAIRNCROSS SPECIMEN AND PHOTO.

epimorphs of 'angel-wing' calcite, forming interlocking plates. In the Musina district, prehnite was regarded as a typical mineral of the Musina copper deposits and a common accessory in these deposits in fault zones in the district. Distinct crystals of prehnite have not been found at the mines. The colour of the rare orange prehnite found at the N'Chwaning II mine in the Kalahari manganese field in April 2000 is caused by traces of manganese in the crystal structure (Cairncross *et al.*, 2000). The crystals are acicular, up to 2 cm in length, and are associated with dark red, iron-rich inesite, calcite and datolite.

Figure 606 The town of Cradock in the Eastern Cape. The hills on the horizon contain Jurassic-age dolerite dykes that host prehnite. The Cradock district is reputed to be the area where prehnite was first discovered in South Africa. BRUCE CAIRNCROSS PHOTO, 2018.

Figure 607 Columns of prehnite, 11.5 cm. Cradock district, Eastern Cape, South Africa. JOHANNESBURG GEOLOGICAL MUSEUM SPECIMEN, BRUCE CAIRNCROSS PHOTO.

Figure 608 Dolerite hills on Bekkerskloof farm east of Cradock, where an early twentieth-century gold mine operated. This historical site yielded the prehnite specimen in figure 607. BRUCE CAIRNCROSS PHOTO, 2018.

Figure 611 A plate of interlocking prehnite, 28 cm (largest crystal 21 cm). Zoetwater, Calvinia district, South Africa. WARREN TAYLOR RAINBOW OF AFRICA COLLECTION, MARK MAUTHNER PHOTO.

Figure 614 Translucent prehnite crystals. Marlin Norite quarry, Belfast, South Africa. Field of view 6 cm. BRUCE CAIRNCROSS SPECIMEN AND PHOTO.

Basalts forming the Goboseb Mountains to the west of the Brandberg in **Namibia** are famous for prehnite that occurs in geodes and cavities (Cook, 1999). Prehnite forms here as attractive, sea-green, rounded aggregates and spheres, up to 10 cm in diameter and larger, which are found isolated or coalesced as plates (Cairncross and Bahmann, 2006b; 2007; Von Bezing *et al.*, 2008; 2014; 2016). Beautiful geodes lined with prehnite are also collected. Associated minerals include amethyst, analcime, calcite, clear quartz, and, rarely, epidote and babingtonite. Some amethyst crystals have spheres of prehnite attached. Attractive, smooth, green, botryoidal aggregates of prehnite come from the Karasburg district and from Tubussis.

In **Zimbabwe**, prehnite is found as green botryoidal masses in some basalt vugs and as a rock-forming mineral in some low-grade metamorphic rocks and igneous granites. It is reported from the Scheelite King mine.

Figure 615 A view of Tafelkop in the distance, west of the Brandberg, Namibia. The hills consist of basalt that hosts prehnite, quartz and other collectable minerals. BRUCE CAIRNCROSS PHOTO, 2017.

Figure 616 A specimen of green prehnite and cream analcime, 18.5 cm. Goboseb Mountains, Namibia. BRUCE CAIRNCROSS SPECIMEN AND PHOTO.

Figure 617 Bright green prehnite on quartz with calcite. Goboseb Mountains, Namibia. Field of view 5.4 cm. BRUCE CAIRNCROSS SPECIMEN AND PHOTO.

Figure 618 A prehnite-lined geode with quartz, 12.2 cm. Goboseb Mountains, Namibia. BRUCE CAIRNCROSS SPECIMEN AND PHOTO.

Figure 619 A large 22.6-carat (2.1-cm) faceted prehnite. Copper Valley, Goboseb Mountains, Namibia. WARREN TAYLOR RAINBOW OF AFRICA COLLECTION, MARK MAUTHNER PHOTO.

Pyrite ◆ FeS$_2$

Pyrite crystallizes in the cubic system, has a hardness of 6 to 6.5, specific gravity of 5, a green-black streak, and metallic lustre. Cubic crystals are common as are octahedral and pyritohedral forms and combinations thereof. Pyrite's distinctive brassy gold colour has earned it the common name 'fool's gold'. It is used as a source of sulphur, which in turn is used to manufacture sulphuric acid, fertilizers, soap, detergents, matches, gunpowder, fireworks and vulcanized rubber. Pyrite is the most widespread and abundant sulphide mineral in all types of rocks and mineral deposits. It is relatively widespread as disseminated grains in many base metal deposits and in certain sedimentary rocks such as banded iron formations. In economic metal deposits, pyrite is usually the most common sulphide present, associated with, for example, galena, chalcopyrite, arsenopyrite and pyrrhotite.

Octahedral crystals of pyrite, up to 4 cm in length, have come from the Rooiberg tin mine, and specimens are known from the other tin mines of the Bushveld Complex in **South Africa**. In the Murchison range, crystals up to 2 cm in diameter occur in emerald-bearing biotite schist. Good-quality specimens are also recorded from the Pilgrim's Rest goldfields. Pyrite is the most common sulphide in the Kalahari manganese field, although it is only sporadically found there as attractive specimens. Composite crystals up to 20 cm in diameter have been found at the Gloria and Wessels mines (Cairncross and Beukes, 2013). Several hundred specimens of pyrite were found in association with calcite, baryte and marcasite in 1990 in an adit in the N'Chwaning II mine. Small geodes containing cubes, octahedrons and modified cubes of pyrite occur in seams (ranging in length from a few millimetres to several centimetres) in the central portion of the Mamatwan opencast workings, in association with massive calcite and chalcedony. Very attractive specimens consist of millimetre-sized cubes and octahedrons sprinkled on white baryte crystals.

Figure 620 Octahedral pyrite crystals with scattered grey cassiterite, 7.5 cm. Rooiberg mine, South Africa. BRUCE CAIRNCROSS SPECIMEN AND PHOTO.

Figure 621 Close-up view of cubic pyrite crystals. N'Chwaning II mine, South Africa. Field of view 2.8 cm. BRUCE CAIRNCROSS SPECIMEN AND PHOTO.

◄ **Figure 622** Sub-millimetre pyrite crystals, cascading over calcite. Palabora mine, South Africa. Field of view 1.2 cm. BRUCE CAIRNCROSS SPECIMEN AND PHOTO.

Figure 623 ➤
A semi-rounded, fine-grained pyrite concretion dredged from the Richards Bay harbour, South Africa, 4.6 cm. BRUCE CAIRNCROSS SPECIMEN AND PHOTO.

Pyrite is dispersed throughout the Witwatersrand gold-bearing conglomerates and is mined as a source of sulphur at the Tau Lekoa, Kopanang and Great Noligwa (Vaal Reefs) and Lorraine mines, and at the New Consort mine in the Barberton district. The pyrite in the matrix of the Witwatersrand conglomerates occurs as well-rounded and subrounded grains, leading to the term 'buckshot pyrite'. Pyrite is common in coal seams and often forms diagenetic concretions in rocks.

Pyrite has been mined commercially at some base metal deposits, as at the Otjihase mine (Goldberg, 1976), **Namibia**, and is particularly abundant at the Matchless mine and Gorob deposits. At Kopermyn, pyrite occurs together with chalcopyrite and covellite. Pyrite was fairly abundant at the Tsumeb mine, with octahedral crystals up to 4 cm in diameter. It is disseminated in the Haib porphyry copper deposits and is common in the orebody at Rosh Pinah, both located in southern Namibia.

Much **Zimbabwean** pyrite occurs as dispersed grains in iron-formation layers. At Kanyemba (Kadoma district) folded chert and ironstone contain solid pyrite lenses up to 10 m thick. Pyrite is a coal-mining by-product at the Wankie Colliery. It is the main ore mineral at the Iron Duke pyrite mine, Mazowe district.

In **Botswana**, pyrite is found in graphite schists at Moshaneng, and in nickel deposits in the Tati schist belt and at Phoenix and Selkirk. It is a common constituent in other base-metal sulphide deposits and a common accessory mineral in coal seams.

In **Eswatini**, pyrite is disseminated in gold and sulphide deposits in the Pigg's Peak and Forbes Reef districts. Crystals occur in conglomerates in the Kubuta district and Mahlangatsha area. Pyrite is found in the country's coal deposits and associated shales.

Figure 624 A 15-cm pyrite specimen, largest crystal 9 cm on edge. Witwatersrand goldfield, South Africa. BRUCE CAIRNCROSS SPECIMEN AND PHOTO.

Pyrolusite ◆ Mn^{4+}O$_2$

Pyrolusite crystallizes in the tetragonal system, has a hardness of 2 to 6.5, specific gravity of 5.06, a black streak, and metallic to dull lustre. Pyrolusite is found as tiny, shiny black, metallic crystals or, more commonly, in solid black layers. It is often the mineral that forms dendrites. If found in large enough quantities and in sufficient concentrations, pyrolusite can be exploited for its manganese content. It is common in manganese ore deposits and in manganese-rich shales.

At the opencast Ryedale manganese mine between Ventersdorp and Krugersdorp, **South Africa**, seams of columnar pyrolusite crystals occur in well-bedded but weathered shale of the Karoo Supergroup. Small cavities in these seams are lined with brilliant silvery pyrolusite crystals. Pyrolusite is found at mines in the Kalahari and Postmasburg manganese fields (Cairncross *et al.*, 1997); a few specimens of small pyramidal crystals have come from the Wessels mine, and radial aggregates of pyrolusite were found in jasper at Langdon-Annexe; similar radially orientated needles up to a few centimetres on edge occurred in pockets at the old Black Rock mine.

Figure 625 Three different habits of pyrolusite from the Ryedale mine, South Africa: **A** concretionary masses with yellow goethite, field of view 7 cm; **B** tightly packed columnar crystals, field of view 4.5 cm; **C** cavities lined with brilliantly lustrous pyrolusite crystals, field of view 3 cm. BRUCE CAIRNCROSS SPECIMENS AND PHOTOS.

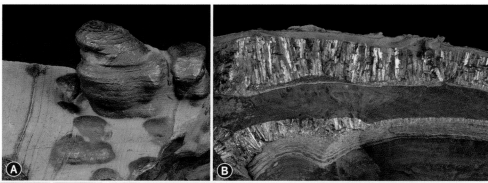

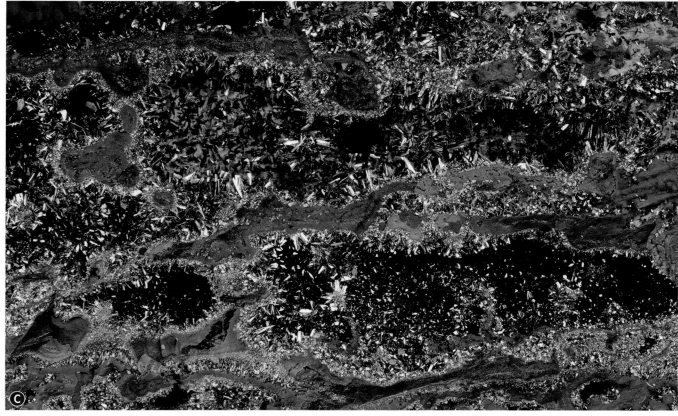

◄ **Figure 626** The open pit at Ryedale mine, with the horizontal sedimentary layers visible. BRUCE CAIRNCROSS PHOTO, 2010.

Figure 627 ➤
End-on view of the radial growth pattern of a 5.1-cm (2.5-cm diameter) columnar pyrolusite specimen. Postmasburg area, South Africa. BRUCE CAIRNCROSS SPECIMEN AND PHOTO.

Manganese mineralization occurs in the Vryheid district in KwaZulu-Natal, where manganese has been leached out of a source rock and concentrated and precipitated on the surface as secondary minerals, such as pyrolusite. On the farms Dipka and Dwaalhoek on the Bivane River, wad (manganese oxides), psilomelane and pyrolusite occur as seams, veins and botryoidal masses, some of which display an attractive shiny black surface texture (Astrup and Hammerbeck, 1998). On the farm Goedgeloof, shales contain concretions and seams of manganese ore.

In **Namibia**, pyrolusite is common in the Otjosondu manganese field and the Kombat mine. Tiny (1-mm) silver-coloured crystals occur as pseudomorphs after siderite in some of the Erongo pegmatites. Pyrolusite was common in iron ore mined on the farm Eisenberg 78 in the Otjiwarongo district (Schneider, 1992b).

Pyrolusite is one of the main manganese minerals exploited in **Zimbabwe** at localities such as the Dan workings (Kwekwe district), Hedgehog (Gweru district) and Morocco (Makonde district). In Mazowe district, it forms attractive reniform coatings and crusts at Gadzema, Shashani Ranch and Dunstable kaolin mine.

Pyrolusite is reported from **Botswana** in an area 3 km south of Ramotswa village and south-east of Lobatse.

Pyrolusite and other manganese oxides are found in **Eswatini** at the old Ngwenya iron mine.

Pyromorphite ◆ $Pb_5(PO_4)_3Cl$

Pyromorphite crystallizes in the hexagonal system, has a hardness of 3.5 to 4, specific gravity of 7.04, white streak, and resinous to sub-adamantine lustre. Pyromorphite is highly prized by collectors, as the crystals are vibrant yellow-green in colour, commonly hexagonal, barrel-shaped and sharp-edged. It is a secondary mineral, usually associated with lead deposits.

Pyromorphite is widely distributed as micro-crystals up to several millimetres in size in the oxidized ore body at the Argent mine near Delmas, Gauteng, and the defunct Edendale mine east of Pretoria, **South Africa** (Atanasova et al., 2016). The crystals vary from yellow to green, clear to opaque, and occur in a number of forms from acicular to hexagonal. In the Marico district, near Ottoshoop in the North West, pyromorphite crystals recovered in the early 1920s measured up to 1 cm in length. They are prismatic and generally aggregated in

Figure 628 Clusters of pyromorphite crystals. Edendale mine, South Africa. Field of view 3 mm. WOLF WINDISCH SPECIMEN AND PHOTO.

parallel or divergent groups. Pyromorphite also occurred as botryoidal masses and crusts ranging in colour from yellow to grey-green to dark olive green. Crystals have also been reported from lead deposits in the Pretoria, Belfast and Lydenburg districts. Small, attractive green crystals, isolated on gossanous matrix, were once found at the Broken Hill mine, Aggeneys. Small green crystals occur at Moore's Hill, Waterval Onder.

In **Namibia**, pyromorphite is known from the farm Uitsab 654, the Tsumeb and Kombat mines, and from Berg Aukas.

Pyromorphite comes from the Midway and Matabele mines in the Masvingo district, **Zimbabwe**, and from the Copper King gossan, where it occurs with mimetite and anglesite. At the Old West mine, 1.8 km north-west of Penhaloga village, north of Mutare, pyromorphite crystals are found in the oxidized portion of the galena orebody.

Figure 629 **A** Mixture of yellow-green pyromorphite and red vanadinite, field of view 2.4 cm; **B** discrete crystals of pyromorphite on weathered goethite matrix, 4 cm. North West, South Africa. BRUCE CAIRNCROSS SPECIMENS AND PHOTOS.

Figure 630 Two specimens of pyromorphite crystals from Argent, South Africa: **A** green crystals on matrix, collected in 1969, 6.7 cm; **B** a micromount specimen, field of view 3.1 mm. BRUCE CAIRNCROSS SPECIMEN A AND PHOTO, WOLF WINDISCH SPECIMEN B AND PHOTO.

Pyrope ◆ $Mg_3Al_2(SiO_4)_3$

Pyrope crystallizes in the cubic system, has a hardness of 7 to 7.5, specific gravity of 3.5 to 3.8, a white streak, and vitreous lustre. Pyrope is a garnet that forms attractive pink, pink-red, orange-red to purple-red dodecahedral crystals. Rhodolite is a red-violet variety of pyrope that is used as a gemstone. Pyrope occurs in ultramafic rocks, kimberlites and in alluvium derived from weathered kimberlite.

Pyrope is found in alluvial gravels from diamond diggings in the Free State, Northern Cape and North West, **South Africa**. It is commonly associated with the diamonds in the kimberlite pipes and in the alluvial gravels of the Orange and Vaal rivers.

In **Namibia**, pyrope occurs in the kimberlites that are clustered in the Gibeon-Brukkaros area.

Pyrope occurs in the Selukwe and Colossus kimberlite pipes, **Zimbabwe**. Rhodolite is exploited in the Mutoko and Beitbridge areas and in the Bubi district. Jack's Luck mine (Karoi district) was another source of gem-quality rhodolite.

Figure 631
A polished slice of harzburgite, backlit, 10 cm. The isolated pink crystals with black rims are pyrope garnet, set in a matrix of orthopyroxene and olivine. Letšeng diamond mine, Lesotho. MESSENGERS FROM THE MANTLE COLLECTION, BRUCE CAIRNCROSS PHOTO.

Pyrophyllite ◆ $Al_2Si_4O_{10}(OH)_2$

Pyrophyllite crystallizes in both the triclinic and monoclinic systems, has a hardness of 1 to 2, specific gravity of 2.65 to 2.9, a white streak, and dull to pearly lustre. Pyrophyllite is a low-grade metamorphic mineral occurring as tiny, mica-like crystals, which are usually white, cream or golden. When pyrophyllite forms thick rock-like layers and can be economically mined, it is better known by its common name 'wonderstone' (Nel et al., 1937), a reference to its many properties, including excellent thermal stability, which keeps it from expanding when heated. Ferripyrophyllite $Fe^{3+}Si_2O_5(OH)$ is the iron analogue of pyrophyllite. Pyrophyllite is a member of the pyrophyllite-talc group and, because of its softness, it is used for stone carvings and can be polished into different colours. Apart from stone carvings and other *objets d'art*, it is used in ceramics and insecticides, and as a filler in cosmetics. More recently, pyrophyllite has been used as a lining in high-pressure synthetic diamond manufacturing apparatus. Pyrophyllite is usually disseminated in low-grade metamorphic rocks.

Figure 632 A block of massive grey pyrophyllite with a zone containing rounded concretions, some of which have weathered out and left cavities, 13.5 cm. Ottosdal, South Africa. BRUCE CAIRNCROSS SPECIMEN AND PHOTO.

Figure 633 Minute golden pyrophyllite crystals on quartz. President Steyn gold mine, South Africa. Field of view 2.1 cm. BRUCE CAIRNCROSS SPECIMEN AND PHOTO.

Figure 634 Ferripyrophyllite from the Amis Schlucht, Brandberg, Namibia. This material has been used for ornamental stone carvings and this specimen has been cut and polished. Field of view 12.5 cm. BRUCE CAIRNCROSS SPECIMEN AND PHOTO.

In **South Africa**, an economic deposit has been quarried on the farm Gestoptefontein 349 IP, close to Ottosdal, west of Klerksdorp, North West, where very ancient Archaean volcanic ash deposits in the Dominion Group have altered over time to form thick beds of grey-coloured pyrophyllite (Astrup and Horn, 1998; Agangi et al., 2020; 2021).

The mineral is recorded as yellow micro-crystals on quartz from the Witwatersrand gold mines, and as small crystals with amesite and diaspore in aluminous shales from the Postmasburg manganese field. Pyrophyllite is also found at the Masala deposit, 15 km north-east of eMkhondo in Mpumalanga, and occurs near Groblersdal.

A deposit is known at Waaihoek in the Western Cape, 17 km north-west of Worcester.

A well-known deposit of pyrophyllite is found in the Amis Schlucht on the western edge of the Brandberg Mountain, north of Uis, **Namibia** (Von Bezing et al., 2016). Light red-brown layers are exploited for lapidary and stone carvings. Bladed white pyrophyllite is also found at Klein Spitzkoppe.

In **Eswatini**, pyrophyllite is found in the ancient volcanic rocks of the Insuzi Group in the Manzini, Mahlangatsha and Mkopeleli areas (Davies et al., 1964). Several deposits occur in schists in the Gege-Sicunansa area in the Mankaiana district, associated with diaspore and andalusite.

Figure 635 Blades of radiating pyrophyllite, 6.6 cm. Klein Spitzkoppe, Namibia. DEBBIE WOOLF SPECIMEN AND PHOTO.

Pyrrhotite ◆ Fe$_{1-x}$S

Pyrrhotite crystallizes in the monoclinic system, has a hardness of 3.5 to 4.5, specific gravity of 4.53 to 4.77, a grey-black streak, and metallic bronze-yellow to reddish lustre. Pyrrhotite is sometimes confused with pyrite, both having a brassy, metallic colour and lustre. Unlike pyrite, which forms cubic and octahedral crystals, pyrrhotite usually forms flat, platy crystals and tends to tarnish an iridescent red-brown. This sulphide species is often found in mafic and ultramafic rocks, associated with other sulphides such as arsenopyrite and pyrite.

In **South Africa**, pyrrhotite occurs in metallic ore deposits such as at Bon Accord, Musina, Okiep, the Witwatersrand goldfield, Broken Hill-Aggeneys, the Bushveld Complex and others, but it is usually found only in massive form. Wafer-thin tabular crystals up to 5 cm in diameter, the finest examples of pyrrhotite crystals in southern Africa, were collected at the Mponeng gold mine in the Carletonville district in the late 1990s and again in early 2003 (Kershaw *et al.*, 2003). They were associated with faults and cavities encountered during the mining of gold-bearing reefs, and were sometimes partially coated with galena and/or sphalerite. The crystals were scattered singly on matrix or clustered together in rosette-like aggregates. Platy crystals and aggregates up to 5 mm in diameter have been found in the Kusasalethu gold mine, associated with quartz, pyrite, calcite and kerogen.

Pyrrhotite is found at the Kobos copper mine 55 km south-west of Rehoboth, **Namibia**. It also occurs at the Matchless and Rosh Pinah mines and, with copper deposits, in the Otjiwarongo district. The Navachab gold mine also has pyrrhotite present in the orebodies. Euhedral crystals are reported from the Namib lead mine.

Pyrrhotite is frequently associated with nickel deposits in **Zimbabwe**, for example at the Empress nickel mine and Trojan mine, its distribution somewhat similar to that of pentlandite.

Pyrrhotite is found in **Botswana** in the Tati schist belt at the Phoenix and Selkirk nickel deposits, with chalcopyrite and pentlandite. In the Selebi-Phikwe district, gneiss and amphibolite contain pyrrhotite with other sulphides. There are many other small, scattered deposits in the region, all of which are associated with either serpentinites or amphibolites. Pyrrhotite occurs with chalcopyrite in Usushwana gabbro at Mhlanbanyati, **Eswatini**. It is also reported from a scheelite occurrence in the Forbes Reef area, in biotite granite in the vicinity of the waterfall on the Malolotsha River.

Figure 636
Massive pyrrhotite forming the bulk of the specimen, with brassy chalcopyrite, pentlandite and hornblende, 5.8 cm. Selebi-Phikwe mine, Botswana. BRUCE CAIRNCROSS SPECIMEN AND PHOTO.

Figure 637 Pyrrhotite specimens: **A** 4 cm; **B** 6.1 cm. Mponeng gold mine, Witwatersrand goldfield, South Africa.
BRUCE CAIRNCROSS SPECIMENS AND PHOTOS.

Quartz ◆ SiO$_2$

Quartz crystallizes in the trigonal system, has a hardness of 7, specific gravity of 2.66, white streak, and vitreous lustre. Although hexagonal crystals are the most common, quartz occurs in other habits, such as finely banded layers of agate, amorphous opaline quartz, and as a replacement mineral in 'tiger's eye'. It has no cleavage, but displays a typical conchoidal fracture pattern. Quartz is one of the most common minerals on Earth, as a component in many sedimentary, igneous and metamorphic rocks. Sandstone, for example, consists primarily of detrital quartz grains. Granite and granitic pegmatites contain abundant quartz, and quartz is present in gneisses and some schists.

Quartz occurs in a great diversity of colours, and several varieties are exploited as gemstones, including amethyst, citrine, 'tiger's eye' and carnelian.

Quartz is the main source of silica and silicon. Building sand and glass contain silica. Silicon chips are widely used in the computer and electronics industries.

Variety	Colour / Features
Agate	multicoloured
Amethyst	purple
Aventurine	green
Carnelian	orange-red
Chalcedony	multicoloured
Chrysoprase	apple-green
Citrine	yellow
Jasper	red
Milky quartz	milky white
Rose quartz	pink
Smoky quartz	grey, black, transparent
Tiger's eye, pietersite	striated brown, blue

Figure 638 A selection of different coloured tumbled stones: virtually all varieties of quartz – brown, blue and red (heat-treated) tiger's eye; blue agate; green chrysoprase; agate; red jasper; and rose quartz – as well as dark blue sodalite (not quartz). BRUCE CAIRNCROSS SPECIMENS AND PHOTO.

Figure 639 Multicoloured quartz crystals from the well-known 'cactus quartz' locality at Boekenhouthoek, South Africa. The yellow specimen (back left) is 6.5 cm. BRUCE CAIRNCROSS SPECIMENS AND PHOTO.

Figure 640 Two large quartz specimens from the Becker's pegmatite/Otjua mine, Namibia: **A** on public display at the Kristall Galerie in Swakopmund; **B** photographed at the Tucson Gem and Mineral Show. BRUCE CAIRNCROSS PHOTOS.

Figure 641 One of the large quartz crystals from Jan Coetzee mine, South Africa. BRUCE CAIRNCROSS SPECIMEN AND PHOTO.

Figure 642 A 28-cm quartz crystal cluster from the Messina mine, South Africa. BRUCE CAIRNCROSS SPECIMEN AND PHOTO.

Quartz crystal (general)

Well-formed crystals of quartz, some up to a metre or more in length, have been found in the Okiep copper district in Namaqualand, **South Africa**, where several, relatively large copper mines operated in the past. The crystals occur in cavities in the orebodies and are often doubly terminated. Associated minerals include axinite, baryte, calcite, chalcocite, chalcopyrite, chlorite, fluorite, orthoclase, pumpellyite and stilbite. At the best-known locality, Jan Coetzee mine, hundreds, if not thousands, of quartz crystals were found in a large cavity that was opened during mining in 1966 (Von Bezing and Kotze, 1993). Many are dark smoky quartz and have green chlorite, calcite and attractive small yellow baryte crystals attached. The Nababeep West mine has also produced fine quartz crystals, some over a metre long (sometimes used by farmers as fence posts).

Figure 643 A 22-cm vug in a massive quartz vein in dolomite. Kuruman district, South Africa. BRUCE CAIRNCROSS SPECIMEN AND PHOTO.

Figure 644 Quartz from South Africa's Witwatersrand goldfield. **A** Quartz with oxidized brown pyrrhotite. Kusasalethu gold mine. Field of view 5 cm. **B** A mass of elongate quartz crystals. Kopanang gold mine. Field of view 4.5 cm. **C** Quartz and minor pyrrhotite, 16 cm. Hartebeesfontein mine.
BRUCE CAIRNCROSS SPECIMENS AND PHOTOS.

Figure 645 A cluster of quartz crystals collected from dolomite outcrop north of Johannesburg, 5 cm. Hennops River, South Africa. BRUCE CAIRNCROSS SPECIMEN AND PHOTO.

In the Reivilo district in the North West, transparent crystals up to 15 cm long occur in cavities in dolomite, together with galena, sphalerite and grey dolomite crystals. There are similar occurrences in quartz veins in dolomites in Mpumalanga in the Emalahleni-Middelburg area and in the Magaliesberg, north of Johannesburg. A popular collecting site for clear quartz is near the Hennops River on the road to Hartebeespoort Dam, north of Johannesburg.

In the Kalahari manganese field, milky quartz is relatively common as crystals forming drusy coatings that are particularly attractive on black manganite crystals. In many cases, these coatings are found on well-developed calcite and baryte crystals in pockets associated with alteration zones at the Mamatwan mine, as well as at Langdon-Annex. Encrustations of quartz and chalcedony on rhodochrosite, manganite, pyrolusite and calcite have been found from other manganese mines. Large calcite scalenohedrons overgrown with quartz from N'Chwaning II are especially attractive.

Brilliant, clear crystals with a vitreous lustre, up to 5 cm in length, were found at the Wessels mine. These crystals were associated with pectolite, hydroxyapophyllite-(K) and sugilite. Some have inclusions of sugilite crystals; in others the sugilite forms the phantom faces. Phantom crystals are formed by earlier stages of crystallization that are outlined by thin veneers of another mineral, in this case sugilite, that outline the shape of the quartz before further growth takes place (see figure 704b for a good example of phantom growth).

Very attractive red and orange quartz has been found in pegmatites along the Orange River. The colour is caused by inclusions of hematite forming cloudy red and orange phantoms in the quartz. These can be associated with fluorite and, very rarely, baryte.

Weathered dolerite in certain regions of KwaZulu-Natal have produced attractive doubly terminated 'Herkimer' quartz crystals, up to 20 cm, some with cavernous cavities and brown clay. These have been found in the Durban North/KwaMashu area, Kloof, north-west of Durban, and near Mooi River.

Figure 646 Examples of quartz crystals from KwaZulu-Natal, South Africa: **A** Transparent quartz with inclusions of brown clay, 2.4 cm. Mooi River. **B** Doubly terminated quartz, 4.5 cm. Ecubazini, Pietermaritzburg district. **C** Quartz specimen, 4.4 cm. Giant's Castle, Drakensberg. BRUCE CAIRNCROSS SPECIMENS AND PHOTOS.

Figure 647 Transparent quartz, 4.6 cm. Makhanda district, Eastern Cape, South Africa. BRUCE CAIRNCROSS SPECIMEN AND PHOTO.

Figure 648 A cluster of glassy quartz extracted from a cavity in the Drakensberg basalts, 5.5 cm. Butha-Buthe, Lesotho.
BRUCE CAIRNCROSS SPECIMEN AND PHOTO.

Figure 649 Glassy quartz crystal with numerous fluid and vapour bubble inclusions, 3.4 cm. Rosh Pinah mine, Namibia.
BRUCE CAIRNCROSS SPECIMEN AND PHOTO.

Good specimens of crystal quartz come from many of the pegmatites in the Karibib, Omaruru and Swakopmund districts, **Namibia**. Klein Spitzkoppe and Erongo have also produced beautiful specimens. The Gamsberg pegmatites have sporadically produced large, ice-white, deeply etched crystals, some with multiple sceptre terminations, called 'cathedral quartz'. On the farm Kos in the Gamsberg region, crystals of over a metre in length were found, together with amethyst and sceptered quartz.

The Karoo basalts at Tafelkop, west of the Brandberg, are famous for fine-quality, clear quartz specimens, as well as smoky varieties and amethyst (Jahn *et al.*, 2006; Von Bezing *et al.*, 2008; 2014; 2016). These have been known since the mid-1950s (Cairncross and Bahmann, 2006b). Thousands of specimens have been dug from the weathered alluvium and unweathered basalt, some as composite groups, but most as single, loose crystals in the 5–10 cm range (although crystals over 35 cm long are known). Matrix specimens are rare. Doubly terminated crystals are common, and beautiful sceptres may be found. Many, if not most, of these crystals are highly included, largely by brown clay, and there are fluid inclusions with gas bubbles up to a few centimetres in diameter that move in the crystal when it is tilted. Some of the most attractive pieces contain phantoms made up of combinations of amethyst and smoky quartz varieties.

In the extreme south of Namibia, quartz that is often coloured blood-red by the inclusion of hematite, and amethyst are found. On the farm Doornfontein 316, 60 km west of the town of Rehoboth, quartz veins cut across schist, marble and quartzites. In places, the veins have cavities lined with transparent quartz crystals that have piezoelectric properties.

Well-formed, giant quartz crystals over 2 m long and 2 m in diameter come from the Otjua pegmatite in the Karibib district. Very unusual giant quartz crystals occur on the farm Verloren in the Hakos Mountains west of Rehoboth, where crudely formed quartz crystals several tens of metres long outcrop along a prominent ridge. Associated with the quartz are equally large crystals of other minerals, such as dolomite and calcite rhombs up to 1 m long, stellate groups of tremolite as long as 20 m and specularite plates 30 cm in diameter.

Clear quartz crystals are the most common mineral in pegmatites throughout **Zimbabwe**. The Karoi district in the north-west has many sites, as does the north-eastern region. Quartz veins are very common in granites and these often contain attractive quartz crystals. Milky quartz is found in the Mutoko district pegmatites and north-west of Mwami. Quartz is common in the many pegmatites in the Karoi district. Some interesting quartz with green 'phantoms' was once extracted from an unspecified locality in Zimbabwe.

In **Botswana**, quartz crystals come from the Selebi-Phikwe area. There are many quartz veins in rocks in the Tati district, near Ramokgwebana and north of Francistown.

Crystals of up to 35 cm long come from the Pigg's Peak district, **Eswatini**. Clear quartz crystals were found at the defunct Devil's Reef gold deposit in the north-west (Jones, 1962). Some have inclusions of red hematite phantoms and tiny silver hematite, as well as spheres of minute, yellow goethite needles.

Quartz crystals are found in **Lesotho** in the same basalts that produce agate and chalcedony in neighbouring South Africa. These can take the form of delicate, drusy stalactites or clusters of brilliantly lustrous, glassy quartz.

Quartz variety agate

Agates are characterized by multiple thin laminations of chalcedony (micro-crystalline quartz) that define distinct banding. Many agates are found as amygdales in basalts and other volcanic rocks.

In **South Africa**, the Drakensberg basalts that outcrop in KwaZulu-Natal, the Free State and Eastern Cape contain many agates (Gliozzo *et al.*, 2019). In northern KwaZulu-Natal, in the Jozini-Pongolapoort area, there are grey-blue agates up to a metre in diameter that weigh many kilograms. Many alluvial diamond-digging dumps at Bloemhof, Barkly West and Lichtenburg contain a variety of agates. Agates also occur in present-day alluvial sediments of the Orange (Gariep), Vaal, Limpopo and Caledon rivers. Moss agates, formed by dendrites in chalcedony, come from the Lebombo Mountains and the Lichtenburg diamond diggings.

Agates are plentiful on some of **Namibia's** beaches, particularly at Lüderitz at the appropriately named Agate Beach. A famous deposit of blue-lace agate occurs in veins associated with dolerite in southern Namibia, south-west of Karasburg, on the farm Ysterputz 254. North of the Ugab River, brown, grey, red and yellow alluvial agates are eroded out of basalt. Diamond-bearing gravels also contain abundant agates.

Figure 650 Two views of the Drakensberg Mountains, South Africa: **A** Alpine Heath; **B** Golden Gate. These are a source of various minerals, including quartz variety agate. In both instances, the mountains on the horizon are basalts that host the minerals. BRUCE CAIRNCROSS PHOTOS, A 2014; B 2015.

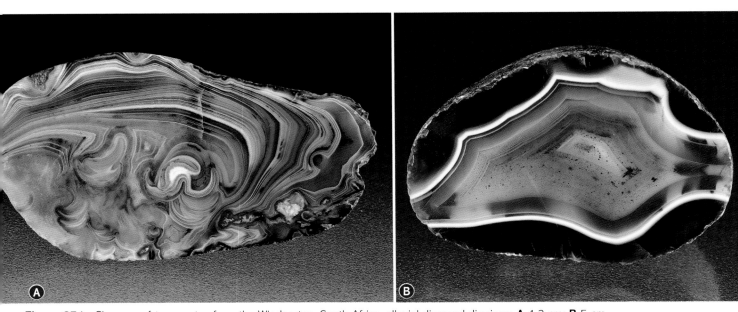

Figure 651 Close-up of two agates from the Windsorton, South Africa, alluvial diamond diggings: **A** 4.2 cm; **B** 5 cm.
BRUCE CAIRNCROSS SPECIMENS AND PHOTOS.

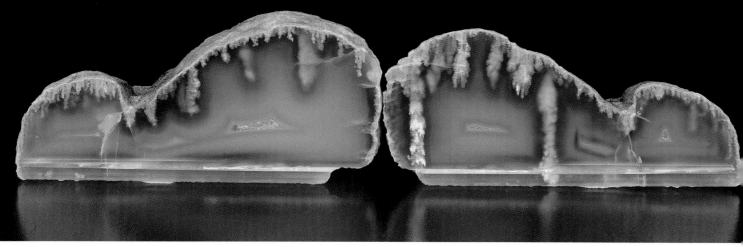

Figure 652 Cut and polished halves of agate from Jozini, KwaZulu-Natal, South Africa. Right hand piece is 8.5 cm. MASSIMO LEONE SPECIMENS, BRUCE CAIRNCROSS PHOTO.

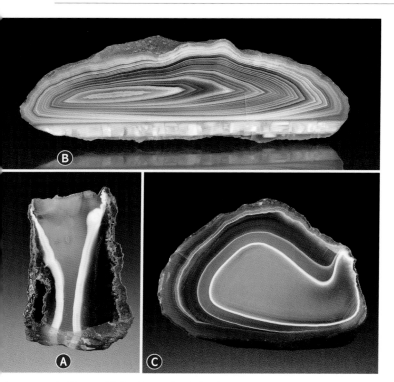

▲ **Figure 653** Cut and polished agates: **A** 5.5 cm; **B** 10.5 cm; **C** 9.7 cm. Bobonong district, Botswana. BRUCE CAIRNCROSS SPECIMENS AND PHOTOS.

Figure 654 Blue lace agate: ➤ **A** 9.7 cm; **B** polished cabochon, 6.2 cm. Ysterputz, Karasburg district, Namibia. BRUCE CAIRNCROSS SPECIMENS AND PHOTOS.

Agates fill amygdales and geodes in Karoo basalts, notably in the Tsholotsho, Bumi Hills and Featherstone areas, **Zimbabwe** (Anderson, S.M., 1980). Being resistant to weathering, they are liberated when basalts weather and crumble, accumulating in alluvium and other weathered debris. Most Zimbabwean agate is grey or blue-grey to pink. Agates have been worked at the Dunkirk, Pimbi and New Brooklyn deposits in the Charter district. They are found together with amethyst, jasper, unakite, quartz and fuchsite in the Nyamandhlovu district (at the Mazibope property and at Chikodzi and Madombe). In the Lupane district, agates occur with green jasper and rose-pink quartz. Agates are also found in the Hwange-Bulawayo area and the Save-Limpopo region.

The famous pink and cream **Botswana** agates come from present-day alluvial deposits in the Bobonong district (Zenz, 2005). Pink varieties are the most sought after, but other colour variants are also popular. These are found in the Tuli region, several kilometres north of Pontdrif, which is on the South African/Botswana border, west of Beitbridge.

Agates come from the southern region of **Eswatini**, originating in Lebombo basalt and rhyolites.

Agate is relatively common in certain geological environments in **Mozambique**, notably associated with volcanics such as the Lebombo Mountains in the Maputo Province that borders

Figure 655 A typical grey-brown agate from Lesotho, cut in half and polished to show the intricate banding, 5.2 cm. BRUCE CAIRNCROSS SPECIMEN AND PHOTO.

Figure 656 General view of KwaMhlanga village, Boekenhouthoek, South Africa. This region has been the source of a variety of multi-coloured quartz specimens, including amethyst. BRUCE CAIRNCROSS PHOTO, 2008.

Figure 657 All the specimens collected at Boekenhouthoek, South Africa, and surrounding areas are via informal diggings on outcrop, such as this one. Quartz veins are found on surface and then excavated in search of crystals. BRUCE CAIRNCROSS PHOTO, 2008.

South Africa and Mozambique. Similar deposits are found in the Canxixe-Doa area of Sofala Province, and the volcanics in the Tete Province, where some geodes occur up to 30 cm in diameter (Lächelt, 2004). In general, the agates display a wide range of colours, including white, grey, pink and red. The Gaza Province is a source of alluvial agates, and alluvial deposits occur on the Zambezi River in the vicinity of Chemba, Sofala Province.

Agates are found in many of the streams in **Lesotho** (Cairncross and Du Plessis, 2018). As in South Africa, they are weathered from the Drakensberg basalts that make up much of the country. Fine yellow-and-blue banded examples come from the upper Sani Pass.

Quartz variety amethyst

Amethyst is found at many localities in southern Africa. It is a mauve to purple variety of quartz, its colour the result of Fe^{3+} ions in its atomic lattice.

Excellent-quality amethyst-lined geodes as well as loose crystals are found in basalt and Lebombo rhyolites in the Jozini area of northern KwaZulu-Natal, **South Africa**.

Some of the best amethyst found in South Africa occurs at Kwaggafontein, Boekenhouthoek and Mathys Zyn Loop in the KwaNdebele area, north-east of Pretoria, on the road to Groblersdal (Ehlers, 2003; Cairncross *et al.*, 2004). It is dug from quartz veins in the Bushveld granophyre, and ranges in colour from light to deep purple. Discovered in 1986, it disappeared off the market, resurfacing en masse years later, in 2001. The mineral's distinctive habit of forming prisms that, except for the termination, are coated with a myriad tiny, drusy crystals, has given rise to the name 'cactus quartz'. It is also called 'spirit quartz' by local diggers, perhaps because some crystals are the colour of methylated spirits. A prodigious quantity of this amethyst and other quartz varieties from this locality has entered collections around the world, all collected by the artisanal diggers.

Figure 658 A selection of variously coloured amethyst:
A 13.2 cm; **B** 5.5 cm; **C** 10.3 cm; **D** 9.2 cm; **E** 4.5cm. Boekenhouthoek, South Africa.
BRUCE CAIRNCROSS SPECIMENS AND PHOTOS.

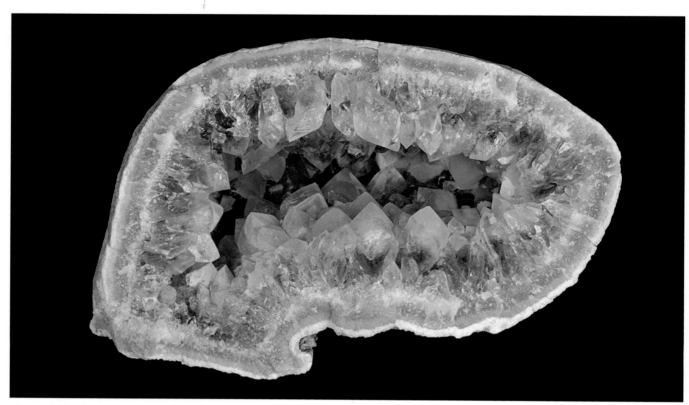

Figure 659 A quartz-amethyst-lined geode, 10.5 cm. Jozini area, South Africa. BRUCE CAIRNCROSS SPECIMEN AND PHOTO.

Figure 660 A large plate of amethyst, some with orange-red iron staining, 42 cm. Boekenhouthoek, South Africa. DEPARTMENT OF GEOLOGY, UNIVERSITY OF JOHANNESBURG COLLECTION, BRUCE CAIRNCROSS PHOTO.

Figure 661 Amethyst from the Kalahari manganese field, South Africa, 6.5 cm. Amethyst is relatively rare from the manganese mines. BRUCE CAIRNCROSS SPECIMEN AND PHOTO.

Figure 662 A 16-cm cluster of amethyst on matrix. Goboboseb, Namibia. DESMOND SACCO SPECIMEN, BRUCE CAIRNCROSS PHOTO.

Amethyst is also found near Pretoria at Hammanskraal and, in the North West, near the Hennops River. Amethyst veins, associated with smoky quartz and citrine, are found in Archaean granite close to Bryanston in the northern suburbs of Johannesburg, and in the koppies around Leeuwkop. It also occurs further north in the granites of the Bushveld Complex near Driehoek station, 30 km west of Modimolle, in Limpopo.

In the Northern Cape, high-quality amethyst crystals are found about 15 km south of Pofadder. Other areas are the Keimoes, Kakamas and Augrabies districts, Vuurdoodberg near Goodhouse on the Orange (Gariep) River, and the farm Vaal Koppies in the Gordonia district. South of Kakamas, small amethyst crystals line fractures and joints in pegmatites. Amethyst, combined with red hematite-included quartz, is sourced from pegmatites in the Onseepkans area. Drusy lavender-purple amethyst was collected from the Kalahari manganese field.

In **Namibia**, superb amethyst crystals, some over 20 cm long – with deep purple amethystine colours that are often combined with dark smoky grey layers, brown clay inclusions and fluid inclusions – come from Karoo-age basalts that make up the Goboboseb Mountains west of the Brandberg, and further south towards the Messum crater (Cairncross and Bahmann, 2006b; Von Bezing et al., 2014; 2016). Doubly terminated crystals are common, as are sceptres of amethyst on stems of both clear and smoky quartz. Unusual reverse sceptres (the terminations taper upwards instead of bulging outwards) are occasionally found. At Sarusas, on the Skeleton Coast, amethyst-lined geodes weather out from the basalt, some crystals up to 5 cm in diameter (Schneider and Seeger, 1992d). Good-quality amethyst comes from the farm Rooisand in the Gamsberg region. Quartz veins containing amethyst occur in granite on the farm Otjipetekera Süd 97 in the Omaruru district. In the Grootfontein district in the north, quartz-calcite veins on several farms are hosted in marble and limestone. Some facet-grade amethyst is found in these veins.

Figure 663 A view of Tafelkop, Goboboseb Mountains, Namibia, with a *Welwitschia mirabilis* in the foreground. The flat-topped hills are composed of basalt that hosts the minerals for which the region is most famous. BRUCE CAIRNCROSS PHOTO, 1999.

Figure 664 Vermiform reverse amethyst sceptres on quartz, 4 cm. Goboboseb, Namibia. BRUCE CAIRNCROSS SPECIMEN AND PHOTO.

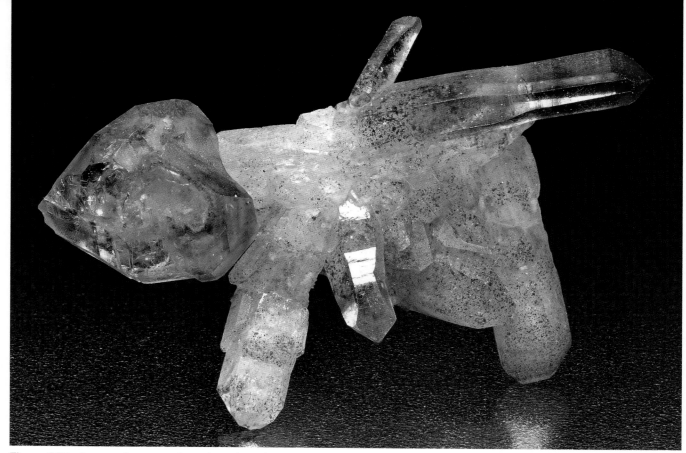

Figure 665 A somewhat unusual specimen from Goboboseb, Namibia. A series of intergrown quartz, amethyst and smoky quartz crystals, some at right angles, and a sceptre (left) forming the 'head' of the specimen. The tiny red crystals inside the quartz are hematite, 5.5 cm. BRUCE CAIRNCROSS SPECIMEN AND PHOTO.

Figure 666 Doubly terminated amethyst: **A** has a sceptre on the upper termination, 7 cm; **B** 8.5 cm. Goboboseb, Namibia.
BRUCE CAIRNCROSS SPECIMENS AND PHOTOS.

In **Zimbabwe**, amethyst lines cavities in Karoo basalts in the Tsholotsho and Hwange districts and is often crystallized on an agate base. Amethyst is also found in pegmatites in the Karoi region. Some noteworthy localities are: Manzinyama (Nyamandhlovu district), where amethyst occurs as transparent mauve-pink crystals lining geodes, and the Coronet and Pat pegmatites (Karoi district), where amethyst was mined with mica, beryl, gem quartz, garnet and aquamarine. Doubly terminated crystals come from the Mutoko district. Amethyst is reported from pegmatites and basalts in the Beitbridge district, Featherstone and Mutoko areas, and the Bumi district.

In recent times (2018 onwards), excellent amethyst specimens have been collected from the Chiredzi district, Masvingo. Many of these specimens have well-formed sceptres, some doubly terminated, that consist of deep purple on white hexagonal quartz stems. Specimens can be over 20 cm in length and comprise individual crystals and multiple interlocking crystals.

Amethyst is reported from the Lebombo Mountains, **Mozambique** (Lächelt, 2004). Specimens are not gem quality, i.e. dark purple and transparent, but rather light mauve and of interest to collectors, not jewellers.

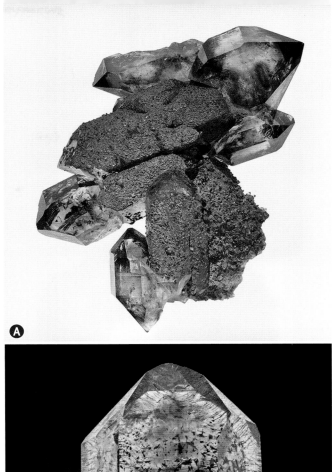

A

B

Figure 668 Amethyst specimens, backlit: **A** cluster of doubly terminated crystals, 7 cm; **B** close-up, showing red hematite inclusions, field of view 4 cm. Chiredzi district, Zimbabwe.
PAUL BOTHA SPECIMENS, BRUCE CAIRNCROSS PHOTOS.

Figure 669 A 2.2-carat faceted amethyst from an undisclosed locality in Zimbabwe. BRUCE CAIRNCROSS SPECIMEN AND PHOTO.

Figure 667 A sceptred quartz-amethyst, 8.5 cm. Chiredzi district, Masvingo, Zimbabwe. BRUCE CAIRNCROSS SPECIMEN AND PHOTO.

Quartz variety aventurine

Aventurine is an apple-green variety of chalcedony, in which minute flecks of chrome-bearing muscovite (fuchsite) create a sparkling appearance. It is popular as a lapidary material.

Aventurine is found in **South Africa** on the farm Santor in the Soutpansberg, at Gravelotte and Leydsdorp in Limpopo, and in the Barberton district in Mpumalanga.

Aventurine occurs in **Zimbabwe** as attractive green quartzite (metamorphosed sandstone) that contains minute green fuchsite crystals. These impart a glittering sheen to the rock, making it a sought-after lapidary material. The best-known deposits are in the northern Beitbridge district, between Gwanda and Beitbridge, where the appropriately named Greenhill mine exploited the mineral. At Jopempe, veins of aventurine, 3–15 m wide, accounted for about 30,000 tonnes of reserves. Good-quality aventurine came from the Altitude mine in the Masvingo district.

Figure 670 A solid specimen of aventurine, 9.6 cm. Barberton district, South Africa. BRUCE CAIRNCROSS SPECIMEN AND PHOTO.

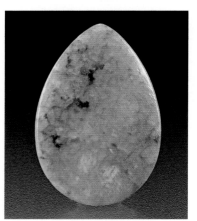

Figure 671 Polished cabochon of aventurine, 3.2 cm. The green fuchsite is clearly visible between the quartz grains. Mashonaland West, Zimbabwe. BRUCE CAIRNCROSS SPECIMEN AND PHOTO.

Figure 673 Naturally fractured carnelian, showing the characteristic conchoidal fracture patterns. The small, hollow centre is lined with drusy quartz, 5.8 cm. West coast, Namibia. BRUCE CAIRNCROSS SPECIMEN AND PHOTO.

Quartz variety carnelian

Carnelian is an attractive orange variety of chalcedony. The colour comes from impurities of iron oxide.

Carnelian occurs in the Hay and Prieska districts of the Northern Cape, **South Africa**. It is relatively common in the alluvial diamond deposits around Kimberley, Barkly West and Lichtenburg. Beautiful, lapidary-quality carnelian is found in Limpopo, in the Musina and Tshipise districts. Carnelian also occurs in northern KwaZulu-Natal, where it weathers out from Lebombo volcanics. It is found in the alluvium of rivers that drain these mountains.

In **Zimbabwe**, carnelian is sometimes found in basalt. It occurs either as layers or bands in amygdales and vugs, or as alluvial specimens weathered out from the basalt. In the Beatrice-Featherstone region, carnelian and agate are found together.

Carnelian is found along the Skeleton Coast, **Namibia**, with particularly rich pockets at Sarusas.

Carnelian is known from south-eastern **Eswatini** in Lebombo rhyolites and associated volcanic rocks. Deposits similar to those in South Africa occur south of Eswatini's northern border.

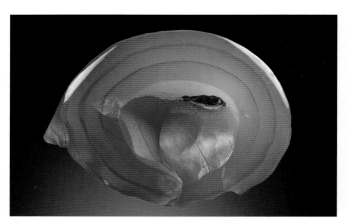

Figure 672 A quartz variety carnelian polished on one side, and showing concentric growth rings, 7.6 cm. Skeleton Coast, Namibia. BRUCE CAIRNCROSS SPECIMEN AND PHOTO.

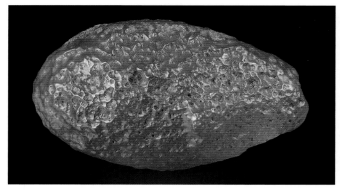

Figure 674 Carnelian displaying remnant surface patterns inherited from the basalt host rock, 3.8 cm. Sarusas, Namibia. BRUCE CAIRNCROSS SPECIMEN AND PHOTO.

Quartz variety chalcedony

Chalcedony is micro-crystalline quartz. It is usually found in shades of white, grey, black or blue, filling fissures and cavities in rocks. Chalcedony often lines amygdales and geodes in Karoo basalts. Mtorolite is an attractive green variety of chalcedony, its colour imparted by chromium. It is used as an ornamental stone and is favoured by lapidarists.

Chalcedony is common in **South Africa** in the alluvial sediments along the Vaal and Orange rivers, and in the Drakensberg basalts and rhyolites. Beautiful blue chalcedony has been extracted from the mines of the Kalahari manganese field. Some striking specimens are associated with red rhodochrosite and black manganite crystals.

In **Namibia**, very high-quality, jewellery-grade, semi-transparent, blue chalcedony occurs as veins and nodules hosted in marble at two farms, Otjoruharui 251 and Troye 253, approximately 150 km north-east of Okahandja (Von

Bezing *et al.*, 2008). When polished, it produces stones of a stunning, limpid blue colour and translucency. Another deposit of attractive blue chalcedony in marble is found 16 km north-east of Karibib on the farm Dobbelsberg 99. Attractive yellow, cream and blue stalactites and botryoidal layers in cavities in jasper occur near Rössing Siding.

Mtorolite, a variety of chalcedony, occurs in narrow veins in the rocks of the Great Dyke, close to Mutorashanga, **Zimbabwe**. At the Jester prospect in the Guruve district, over 7,000 tonnes was exploited (Anderson, S.M., 1980). A smaller

Figure 675 Botryoidal chalcedony on goethite. Gamsberg mine, South Africa. Field of view 2.5 cm. ALLAN FRASER SPECIMEN, BRUCE CAIRNCROSS PHOTO.

Figure 677 Botryoidal chalcedony, 8.2 cm. N'Chwaning II mine, Kalahari manganese field, South Africa. BRUCE CAIRNCROSS SPECIMEN AND PHOTO.

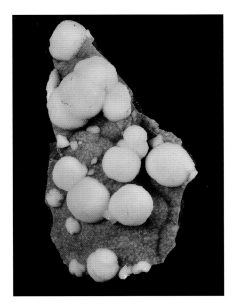

Figure 676 Two specimens of blue chalcedony: (left) natural, 6.7 cm; (right) cut and polished. Otjoruharui farm, Namibia. BRUCE CAIRNCROSS SPECIMENS AND PHOTO.

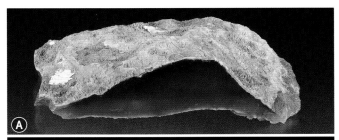

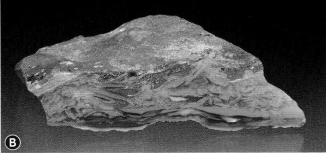

Figure 678 Two specimens of green chalcedony variety mtorolite: **A** 12.7 cm; **B** 7.5 cm. The green colour is caused by traces of chromium. Mtoroshanga, Makonde district, Mashonaland West, Zimbabwe. BRUCE CAIRNCROSS SPECIMENS AND PHOTOS.

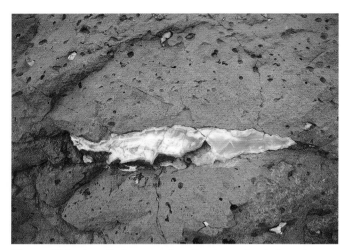

Figure 679 A cavity in a grey basalt outcropping in Lesotho. The right side is filled with blue-grey chalcedony and the left side with banded agate. The surrounding hollows in the rock are unfilled vesicles. HERMAN DU PLESSIS PHOTO, 2016.

deposit was worked at Spruit in the Mutare area. Some light blue chalcedony is found in the Tsholotsho district.

Green chalcedony comes from the north-eastern region of **Eswatini**.

Chalcedony is relatively common in the basalt from **Lesotho** (Cairncross and Du Plessis, 2018). It occurs as pale blue-grey masses usually lining or filling vugs.

Quartz variety chrysoprase

Chrysoprase is an apple-green variety of massive quartz. The vibrant green colour is caused by trace amounts of nickel. Chrysoprase can form by deep weathering of nickel-rich serpentinites.

Chrysoprase is found in **Namibia** near Rehoboth and in the Karasburg district. It is known from the Kaokoveld, 45 km north-west of Otjovazandu.

Good-quality chrysoprase comes from the eastern regions of **Zimbabwe**.

Figure 680 Polished slice of chrysoprase, 8.8 cm. Zimbabwe. DEPARTMENT OF GEOLOGY, UNIVERSITY OF JOHANNESBURG SPECIMEN, BRUCE CAIRNCROSS PHOTO.

Quartz variety citrine

Citrine is the yellow variety of quartz. The name usually refers to transparent crystals that can be faceted into gemstones. The yellow colour is caused by trace ferric iron contamination. Citrine is used extensively in the gemstone industry, and because natural citrine is rare, most material is obtained by heating amethyst or smoky quartz to produce commercially derived citrine. Laboratory-grown citrine is readily used in the jewellery trade. Some crystals purported to be citrine are yellow, iron-stained quartz.

In **South Africa**, citrine is found in the Free State, on the farm Jagersfontein 14. Yellow quartz from the KwaNdebele area has been referred to as citrine, but most is iron-stained quartz. Some of these crystals are beautiful with yellow iron-staining but are not true citrine.

Citrine is found in tantalite-bearing pegmatites south of the Rubikon mine, **Namibia**, on the farms Okongava Ost 72 and Otjimbingwe 104 (Schneider and Seeger, 1992d).

Veins containing citrine are found in granites in the Harare South region, **Zimbabwe**, and due west and north of Marondera (Anderson, S.M., 1980). Some deposits have produced significant quantities of citrine, for example, 21,547 kg from the deposit at Ulva.

Figure 681 An example of yellow, iron oxide/hydroxide-stained quartz incorrectly referred to as citrine, 7.4 cm. Boekenhouthoek, South Africa. BRUCE CAIRNCROSS SPECIMEN AND PHOTO.

Quartz variety jasper

Jasper is a variety of chert that has been coloured bright red by the presence of ferric (Fe^{3+}) iron. It is hard and breaks with a conchoidal fracture pattern. Jasper is a common constituent of iron-rich sedimentary rocks known as banded iron formations. As jasper is both colourful and hard, it has been used extensively as an ornamental stone.

The extensive banded iron formations that occur in the Northern Cape, **South Africa**, have long been a source of lapidary-grade jasper. Small jasper occurrences are found in some of the Archaean greenstone belts, such as in the Barberton mountainland (McIver, 1966; Macintosh, 1990; Cairncross and Dixon, 1995; Snyman, 1998). Some of the jasper is uniform red, while another variety, brecciated jasper, has an attractive play of white, cream and red that fills the voids between fragments of jasper.

Namibia has scattered jasper deposits in the Swakopmund-Karibib districts (Schneider and Seeger, 1992d).

In **Zimbabwe**, jasper is known to occur in some of the Archaean banded iron formations.

Quartz variety milky quartz

Milky quartz is translucent/opaque to white, hence milky. This variety is abundant and occurs in virtually any geological environment in southern Africa, but is common in quartz veins and granitic pegmatites. Some of the finest, large, well-formed milky quartz crystals were collected in Namaqualand, **South Africa** (Böllinghaus *et al.*, 2007). Milky quartz crystals,

Figure 683 Milky quartz crystal with smoky quartz attached, 12.8 cm. Steinkopf district, Northern Cape, South Africa. BRUCE CAIRNCROSS SPECIMEN AND PHOTO.

Figure 682 A block of jasper weighing several hundred kilograms. Northern Cape, South Africa. BRUCE CAIRNCROSS PHOTO.

Figure 685 Jasper disrupted by veins of white quartz and pale yellow-green andradite garnet, 13 cm. Griquatown district, South Africa. BRUCE CAIRNCROSS SPECIMEN AND PHOTO.

Figure 684 Lustrous, white milky quartz crystals: **A** 7.5 cm; **B** 6.8 cm. Boekenhouthoek, South Africa. BRUCE CAIRNCROSS SPECIMENS AND PHOTOS.

Figure 686 A very large slab of milky quartz crystals used as a garden feature. Messina mine, South Africa. BRUCE CAIRNCROSS PHOTO.

resembling small spinning tops, occur in an alluvial deposit between Hluhluwe and Umfolozi in KwaZulu-Natal. The crystals are small (under 1 cm), and consist of two joined hexagonal pyramids with no intervening hexagonal prism.

Quartz variety rose quartz

Rose quartz is a popular variety of quartz due to its attractive pink hue. It is found as massive forms as well as rose-pink crystals, which are much less common. In southern Africa, euhedral rose quartz crystals are unknown.

In **South Africa**, rose quartz is found as massive lumps and veins in pegmatites near Kenhardt and Keimoes and at Riemvasmaak in the Gordonia district in the Northern Cape (Hugo, 1970). High-quality, lapidary-grade rose quartz occurs in a pegmatite on the farm Jakkalswater, associated with beryl, and was mined at the Sleight's mine. Some pegmatites in the Goodhouse-Wolftoon area in the Orange River valley contain light to deep pink rose quartz.

Rose quartz is found in granites in Limpopo, on the farms Barend 523 MS and Palm Grove 14 MT in the Soutpansberg district, and at Selatidrift near Gravelotte. It also occurs at some tin mines in the Bushveld Complex. Pegmatitic rose quartz varies from very pale pink to light red on the farms Witkop 507 MT and Assegai 143 HT in the eMkhondo district in Mpumalanga.

Rose quartz is known from Rössing Siding, east of Swakopmund, **Namibia**. It is found in pegmatites in southern Damaraland and the Karibib and Karasburg districts (Schneider and Seeger, 1992d). Deep pink, gem-quality rose quartz came from the Bella Rosa rose quartz mine in the Karasburg district, where the Mickberg rose quartz mine is also located. South of Uis and east of Cape Cross, pegmatites contain beautiful translucent rose quartz. Large veins of rose quartz outcrop on the road near the entrance of the Ai-Ais resort.

Good-quality rose quartz comes from pegmatites in the Karoi and Mutoko districts and from the Beitbridge areas, **Zimbabwe**. In the late 1960s and early 1970s, gem-quality pink quartz was extracted from geodes in basalts at the Nyamandhlovu workings in the Manzinyama district.

Figure 687 A vein of rose quartz exposed close to Ai-Ais, Namibia. BRUCE CAIRNCROSS PHOTO, 1979.

Figure 688 Rose quartz and white milky quartz, with polished surface, 12 cm. Northern Cape, South Africa. BRUCE CAIRNCROSS SPECIMEN AND PHOTO.

Figure 689 Deep pink rose quartz, partially backlit to show the translucency, 21.7 cm. Unspecified pegmatite, Namibia. BRUCE CAIRNCROSS SPECIMEN AND PHOTO.

Quartz variety smoky quartz

The grey-brown-black colour in smoky quartz may be due to radiation from natural sources, and/or traces of Al^{3+} in its crystal lattice.

Well-crystallized smoky quartz occurs in **South Africa** in many granitic terrains, such as near the Nantes Dam on Paarl Mountain in the Western Cape. Beautiful crystals have been found in vugs in the granites of Bushveld Complex tin mines, notably at Zaaiplaats, where large crystals were associated with calcite, cassiterite and fluorite. Good-quality smoky quartz has been found in pegmatites in the Northern Cape, especially the Noumas pegmatites, at Angelierspan No.1, Witkop, and the Jakkalswater-Uranoop pegmatites (Hugo, 1970; 1986). In 1966, the Jan Coetzee mine north of Springbok produced spectacular large specimens (Von Bezing and Kotze, 1993). Smoky quartz crystals are occasionally found at the Hennops River area west of Pretoria.

Smoky quartz crystals occur in cavities in Karoo basalts at Tafelkop, west of the Brandberg, **Namibia**. Smoky quartz is also found in pegmatites in the Gamsberg region and the Karibib, Usakos and Swakopmund districts. At Neu Schwaben, large, sceptred smoky quartz crystals are found. The Becker's pegmatite located on the farm Otjua has crystals 40–50 cm in diameter, along with doubly terminated crystals up to 2 m long. Some of the Becker's pegmatite smoky quartz is gem quality. Beautiful, dark grey to black, transparent smoky quartz comes from the Klein Spitzkoppe pegmatites.

Smoky quartz comes from granites south of Harare and the Marondera region, **Zimbabwe**. It also occurs in the Karoi district pegmatites.

Figure 690 Dark smoky quartz cluster, 9.5 cm. Boekenhouthoek, South Africa. BRUCE CAIRNCROSS SPECIMEN AND PHOTO.

Figure 691 A smoky quartz sceptre, 7 cm. Tiny purple fluorite crystals are visible in milky quartz. Neu Schwaben, Namibia. BRUCE CAIRNCROSS SPECIMEN AND PHOTO.

Figure 692 Smoky quartz crystals from Jan Coetzee mine, South Africa: **A** 12.9 cm; **B** 31.2 cm. Associated minerals are yellow baryte, calcite and red hematite. BRUCE CAIRNCROSS SPECIMEN AND PHOTO A; RAINBOW OF AFRICA COLLECTION, MARK MAUTHNER PHOTO B.

Figure 693 Smoky quartz sceptre quartz on matrix, 19.5 cm. Goboboseb Mountains, Namibia. DESMOND SACCO SPECIMEN, BRUCE CAIRNCROSS PHOTO.

Quartz variety tiger's eye / pietersite

'Tiger's eye' forms when groundwater containing dissolved silica infiltrates the relatively brittle crystals and fibres of weathered crocidolite (Gutzmer *et al.*, 2003). Thick (over 10-cm) slabs that can be fashioned into large ornaments are the most sought after. The colours of 'tiger's eye' are dependent on the degree and intensity of weathering. True 'tiger's eye' is yellow-brown, but grey-green and yellow-green varieties can also form. The blue variety forms by silicification of unweathered blue crocidolite. Red 'tiger's eye' is produced by heating yellow-brown 'tiger's eye' to a temperature of 400°C.

'Tiger's eye' outcrops in the Griqualand West region of **South Africa**, where mining has been undertaken at Niekerkshoop and at several other places north-east of Prieska. However, most seams today are thin, yielding rough material for beads, small stone eggs and spheres.

Figure 694 A large smoky quartz crystal with white feldspars attached, 24.6 cm. Erongo Mountains, Namibia. BRUCE CAIRNCROSS SPECIMEN AND PHOTO.

Figure 695 Brown and blue varieties of 'tiger's eye', left specimen 9.2 cm. All have had a least one surface cut and polished. Griquatown district, South Africa. BRUCE CAIRNCROSS SPECIMENS AND PHOTO.

Pietersite is a variety of chalcedony with embedded fibres of amiphiboles. The altered amphibole causes a chatoyancy. It occurs on the farm Hopewell 240, about 40 km northeast of Outjo in **Namibia**. It was discovered in 1962 by Sid Pieters of Windhoek. Elongate crystals have been folded and broken to form very intricate patterns, and the material is a beautiful blue-brown colour due to the mixing of oxidized brown and unoxidized blue fragments. Pietersite has a different genesis to South African 'tiger's eye' and is not quartz, but a variety of chalcedony with fibrous amphibole minerals.

Quartz miscellaneous

Apart from several varieties of quartz, there are others that have unusual habits, various colours, and quartz with inclusions. A variety of quartz known as polyhedral quartz exhibits forms that bear no resemblance to normal quartz crystals. These come from the Seven Oaks farm in the Greytown district of KwaZulu-Natal in **South Africa** (Heron, 1989) and from the old tin mines in **Eswatini** (Mountain, 1942; Cairncross and Dixon, 1995). They are hollow with angular faces that have no crystallographic relationship to the known crystal systems and are believed to form in the spaces between other crystals, such as feldspar, that have dissolved away, leaving the moulds of silica in the form of the polyhedrons.

Figure 696 'Poker chip' quartz, 5.9 cm. Rehoboth district, Namibia.
BRUCE CAIRNCROSS SPECIMEN AND PHOTO.

Figure 697 A 'faden' quartz, 5.6 cm. The white line running through the centre of the specimen is caused by repetitive cracking of the crystal during growth, and the white colour is caused by fluid inclusions. Orange River region, South Africa.
BRUCE CAIRNCROSS SPECIMEN AND PHOTO.

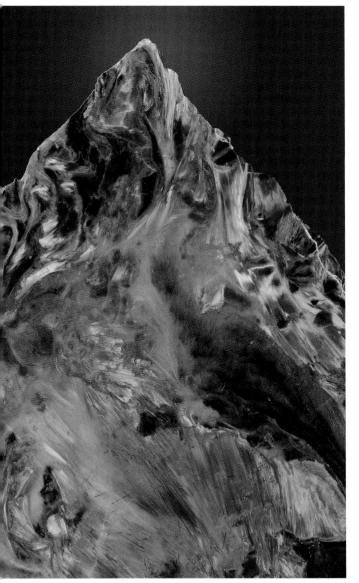

Figure 698 Close-up of a polished slice of pietersite from Hopewell farm, Namibia. Field of view 12 cm. BRUCE CAIRNCROSS SPECIMEN AND PHOTO.

Figure 699 Polyhedral quartz, the centre specimen revealing the typical hollow inside of these angular specimens: (left to right) 3.9 cm, 4.5 cm, and 4 cm. Seven Oaks, Greytown district, KwaZulu-Natal, South Africa. BRUCE CAIRNCROSS SPECIMENS AND PHOTO.

Figure 700 Drusy hollow quartz pseudomorphs after a cubic crystal (pyrite? fluorite?) attached to a milky quartz crystal. Field of view 4.5 cm. Boekenhouthoek, South Africa. BRUCE CAIRNCROSS SPECIMEN AND PHOTO.

Figure 701 Three sub-parallel sceptered quartz crystals, partly coated by goethite, 6.3 cm. Boekenhouthoek, South Africa. BRUCE CAIRNCROSS SPECIMEN AND PHOTO.

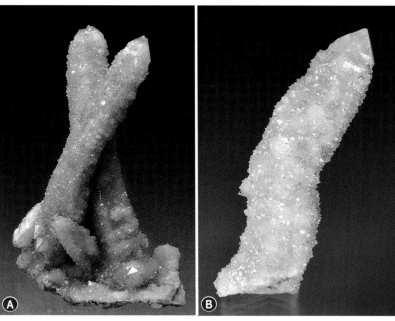

Figure 702 Two distorted quartz crystal specimens, possibly produced by mechanical stress during growth: **A** 10.5 cm; **B** 7.6 cm. Boekenhouthoek, South Africa. BRUCE CAIRNCROSS SPECIMENS AND PHOTOS.

Figure 703 A stalactitic (stalagmitic?) drusy quartz specimen, 6.8 cm. Butha-Buthe, Lesotho. BRUCE CAIRNCROSS SPECIMEN AND PHOTO.

Quartz with inclusions

Because quartz is often one of the last minerals to crystallize, it commonly incorporates and traps earlier formed minerals within it. These can add colour, vibrancy and value to such specimens; a simple quartz crystal is worth far less, aesthetically, scientifically or money-wise, than one with inclusions.

Figure 704 The Messina copper mines in South Africa have been known for decades for an array of included quartz crystals. Some of these are shown here: **A** pale red hematite-included doubly terminated quartz, 14 cm; **B** quartz crystal cut parallel to the long axis, showing many stages of 'phantom' sequential crystallization; there are also several trapped pyrite crystals, 8.5 cm; **C** polished quartz crystal with blue papagoite, 9 cm; **D** two quartz crystals polished on the crystallographic faces revealing red hematite, white kaolinite, light blue ajoite and dark blue papagoite, 6.3 and 5.2 cm. BRUCE CAIRNCROSS SPECIMENS AND PHOTOS.

Figure 705 Quartz from the Orange River region is well known for: **A** its red hematite inclusions that produce spectacular specimens, 5.5 cm; **B** other included minerals such as green chlorite, field of view 2.2 cm. Orange River, South Africa and southern Namibia.
BRUCE CAIRNCROSS SPECIMENS AND PHOTOS.

Figure 706 Quartz with hematite inclusions from Silkaats Nek, south of Brits, South Africa. Field of view 4.4 cm. BRUCE CAIRNCROSS SPECIMEN AND PHOTO.

Figure 707 Clear quartz crystals with inclusions of purple sugilite associated with white pectolite. Wessels mine, South Africa. Field of view 3 cm. BRUCE CAIRNCROSS SPECIMEN AND PHOTO.

Figure 708 Inclusions of silvery hematite and orange goethite in transparent quartz crystals. Artonvilla mine, South Africa. Field of view 2.5 cm. BRUCE CAIRNCROSS SPECIMEN AND PHOTO.

Figure 709
Quartz crystal with chalcopyrite inclusions 1.6 cm. Griquatown district, South Africa. BRUCE CAIRNCROSS SPECIMEN AND PHOTO.

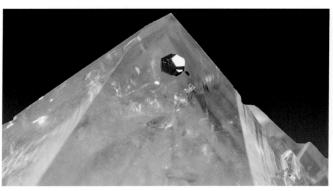

Figure 711 A perfectly formed pyritohedral pyrite crystal trapped in quartz. Boekenhouthoek, South Africa. Field of view 14 mm. BRUCE CAIRNCROSS SPECIMEN AND PHOTO.

Figure 710 A quartz crystal, 4.5 cm, and polished quartz gemstone, 1.69 carats, displaying bright red hematite inclusions. Goboboseb Mountains, Namibia. BRUCE CAIRNCROSS SPECIMENS AND PHOTO.

Figure 713 So-called 'Fenster quartz' – window quartz, or skeletal quartz, where the edges of the crystal grow more quickly than the faces, forming frames or windows: **A** this specimen is also included by brown clay, 15.8 cm. Goboboseb Mountains, Namibia; **B** a typical skeletal quartz from Eswatini, 2.4 cm. BRUCE CAIRNCROSS SPECIMENS AND PHOTOS.

Figure 712 Red and silver hematite and brown goethite inclusions create a multi-coloured crystal, 4.5 cm. Devil's Reef, Eswatini. BRUCE CAIRNCROSS SPECIMEN AND PHOTO.

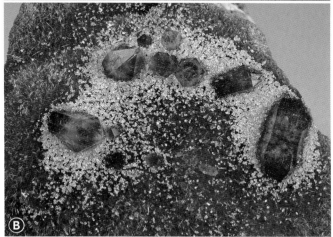

▲ Figure 714
Two examples from the Kaokoveld, northern Namibia: **A** blue shattuckite and green malachite inclusions, 5.1 cm; **B** a myriad of sugary quartz, and some larger crystals on and containing blue shattuckite, field of view 1.5 cm. BRUCE CAIRNCROSS SPECIMENS AND PHOTOS.

Figure 717 ▶
Top: amethyst with large fluid inclusions, 58.08 carats (2.1 cm), Goboboseb Mountains; bottom: tourmaline included in quartz, 21.8 carats (2.2 cm), Gamsberg area. GEMSTONES FROM THE WARREN TAYLOR RAINBOW OF AFRICA COLLECTION, MARK MAUTHNER PHOTO.

▲ Figure 715 Top to bottom: citrine, 9.15 carats (1.5 cm), Sarusas; rose quartz, 20.6 carats (2 cm), Roselis mine, Swakopmund district; amethyst, 16.69 carats (2 cm), Sarusas. GEMSTONES FROM THE WARREN TAYLOR RAINBOW OF AFRICA COLLECTION, MARK MAUTHNER PHOTO.

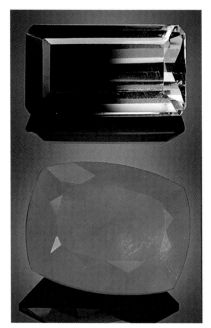

▲ Figure 716 Top to bottom: rose quartz, 18.52 carats (1.9 cm), Mickberg mine, Karasburg district; amethyst, 28.6 carats (2.5 cm), Platveld mine, Erongo Region; rose quartz, 31.9 carats (1.8 cm), Bella Rosa, Karasburg district; smoky quartz, 41.18 carats (2.5 cm), Klein Spitzkoppe. GEMSTONES FROM THE WARREN TAYLOR RAINBOW OF AFRICA COLLECTION, MARK MAUTHNER PHOTO.

◀ Figure 718
Top: bicolour quartz, 32.94 carats (2.6 cm), Klein Spitzkoppe; bottom: chalcedony, 46.73 carats (2.7 cm), Otjoruharui, Okahandja district. GEMSTONES FROM THE WARREN TAYLOR RAINBOW OF AFRICA COLLECTION, MARK MAUTHNER PHOTO.

Rhodochrosite ◆ $Mn^{2+}CO_3$

Rhodochrosite crystallizes in the hexagonal system, has a hardness of 2.5 to 4, specific gravity of 3.7, white streak, and a vitreous to pearly lustre. Rhodochrosite is relatively rare, but the mineral – particularly the famous blood-red crystals that have come from South Africa's Kalahari manganese field – is one of the most flamboyant. It is a manganese carbonate and forms from the oxidation of primary manganese ore. While the ore is typically black, rhodochrosite crystals are cranberry-red or deep pink or colourless to pink, orange, light red or dark red to brown-red and have a variety of habits. In itself rhodochrosite has no economic value, but the best specimens command extremely high prices on the collectors' market. Rhodochrosite occurs as a secondary mineral associated with manganese deposits.

Rhodochrosite was the mineral that brought international fame to the mines of the Kalahari manganese field in **South Africa**, when the first specimens were found in opencast workings at the Hotazel mine in the early to mid-1960s. The specimens that emerged in 1977 from the N'Chwaning I mine were the mineral sensations of the 1977 international gem and mineral shows (Wilson and Dunn, 1978; Von Bezing *et al.*, 1991; Cairncross and Beukes, 2013; Cairncross *et al.*, 2017). Noteworthy discoveries included dark red spheres and dark, intense red, often transparent scalenohedrons, up to 7 cm in length, on a matrix of brilliant jet-black manganite crystals.

Truly exceptional, undamaged rhodochrosites are not that common, particularly the scalenohedral form, whose terminations cleave off very easily. Some gem-quality material has been faceted from fragments yielding stones up to 60 carats. Most of the high-quality rhodochrosite was extracted from N'Chwaning I mine. Specimens from Hotazel and N'Chwaning I mines typically occur with chalcedony, gypsum, manganite, drusy quartz and todorokite.

Apart from its colour variations, which range from wine-red to pale pink, the rhodochrosite occurs in a variety of shapes and sizes, from 'dog-tooth' (scalenohedral) varieties to smooth spheres and 'wheat-sheaf' bundles of crystals.

Small opaque pink rhombohedrons and somewhat larger crystals up to 3 cm on edge have come from the adjacent Wessels mine.

Drusy pink rhodochrosite was the most common mineral found associated with the 2006 N'Chwaning I mine shigaite specimens (see shigaite). Most have drusy, pale pink to pink interlocking rhodochrosite crystals, one to three millimetres long. Some specimens have translucent pink crystals up to a centimetre or slightly longer, but these tend to be isolated on the drusy matrix. One variation of the pink rhodochrosite is thin, prismatic needle-like crystals up to two centimetres, which are intergrown with each other. These crystals are transparent, pale red and bicoloured, almost always with a darker, red-brown colour at the termination of the crystals. In early 2006, a very small discovery of rhodochrosite was made at N'Chwaning II mine, probably fewer than three or four dozen specimens. The difference from the classic Kalahari rhodochrosite is that this small pocket yielded pink, curved rhombohedrons, up to a centimetre. Matrix specimens coated with these crystals were collected, the largest being 12 cm.

Figure 719 Scalenohedral 'dog's tooth' rhodochrosite: **A** 4 cm; **B** 8.2 cm. N'Chwaning I mine, South Africa. DESMOND SACCO SPECIMENS, BRUCE CAIRNCROSS PHOTOS.

Figure 720 Clusters of bright red rhodochrosite, 9.5 cm. N'Chwaning I mine, South Africa. DESMOND SACCO SPECIMEN, BRUCE CAIRNCROSS PHOTO.

Figure 721 A glassy, bright red rhodochrosite, 16 cm. N'Chwaning I mine, South Africa. DESMOND SACCO SPECIMEN, BRUCE CAIRNCROSS PHOTO.

Rhodochrosite has also been found at the Broken Hill mine at Aggeneys as small, pale pink cauliform aggregates of clear crystals, associated with balls of siderite. Minute pale pink Fe-Zn-rich rhodochrosite occurs at Houtenbeck (Atanasova *et al.*, 2016).

Rhodochrosite is reported from the Ai-Ais lead mine, **Namibia**, where it is associated with galena and malachite. It has also been reported from the Kombat mine and from Aris, the latter producing tiny off-pink rhombohedral crystals.

Rhodochrosite has been found at the Iona mine (Mutare district) and the Kingsley mine (Bindura district), **Zimbabwe**.

Figure 722 A group of many flat-topped rhodochrosite crystals, 13 cm. N'Chwaning I mine, South Africa. DESMOND SACCO SPECIMEN, BRUCE CAIRNCROSS PHOTO.

Figure 723 A 1960s rhodochrosite specimen from Hotazel mine, South Africa, one of the early specimens to come from the Kalahari manganese field, 7.5 cm. BRUCE CAIRNCROSS SPECIMEN AND PHOTO.

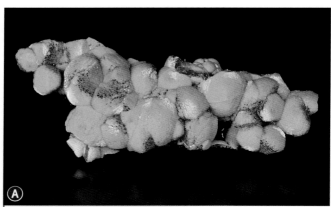

(A)

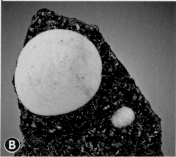

(B)

Figure 724 Platy, pink rhodochrosite on blue chalcedony, with black manganite, 7.6 cm. N'Chwaning I mine, South Africa. BRUCE CAIRNCROSS SPECIMEN AND PHOTO.

Figure 725 Two pink varieties of rhodochrosite, although of very different habits: **A** radiating pink cone-like clusters, 20.1 cm; **B** near-perfect spheres on black manganite. Field of view 7 cm. N'Chwaning I mine, South Africa. DESMOND SACCO SPECIMENS, BRUCE CAIRNCROSS PHOTOS.

Figure 726 Pink rhombohedral rhodochrosite, 5 cm. N'Chwaning II mine, South Africa. BRUCE CAIRNCROSS SPECIMEN AND PHOTO.

Figure 727 Faceted Kalahari manganese field rhodochrosite is rare, and these examples show the different faceted forms, transparency, translucency and colour: **A** 24.4 carats; **B** 2.85 carats (left) and 3.65 carats (right); **C** 7.58 carats; **D** 5.08 carats. BRUCE CAIRNCROSS SPECIMENS AND PHOTOS A AND B; WARREN TAYLOR RAINBOW OF AFRICA COLLECTION, MARK MAUTHNER PHOTOS C AND D.

Figure 728 A 4.2-cm cluster of rhodochrosite mined in 2017. N'Chwaning I mine, South Africa. BRUCE CAIRNCROSS SPECIMEN AND PHOTO.

Figure 729 Elongate, colour-zoned rhodochrosite with shigaite. N'Chwaning I mine, South Africa. Field of view 5.2 cm. BRUCE CAIRNCROSS SPECIMEN AND PHOTO.

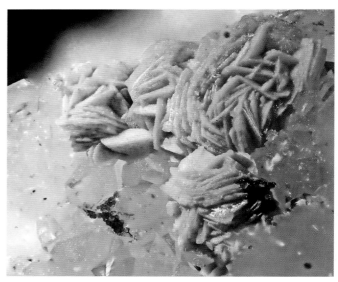

Figure 730 Pale Fe-Zn-rich rhodochrosite on quartz. Field of view 3.8 mm. Houtenbeck mine, South Africa. WOLF WINDISCH SPECIMEN AND PHOTO.

Figure 731 Pink rhodochrosite coating two scalenohedral calcite crystals and present as radiating clusters. Pyrite forms part of the matrix, and tiny pyrite crystals are scattered about the rhodochrosite, 5.2 cm. Broken Hill mine, South Africa. BRUCE CAIRNCROSS SPECIMEN AND PHOTO.

Riebeckite ◆ $Na_2(Fe^{2+},Mg)_3Fe_2^{3+}Si_8O_{22}(OH)_2$

Riebeckite crystallizes in the monoclinic system, has a hardness of 5, specific gravity of 3.32 to 3.38, a blue-grey streak, and vitreous to silky lustre. It is a member of the amphibole group of minerals, some of which form the so-called asbestiform minerals. Asbestos is a common name for fibrous minerals that have been used for decades as fireproofing agents. The amphibole minerals that fall into this group are actinolite, anthophyllite, crocidolite, grunerite (amosite), riebeckite and tremolite. Asbestos-type minerals have been mined in various parts of the world, including South Africa and Zimbabwe. In southern Africa, crocidolite and amosite have been mined, but these are not officially recognized species. Crocidolite is riebeckite, and amosite is a fibrous variety of grunerite. Chrysotile was previously named clinochrysotile, and it is a member of the serpentine group of minerals. Riebeckite is produced by low-grade metamorphism of banded iron formations.

This amphibole is common in **South Africa** as bright blue crocidolite asbestos. It occurs in a belt from Pomfret in the North West to Prieska in the Northern Cape, a distance of some 300 km, and was worked in several mines in the northern and central parts of this belt. Some deposits have riebeckite displaying cone-in-cone structures where the layers of fibres are no longer even and parallel, but rather have small peaks and hollows resembling small hills and valleys.

In **Namibia**, deposits of weathered crocidolite are found on the farm Groot Aub 267 in the Rehoboth district. A silicified variety of riebeckite, called 'pietersite' is used as a lapidary material (see 'tiger's eye'/pietersite).

Figure 732 Crocidolite asbestos with an isolated cone structure, 12 cm. Pomfret mine, South Africa. BRUCE CAIRNCROSS SPECIMEN AND PHOTO.

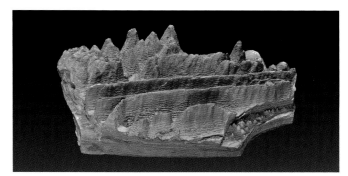

Figure 733 Extreme forms of cone-in-cone structure with sharp pointed pinnacles and valleys in weathered riebeckite, 11.6 cm. Pomfret mine, South Africa. BRUCE CAIRNCROSS SPECIMEN AND PHOTO.

Figure 734 A large plate of riebeckite, 42 cm. Riries mine, Northern Cape, South Africa.
DEPARTMENT OF GEOLOGY, UNIVERSITY OF JOHANNESBURG SPECIMEN, BRUCE CAIRNCROSS PHOTO.

Rutile ◆ TiO$_2$

Rutile crystallizes in the tetragonal system, has a hardness of 6 to 6.5, specific gravity of 4.23, pale brown to yellow streak, and adamantine to submetallic lustre. Rutile can form elongate, acicular crystals or more robust crystals that are often twinned. The mineral is typically dark brown to dark red. It is a titanium oxide and hence resistant to weathering processes. Rutile is one of the main ore minerals of titanium. Rutile is an accessory mineral, occurring in minor quantities in many igneous and metamorphic rocks. Its resistance to weathering results in its accumulating in economic heavy-mineral sand deposits in sandy coastline beaches. These are composed of rutile, ilmenite, zircon, garnet and monazite, with other accessory heavy minerals mixed in the quartz sand.

Rutile is found as crystals in volcanic tuffs in the Pilanesberg Complex and at the Glenover phosphate mine, **South Africa**. It is mined from sand dunes at Richards Bay on the north-east coast of KwaZulu-Natal, and at Namakwa Sands on the west coast. High-energy ocean waves rework the beach sand and concentrate the rutile and ilmenite in economic deposits. Large deposits are also found at Umgababa. Beautiful, tiny acicular red rutile associated with clear quartz, comes from the Cape Lime Company's dolomite quarry near Vredendal in the Western Cape.

Single and twinned crystals of rutilated quartz over 2 cm come from the farm Rooisand in the Gamsberg region of **Namibia**. Geniculate (so-called 'elbow twinned') rutile is common. The Onganja copper mine is famous for large red rutile crystals up to several centimetres in length (Cairncross and Moir, 1996). Euhedral crystals come from the farm Eusgaubib 31, west of Windhoek, where rutile is associated with titanite. Quartz veins with gold and silver mineralization also contain rutile. These outcrop on the farm Duruchaus 612 in the Rehoboth district (Von Bezing et al., 2016). One of the best-known rutile occurrences in Namibia is at Giftkuppe on the farms Kanona Ost 81 and Erongo Ost 32 in the Omaruru district, where veins of rutile in granite were mined commercially. Euhedral crystals of rutile collected from these veins measured up to 20 cm in length and 7 cm in diameter.

Rutile occurs in the Mwami pegmatite region, **Zimbabwe**, usually as tiny needle-like crystals included inside quartz.

Minor quantities of rutile are found in the Halfway Kop kyanite deposit, **Botswana**.

Rutile is found in schists at Mhlabane Mountain, Gege district, **Eswatini**, together with zircon and magnetite. It is reported from a biotite gneiss near the Malolotha River waterfall in the Forbes Reef area and is also found in some alluvial gravels.

Economic deposits of rutile and associated heavy minerals occur along the southern coastline of **Mozambique**.

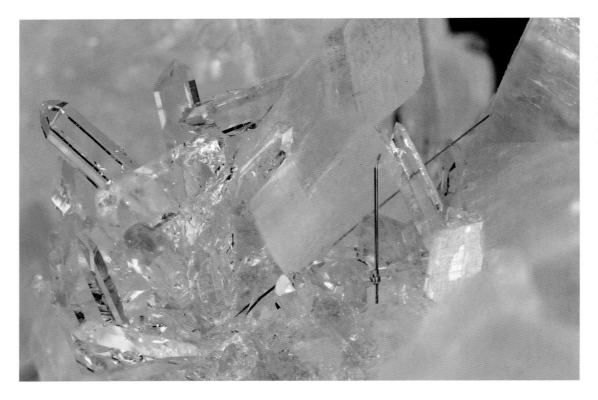

Figure 735
Acicular red rutile crystals with quartz and calcite. Vredendal district, South Africa. Field of view 1.2 cm.
BRUCE CAIRNCROSS SPECIMEN AND PHOTO.

Figure 736 Rutile crystal, 3 cm. Khomas Hochland, Namibia. BRUCE CAIRNCROSS SPECIMEN AND PHOTO.

Figure 737 Rutile crystals on pseudomorphs of goethite after siderite, 5.7 cm. Gamsberg area, Namibia. BRUCE CAIRNCROSS SPECIMEN AND PHOTO.

Figure 738 Rutile crystals on quartz, 5.1 cm. Gamsberg area, Namibia. BRUCE CAIRNCROSS SPECIMEN AND PHOTO.

Scheelite ◆ CaWO$_4$

Scheelite crystallizes in the tetragonal system, has a hardness of 4.5 to 5, specific gravity of 6.1, white streak, and vitreous to adamantine lustre. Like ferberite, scheelite is of economic importance and is mined for tungsten. Crystals are typically heavy because of their tungsten content. They are usually colourless, but can be an attractive yellow, orange, green or red. Scheelite fluoresces readily and can be found at night with an ultraviolet light. Scheelite is found in granite, pegmatites, skarns and hydrothermal deposits and in some metamorphic rocks.

Scheelite was an important ore mineral at the Stavoren mine in the Bushveld Complex, **South Africa**, occurring as white to honey-yellow masses that fluoresce a brilliant white. Well-crystallized specimens 2.5–4 cm long (and even up to 9 cm) were relatively common in the early days. Crystals are very rare now, although small masses of scheelite can still be recovered from the old mine dump.

Scheelite occurs at Pilgrim's Rest, Barberton, and in the Murchison mountain range, but only very rarely as euhedral crystals. It is found in Namaqualand in several deposits associated with the metamorphic rocks of the region (Bowles, 1988). It was mined at the Nababeep West Tungsten mine in the Okiep district (Söhnge, 1950), which has several other tungsten deposits.

Scheelite occurs in granites, skarns and pegmatites, as well as Alpine cleft deposits in **Namibia**. Enormous, attractive orange crystals, weighing up to 60 kg, have been found on the farm Kos in the Gamsberg area. Some are transparent and have been faceted into fancy gemstones, including a 58-carat gem. Beautiful, green copper-bearing crystals, up to 20 cm on edge and sometimes transparent, have come from the farm Natas 220, 120 km south-west of Windhoek. Pegmatites in amphibolite and schist in the Warmbad district contain scheelite (Diehl, 1992b). The Brandberg West mine had scheelite, as did the Krantzberg mine in the Erongo region (Pirajno and Schlögl, 1987). The farms Otjua 37 and Schönfeld 92, approximately 30 km north of Omaruru, contain a skarn deposit with scheelite in marble and calc-silicates.

Figure 739 Scheelite crystal, 3 cm: **A** viewed under normal light and **B** fluorescing blue under 365 nm long-wave ultraviolet light. Northern Cape, South Africa. BRUCE CAIRNCROSS SPECIMEN AND PHOTOS.

Figure 740 **A** Scheelite, 3.7 cm; **B** fluorescing green under 365 nm long-wave ultraviolet light. Natas mine, Namibia. BRUCE CAIRNCROSS SPECIMEN AND PHOTOS.

Figure 741 Scheelite, 7.8 cm. Natas mine, Namibia. WARREN TAYLOR RAINBOW OF AFRICA COLLECTION, MARK MAUTHNER PHOTO.

Figure 742 Two fine, faceted scheelite gemstones: **A** 9.48 carats (1.7 cm); **B** 38.13 carats (1.8 cm). Natas mine, Namibia. WARREN TAYLOR RAINBOW OF AFRICA COLLECTION, MARK MAUTHNER PHOTOS.

Scheelite and ferberite have been mined at over 300 localities, including the Bulawayo area and north and north-east **Zimbabwe** (Anderson, 1979). These two minerals may be found together in the same deposits, but are sometimes mutually exclusive. Scheelite is found in ancient greenstone-belt rocks. It was mined extensively in the north/north-east at the Alton, Hill Top and Scheelite King mines in the Mazowe district. The scheelite occurred in pods in a granite skarn, together with garnet, quartz, calcite and hornblende. The Insiza district in the Bulawayo region had several important scheelite mines, including Clan Mac, AMP, Fernando, Fit, Big Dipper, Lioness and Syd Kom. The last two mines produced over 1,000 tonnes of scheelite. At the Beardmore mine in the Bikita district, where the mineral occurred together with epidote, garnet and vesuvianite, 1,400 tonnes of scheelite was extracted. An interesting deposit of scheelite associated with stibnite and gold is known in gneiss at the Grand Parade mine in the Kadoma district. The Dete region is also known to contain scattered tungsten mineralization (Lockett, 1979).

Scheelite is relatively widespread in the Tati schist belt, **Botswana**. It usually occurs in quartz veins, often with epidote.

In **Eswatini**, scheelite is found in tin-bearing gravels near Sinceni and in the Mbabane district. Pale yellow scheelite occurs in lenses up to 15 cm thick, with calcite, albite, rutile and pyrrhotite, in biotite gneiss at Forbes Reef, near the waterfall on the Malolotsha River (Davies, 1964).

Schizolite ◆ $NaCaMnSi_3O_8(OH)$

Schizolite crystallizes in the triclinic system, has a hardness of 5, specific gravity of 3.09, pale pink to brownish-white streak, and a vitreous to sub-vitreous lustre. Schizolite is a rare mineral. It was originally described in 1903 from South Greenland, but, as can sometimes occur in mineralogy, it was 'discredited' as a valid species in 1955. Based on specimens found in the 2000s in the South African Kalahari manganese field, the mineral was named marshallsussmanite. However, further research revealed that marshallsussmanite was in fact the historic schizolite, and this name was reinstated by the International Mineralogical Association, and the name marshallsussmanite was then discredited (Grice *et al.*, 2019). Schizolite is included here, as the finest crystals known have come from southern Africa.

The specimens from **South Africa** were reportedly discovered simultaneously at the Wessels and N'Chwaning II mines in the Northern Cape, which is somewhat unusual, that two separate mines should simultaneously produce the same rare mineral, particularly as few specimens were found. The mineral ranges from pale pink to bright pink and forms tabular crystals. Some are associated with calcite, hydroxyapophyllite-(K) and aegirine. When first discovered, it was thought to be bustamite, another pink mineral, but this was not the case. The Kalahari manganese field schizolite specimens are the finest known, perhaps only equalled by some from Mont Saint-Hilaire, Quebec, Canada.

Schizolite is reported from one other southern African locality, the Aris quarry in **Namibia**. Here, the crystals are small, a millimetre or less, and are bladed, lath-like and colourless, and associated with aegirine and natrolite.

Figure 743 Pink schizolite crystals: **A** with white hydroxyapophyllite-(K), 6.8 cm; **B** 7.4 cm; **C** 6 cm. Kalahari manganese field, South Africa.

BRUCE CAIRNCROSS SPECIMENS AND PHOTOS.

Schorl ◆ $NaFe_3^{2+}Al_6(BO_3)_3Si_6O_{18}(OH)_4$

Schorl crystallizes in the hexagonal system, has a hardness of 7, specific gravity of 3.1 to 3.25, white streak, and vitreous to resinous lustre. Schorl is the iron-rich member of the tourmaline group and perhaps the most common tourmaline species. It forms distinctive jet-black crystals, but can also be found in other colours such as deep red or deep mauve. It is a common accessory mineral in granite and granitic pegmatites and hydrothermal veins. Associated minerals can be quartz, muscovite, alkali feldspar, cassiterite and topaz. Schorl is a common accessory mineral in granite pegmatites, tin-mineralized granites and in some metamorphic rocks.

Schorl is found sporadically in many Northern Cape pegmatites, **South Africa**. It commonly occurs as crude to well-formed striated black crystals up to 25 cm long. In the Straussheim and Angelierspan pegmatites, crystals have been found up to 1 m in length and 15 cm in diameter. Schorl is commonly found embedded in quartz in pegmatites in the Limpopo and Mpumalanga provinces. Schorl tourmaline is a common accessory mineral with cassiterite mineralization in most of the Bushveld tin mines, occurring as radiating stellate groups.

Some schorl crystals from the Erongo Mountains in **Namibia** rival the finest in the world and are certainly the best known from southern Africa and perhaps all of Africa (Cairncross and Bahmann, 2006a; Von Bezing *et al.*, 2008; 2014; 2016). Single, jet-black, prismatic crystals up to 40 cm long, with mirror-like lustres typify the crystals, while some are more matt with highly striated prism faces. There is also a bewildering array of other habits, from dodecahedral-like crystals resembling garnet to crystals with triangular terminations that are a replica of the Mercedes Benz logo. They occur singly or in clusters over 50 cm on edge and stunning specimens result when they are associated with deep blue aquamarine, green fluorite and white feldspar.

Figure 744 A mass of black schorl crystals, 7.6 cm. Orange River region, South Africa. BRUCE CAIRNCROSS SPECIMEN AND PHOTO.

Figure 745 Black schorl crystals embedded in coarse-grained pegmatitic quartz and alkali feldspar. Omapyu pegmatite, Namibia. BRUCE CAIRNCROSS PHOTO, 2014.

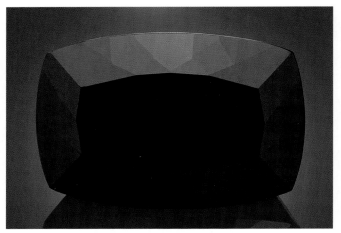

Figure 746 A faceted schorl, 75.65 carats (3.1 cm). Erongo Mountains, Namibia. WARREN TAYLOR RAINBOW OF AFRICA COLLECTION, MARK MAUTHNER PHOTO.

Figure 747 **A** Unusual radiating clusters of schorl. Some of the crystals are dark green and may be different species (see text for details), 9 cm; **B** doubly terminated quartz with schorl, 7 cm. Erongo Mountains, Namibia. BRUCE CAIRNCROSS SPECIMENS AND PHOTOS.

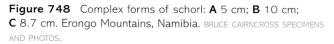

and these require further investigation to determine whether they are in fact schorl and not one of the other tourmaline species. Research suggests that some of these black crystals may not be schorl, but rather the very rare tourmaline foitite or even a new species of tourmaline altogether (Lensing-Burgdorf *et al.*, 2017; Falster *et al.*, 2018). One pocket produced crystals (up to 7 cm in diameter) that have an isometric habit, resembling black garnets. These are either individual floater crystals or aggregates. A handful of unique vermiform schorl was collected during January 2006. These are miniature specimens and consist of contorted finger-like groups of crystals. Most of the farms located in the south-west, west and north-west of the Erongo Mountains have yielded outstanding schorl.

Another interesting feature of the Erongo tourmalines is the internal colour zonation that is revealed when very thin slices are cut either perpendicular or parallel to the *c*-axis of crystals; when backlit, horizontal sections tend to reveal blue colours while vertical slices are mostly red-brown. Multiple coloured zones that are either trigonal or pseudohexagonal in shape occur in the crystals. These are reminiscent of the famous Madagascar liddicoatite, and colours include yellow, orange, blue and green.

Elsewhere in Namibia, schorl is found scattered in many pegmatites.

Schorl is common in pegmatites in **Zimbabwe**. It was particularly abundant at the Patronage lithium pegmatite and is widespread in the Goromonzi district.

Not all Erongo schorl appears as pristine crystals. Some are highly corroded, etched and resorbed and some crystal faces are studded by acicular, grey quartz, muscovite and feldspar. Some terminations are cavernous, arena-like hollows produced by preferential growth of sections of the outer prism faces, accompanied by less rapid crystal growth towards the centre of the termination. Later-stage muscovite and fluorite commonly crystallize in these hollow terminations.

Some specimens are coated by fluorescent hyalite. Other associated species include aquamarine, muscovite, orthoclase and quartz. Some crystals are very dark green and deep red,

Shattuckite ◆ $Cu_5(Si_2O_6)_2(OH)_2$

Shattuckite crystallizes in the orthorhombic system, has a hardness of 3.5, specific gravity of 4.1, blue streak, and dull to silky lustre. It is characteristically blue and often forms felt-like fibrous crystals that aggregate into semi-circular masses. Visually, it can be confused with plancheite, another blue secondary copper mineral with similar colour and habits. Shattuckite forms as a secondary mineral associated with some copper deposits.

In **South Africa**, shattuckite has been reported from the copper deposits in the Musina district, Limpopo, although quantitative analyses to properly identify the species are lacking. Recent analyses of several dark blue inclusions in quartz showed the mineral to be papagoite (Matsimbe, 2019).

Some of the finest shattuckite specimens in the world come from 400 km north-west of Windhoek in the Kaokoveld, Kunene Region, northern **Namibia**. These have been collected from various copper prospects in the region, including Omaue, Kandesi and Okenwasi (Bowell *et al.*, 2013). Specimens are pale blue to vibrant royal blue and are associated with chrysocolla, dioptase, malachite, plancheite and quartz.

Figure 749 Shattuckite from the Kaokoveld, Namibia: **A** soft, velvety shattuckite lining a cavity, 4.4 cm; **B** dark blue shattuckite in a quartz vug, field of view 5.1 cm; **C** unusual sprays of shattuckite that might be pseudomorphic replacement of malachite, 7.2 cm; **D** shattuckite associated with pale blue plancheite and green malachite, 7.2 cm. BRUCE CAIRNCROSS SPECIMENS AND PHOTOS.

Shigaite ◆ $[AlMn^{2+}_2(OH_6)]_3(SO_4)_2Na_2(H_2O)_6\{H_2O\}$

Shigaite crystallizes in the trigonal system, has a hardness of 2, specific gravity of 2.32, pale yellow to white streak, and vitreous to dull lustre. Shigaite forms hexagonal, micaceous minerals that form rosette-shaped aggregates. The colour is characteristically amber to gold-brown and the crystals can be transparent to opaque. Shigaite is found in several deposits worldwide, but the crystals discovered in South Africa are the finest known, hence their inclusion here (Cairncross, 2016b).

One of the first finds of shigaite in **South Africa** was made in 1993 at the Wessels mine, Kalahari manganese field, South Africa. These were lustrous, amber-coloured crystals associated with small pink rhodochrosite crystals, leucophoenicite, gageite and caryopilite. Crystals have a micaceous habit and tend to occur as foliated masses. Most were tiny, micromount specimens, but some aggregates were 0.5–1 cm, and in some exceptional cases, crystals are 2 cm in diameter, although these are rare. However, these early 1990s specimens were superseded by those collected 13 years later at the N'Chwaning I mine.

During the first half of 2006, an area in the old N'Chwaning I mine produced some extraordinary specimens of shigaite associated with rhodochrosite (Cairncross and Beukes, 2013; Cairncross et al., 2017). Shigaite was the

Figure 750 A large 1.5-cm shigaite crystal with smaller ones attached on pink rhodochrosite. N'Chwaning I mine, South Africa.
BRUCE CAIRNCROSS SPECIMEN AND PHOTO.

Figure 751 A cluster of shigaite with colourless baryte, pink rhodochrosite and tiny, transparent orange shigaite. Field of view 2.8 cm. N'Chwaning I mine, South Africa.
BRUCE CAIRNCROSS SPECIMEN AND PHOTO.

premier mineral discovered in some of the cavities. The crystals tend to be bright, lustrous dark amber to red and display a typically micaceous habit. Shigaite crystals vary in size from 1 mm to 2.8 cm, the latter being rare. The smaller crystals are thin, but the larger crystals can be up to several millimetres thick and some have a rosette-like, layered habit.

The smallest crystals usually have the brightest orange colour and are transparent. The shigaite is commonly scattered over the rhodochrosite matrix. The associated mineral assemblage consists of shigaite, baryte, kutnahorite, manganocalcite, pyrochroite, rhodochrosite and soft, fibrous sussexite, much of which is washed off the specimens.

Figure 752 A 2.1-cm shigaite crystal on pale pink rhodochrosite matrix, mined in May 2006. N'Chwaning I mine, South Africa. PAUL BALAYER SPECIMEN, BRUCE CAIRNCROSS PHOTO.

Figure 753 Assorted shigaite crystals, rhodochrosite and white baryte. Field of view 3.2 cm. N'Chwaning I mine, South Africa. BRUCE CAIRNCROSS SPECIMEN AND PHOTO.

Figure 754 A large, composite 2.8-cm shigaite crystal. N'Chwaning I mine, South Africa. BRUCE CAIRNCROSS SPECIMEN AND PHOTO.

Siderite ◆ Fe²⁺CO₃

Siderite crystallizes in the hexagonal system, has a hardness of 4, specific gravity of 3.96, a white streak, and vitreous to pearly lustre. This iron carbonate species can be found in various shades of brown, orange to green rhombohedral crystals. It also forms rounded to subrounded diagenetic concretions in sedimentary rocks. Siderite is a relatively common constituent of the matrix of some sedimentary rocks, such as sandstones, and is a rock-forming mineral in banded iron formations. It can also be found in hydrothermal ore veins and as secondary veinlets and layers in coal seams, sandstones and limestones.

Large specimen-quality siderite crystals are rare in **South Africa**. Small orange crystals at the Clairwood quarries in KwaZulu-Natal are associated with calcite in fissures and fractures in Dwyka tillite, and make attractive specimens. Layers of siderite in sandstone are found in KwaZulu-Natal near Durban, and at Zinguin tunnel at Vryheid. Brown siderite crystals are rare at the Argent mine near Delmas in Gauteng, although some have been found up to several centimetres in length. At the Broken Hill mine at Aggeneys in the Northern Cape, balls of siderite are associated with pale pink rhodochrosite crystals. Honey-coloured crystals of siderite up to 4 mm line cavities in the oxidized part of the orebody.

An entirely different variety of siderite from the Broken Hill mine consists of pseudomorphs of siderite, mixed with sphalerite replacing large calcite crystals. Some are several centimetres in size, and clusters of these reached tens of centimetres. These 'casts' or moulds are invariably completely hollow; the siderite-sphalerite precipitated over the calcite, which then dissolved away, leaving the pseudomorphs.

Figure 755
Brown siderite-sphalerite. These are pseudomorphs after calcite and are hollow, with the original calcite that provided the mould for the brown coating, dissolved away: **A** 12 cm; **B** 8.2 cm. Aggeneys, Northern Cape, South Africa. BRUCE CAIRNCROSS SPECIMENS AND PHOTOS.

Figure 756   Three siderite specimens displaying different colours and habits: **A** yellow crystals with rhodochrosite and grey galena, field of view 2.2 cm; **B** discrete crystals on matrix, field of view 2.8 cm; **C** olive-green siderite, field of view 3.2 cm. Aggeneys, South Africa. BRUCE CAIRNCROSS SPECIMENS AND PHOTOS.

Figure 757 Close-up view of siderite. Coedmore quarry, Durban, South Africa. Field of view 2.4 cm. BRUCE CAIRNCROSS SPECIMEN AND PHOTO.

Figure 758 A giant 27-cm (9.4-kg) crystal of siderite, now altered and replaced by goethite and hematite. Erongo Mountains, Namibia. ULI BAHMANN SPECIMEN, BRUCE CAIRNCROSS PHOTO.

Siderite crystals several centimetres on edge come from pegmatites at Klein Spitzkoppe, **Namibia**, where they are usually associated with quartz. Some have pseudomorphed to goethite (Cairncross, 2005a). Spectacular crystals over 20 cm on edge have been collected from the miarolitic pegmatites of the Erongo Mountains, although these are somewhat oxidized and partly pseudomorphed to goethite (Cairncross and Bahmann, 2006a). They are often associated with schorl, aquamarine and quartz. Beautiful, semi-transparent, lustrous, dark amber crystals of arsenic-rich siderite up to 2 cm on edge were found at the Tsumeb mine. Brown siderite crystals have occasionally been found associated with quartz from the Goboboseb Mountains.

Siderite occurs as nodules and in some Karoo sedimentary rocks in **Zimbabwe**. Crystals of siderite and massive aggregates are found in gangue assemblages at the Vubachikwe arsenic mine in the Gwanda district and at the Silverside copper mine, Makonde district.

Siderite comes from banded iron formations at the Ngwenya iron-ore mine, **Eswatini**. It is also found in the Forbes Reef district as crystalline material in outcrops of ancient sedimentary rocks.

Figure 759 Brown siderite crystals, 7.4 cm. Goboboseb Mountains, Namibia. BRUCE CAIRNCROSS SPECIMEN AND PHOTO.

Figure 760 Siderite gem crystals. Matchless mine, Namibia. Field of view 1.8 cm. BRUCE CAIRNCROSS SPECIMEN AND PHOTO.

Silver ◆ Ag

Silver crystallizes in the cubic system, has a hardness of 4 to 4.5, specific gravity of 10.1, a silver-grey streak, and a metallic lustre. Silver very rarely appears as well-formed cubic crystals, but rather as flat sheets, wire-like threads and arborescent aggregates that resemble fern leaves. Like gold, silver is malleable and ductile, and is used in jewellery and coinage. It is still extensively used in photographic film, silver plating and silverware. Native silver is rare in southern Africa. The metal is usually obtained as a by-product from gold and base-metal mines.

In **South Africa**, native silver has been reported from the Albert mine north of Bronkhorstspruit and the Argent mine near Delmas (Wilson-Moore and Carrington Wilmer, 1893). Arborescent masses enclosed in bornite from these now-defunct mines weighed up to several kilograms. Silver occurs as a minor constituent of the gold ores of the Witwatersrand goldfield and at Barberton. It is a by-product of mining operations at the Palabora, Maranda and Broken Hill mines. Silver was likewise extracted as a by-product of lead mines in dolomites and deposits associated with the Bushveld Complex aureole. Argentiferous galena occurred at the aptly named Silver Hills mine at Hennops River, north of Johannesburg.

Much of the silver produced in **Namibia** originates as a by-product of the mining of base-metal lead-zinc-copper sulphide deposits (Hirsch and Genis, 1992b). The Tsumeb, Namib Lead, Rosh Pinah, Ai-Ais and Otjihase mines all produced silver. However, native silver has been found at the Klein Aub mine and was sometimes associated with native copper at the Onganja copper mine. There was some native silver in the orebodies at

Kombat mine and Oamites. The Tsumeb mine produced the best Namibian silver specimens, but these were found not to be pure silver but an amalgam with mercury (Gebhard, 1999).

Silver from **Zimbabwe** is obtained mostly as a by-product of the gold mining and, to a lesser degree, copper mining industries. Exceptions are the Osage mine in the Kwekwe district (where native silver was mined together with sphalerite, galena and a suite of copper minerals), Jessie copper mine in the Gwanda district, Mangula mine, and the Old West lead mine in the Mutare district.

Most of the silver produced in **Botswana** is a by-product of the Tati gold mining operations. Silver also comes from argentiferous galena at the Bushman mine.

Silver is obtained as a by-product of the gold mining industry in **Eswatini**.

Figure 761 Silver on copper ore, 2 cm. Tsumeb mine, Namibia.
BRUCE CAIRNCROSS SPECIMEN AND PHOTO.

Figure 762 Platy silver on copper ore, 2.7 cm. Kombat mine, Namibia. BRUCE CAIRNCROSS SPECIMEN AND PHOTO.

Figure 763 Silver amalgam with green zincolivenite and yellow ferrilothermeyerite. Tsumeb mine, Namibia. Field of view 2.6 cm.
BRUCE CAIRNCROSS SPECIMEN AND PHOTO.

Smithsonite ◆ ZnCO$_3$

Smithsonite crystallizes in the trigonal system, has a hardness of 4 to 4.5, specific gravity of 4.3 to 4.5, a white streak, and vitreous to pearly lustre. Smithsonite can occur either as well-formed rhombohedral crystals or as botryoidal masses. It resembles calcite but is heavier. Pure smithsonite is white to colourless, but the presence of minute chemical impurities can impart a wide spectrum of colours. Smithsonite is usually associated with zinc sulphide deposits where the oxidation of the primary zinc ore gives rise to this secondary zinc carbonate.

Lead-zinc deposits in the Ottoshoop district, **South Africa**, and the Argent mine have produced rare, small crystals associated with cerussite. Prismatic smithsonite crystals, up to 5 mm long, occurred with hemimorphite, calcite and quartz at the Broken Hill mine at Aggeneys in the Northern Cape. Smithsonite occurs sparsely in the Okiep copper mines, together with quartz crystals, calcite and baryte.

Some of the finest and most colourful specimens of smithsonite known originated from the Tsumeb mine, **Namibia** (Cairncross, 2010a). Colourless to opaque-white crystals of pure smithsonite were plentiful, and the contamination of other metallic elements produced an array of brilliant green and blue cupro-smithsonite; vibrant pink cobalt- and mangan-smithsonite; and bright yellow, orange and caramel cadmium-smithsonite. Large individual crystals up to 7 cm on edge are known, as well as attractive botryoidal masses. Beautiful white smithsonite, associated with willemite and cerussite, came from the Abenab mine.

Figure 764 Various coloured smithsonite from Tsumeb mine, Namibia: **A** yellow cadmium smithsonite, 12 cm; **B** pink cobalt smithsonite, 6.8 cm; **C** pale green copper-rich smithsonite, 4.2 cm; **D** close-up of unusual red smithsonite with glassy cerussite, field of view 5.2 cm; **E** orange smithsonite with white cerussite, 7.2 cm; **F** bright green copper-rich smithsonite, 10.6 cm; **G** pale pink cobalt smithsonite, 9.5 cm. DESMOND SACCO SPECIMENS F & G, BRUCE CAIRNCROSS SPECIMENS A—E AND PHOTOS.

Figure 765 Radiating yellow smithsonite, 3.8 cm. Berg Aukas mine, Namibia.
BRUCE CAIRNCROSS SPECIMEN AND PHOTO.

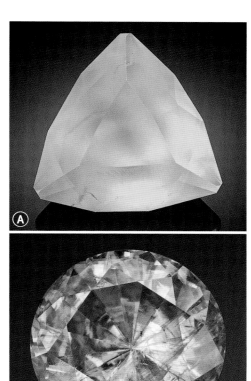

Figure 766 Faceted smithsonite: **A** 48 carats (2.1 cm); **B** 19 carats (1.4 cm). Tsumeb mine, Namibia. WARREN TAYLOR RAINBOW OF AFRICA COLLECTION, MARK MAUTHNER PHOTOS.

Figure 767 Transparent smithsonite crystals, 3.2 cm. Berg Aukas mine, Namibia.
BRUCE CAIRNCROSS SPECIMEN AND PHOTO.

Figure 768 A large smithsonite crystal overgrowing on smaller crystals. The brown matrix is willemite, 3.3 cm. Berg Aukas mine, Namibia. BRUCE CAIRNCROSS SPECIMEN AND PHOTO.

The Berg Aukas mine has also been a premier source of smithsonite, albeit not the same volume of specimens as from Tsumeb (Cairncross, 2021c). Crystals are associated with descloizite and willemite, and range in colour from off-white, olive-green, to yellow and orange.

Two of southern Namibia's mines have also produced collectable smithsonite: Skorpion mine smithsonite is typically pale blue to white and can occur as rice-like crystals or as teardrop-shaped clusters (Von Bezing *et al.*, 2014). It can be associated with hemimorphite and, more rarely, tarbuttite. Small 3–6 mm translucent yellow 'rice grain' smithsonite on weathered gossanous matrix came from the Rosh Pinah lead-zinc mine (Cairncross and Fraser, 2012).

Figure 769 Clusters of translucent smithsonite, 12.3 cm. Berg Aukas mine, Namibia. JIM AND GAIL SPANN COLLECTION, TOM SPANN PHOTO.

Figure 770 The Skorpion mine in southern Namibia has been a source of fine smithsonite specimens such as the ones shown here: **A** 16 cm; **B** 12.5 cm. DESMOND SACCO SPECIMENS, BRUCE CAIRNCROSS PHOTOS.

Figure 771 Bowtie smithsonite on drusy hemimorphite, 3.4 cm. Skorpion mine, Namibia. BRUCE CAIRNCROSS SPECIMEN AND PHOTO.

Figure 772 Two smithsonite specimens on gossanous matrix: **A** 3.2 cm; **B** field of view 3.8 cm. Rosh Pinah mine, Namibia. BRUCE CAIRNCROSS SPECIMENS AND PHOTOS.

Sodalite ◆ $Na_8Al_6Si_6O_{24}Cl_2$

Sodalite crystallizes in the cubic system, has a hardness of 5.5 to 6, specific gravity of 2.14 to 2.4, a white streak, and vitreous to greasy lustre. It is a member of the feldspathoid group. Sodalite is an ornamental stone used extensively for cladding, tiles, carvings, stonework and jewellery. It occurs in some carbonatite and syenite, i.e. silica-undersaturated rocks. The largest deposit in southern Africa is periodically mined in northern Namibia.

Sodalite is found in minor quantities in **South Africa** in the rocks of the Pilanesberg in the North West province.

The best-known sodalite deposit in southern Africa is in **Namibia**, in the Kaokoveld, 10 km west of Swartbooisdrif,

4 km south of the Kunene River (Menge, 1986). The sodalite is hosted in carbonatite and syenite and occurs in dykes 2–4 km long. It usually contains bands or layers of white ankerite (a calcium-iron carbonate), yellow cancrinite (a sodium-calcium silicate) and pale pink analcime (a sodium member of the zeolite group of minerals). It is sporadically exploited commercially for ornamental stonework. Beautiful dark blue masses occur, up to 8 m long.

Attractive sodalite has been extracted in the Mwami and Karoi districts, **Zimbabwe**.

Figure 773 A polished slab of sodalite cross-cut by veins of cream cancrinite, 11.5 cm. Swartbooisdrif, Namibia. BRUCE CAIRNCROSS SPECIMEN AND PHOTO.

Sperrylite ♦ PtAs$_2$

Sperrylite crystallizes in the cubic system, has a hardness of 6 to 7, specific gravity of 10.46 to 10.6, a black streak, and metallic lustre. Sperrylite is a rare sulphide species, which is exploited for platinum. It forms small, heavy, silver, brilliantly lustrous, cubic and octahedral crystals. The Bushveld Complex platinum deposits in South Africa have produced some of the finest sperrylite crystals in the world. Sperrylite occurs in ultramafic pyroxenites mostly in layered complexes such as the Bushveld Complex, South Africa, and the Great Dyke, Zimbabwe.

In **South Africa**, sperrylite crystals and associated rare sulphides braggite, cooperite and laurite are characteristic of the Merensky Reef (Wagner, 1927). Most are microscopic, but in the 1920s on the farm Tweefontein 238KR in the Mokopane district, large, perfect, tin-white crystals with a very high lustre were recovered from a limonitic gossan (Wilson, 2010). One of the largest measured 1.85 x 1.625 x 1.5 cm. Most were cubic, but there were also combinations of cubic and octahedral forms. These crystals have been surpassed by larger crystals from Russia and Canada, but still remain iconic for the species. At Insizwa in the Eastern Cape, sperrylite is fairly common in the massive ore and occurs as tiny crystals up to 0.2 mm on edge associated with pyrrhotite, cubanite and parkerite.

At the Mimosa mine located on the Great Dyke, Zvishavane district, **Zimbabwe**, sperrylite occurs together with nickel and copper sulphides in certain horizons such as the Great Dyke's 'potato reef'.

Figure 774 A large 1.2-cm sperrylite crystal in matrix, 3.4 cm, Tweefontein farm, Mokopane, Eastern Bushveld Complex, Limpopo, South Africa. JIM AND GAIL SPANN COLLECTION, TOM SPANN PHOTO.

Figure 775 A bright silver 4-mm sperrylite crystal in weathered matrix stained in places by green malachite. Tweefontein farm, Mokopane, South Africa. BRUCE CAIRNCROSS SPECIMEN AND PHOTO.

Spessartine ◆ $Mn^{2+}_3Al_2(SiO_4)_3$

Spessartine crystallizes in the cubic system, has a hardness of 7 to 7.5, specific gravity of 3.8 to 4.25, a white streak, and vitreous lustre. Spessartine is a variety of garnet that can be very valuable as a gemstone. Spessartine can form beautiful, gemmy bright orange transparent crystals prized by the jewellery trade. It is found in pegmatites, schists and granite-gneiss.

This garnet, characteristically light brown to reddish-brown to red, occurs sporadically in pegmatites of the Northern Cape, **South Africa**, as well-developed crystals reaching 3.5 cm in diameter. Crystals up to 5 cm come from the Angelierspan pegmatite (Cairncross and Dixon, 1995). Spessartine has also been reported from schists of the Murchison mountain range in Limpopo.

Superb, bright orange, gemstone-quality spessartine was discovered in 1991 in muscovite schists in the north-west corner of **Namibia**, near the confluence of the Marienfluss River south of the Kunene River (Von Bezing *et al.*, 2014). It is found either as very well-formed, sharply defined crystals or as shapeless nodules in the host rock. Faceted stones of over 20 carats – dubbed 'Mandarin Garnet' by the gem trade – have been cut from rough material. Spessartine is also found in the Otjosondu manganese field and the Kombat mine.

Spessartine is found in metamorphosed granite-gneiss and metamorphosed quartzites in the Karoi and Masvingo districts, **Zimbabwe**, and at the Ntabeni beryl pegmatite in the Mutoko district. Good crystals have come from the Good Days beryl-lithium pegmatite in the Mutoko district.

Spessartine is found in crystals up to 1 cm in diameter in a pegmatite 12 km south-west of Mbabane, **Eswatini**, together with the main ore mineral, cassiterite, and a host of other pegmatite species. At Gege, spessartine is found as individual crystals or as layers with iron-ore in quartzite.

Figure 776
Two gem-quality spessartine crystals and a faceted 0.73-carat stone. The left crystal is 1.2 cm. Marienfluss, Namibia. BRUCE CAIRNCROSS SPECIMENS AND PHOTO.

Figure 777 Three faceted spessartine garnets illustrating the different colours: **A** 41.76 carats (2.1 cm); **B** 3.97 carats (9 mm); **C** 4.62 carats (1 cm). Marienfluss, Namibia. WARREN TAYLOR RAINBOW OF AFRICA COLLECTION, MARK MAUTHNER PHOTOS.

Sphalerite ◆ (Zn,Fe)S

Sphalerite crystallizes in the cubic system, has a hardness of 3.5 to 4, specific gravity of 3.9 to 4.1, a white to pale brown streak, and resinous to adamantine lustre. Sphalerite crystals are usually dark orange-amber with a typically resinous, waxy lustre, but may also be brown, black or, rarely, green. Sphalerite is the main ore mineral of zinc, which is used in bronze and brass alloys. It is also a filler in paint and rubber, an oxide in feed additives and soil rejuvenation, and is used in special zinc batteries. Sphalerite is common in some sedimentary and hydrothermal ore deposits, often in association with galena.

At the Nababeep West mine, **South Africa**, sphalerite occurred as euhedral crystals up to 3 cm on edge. Beautiful honey-coloured crystals have come from the Pering mine at Reivilo, south-west of Vryburg. These crystallized in cavities in dolomite and are found with grey dolomite crystals, galena and quartz (Southwood, 1986; Wheatley *et al.*, 1986b). Excellent, dark amber, euhedral crystals up to 3 cm on edge were also collected together with galena and dolomite. Large-

cleavage masses of sphalerite are mined at a similar type of deposit at Bushy Park, near Griquatown. Sphalerite occurs at the Maranda mine in the Murchison area, Limpopo. The Gamsberg and Broken Hill deposits situated close to Aggeneys in the Northern Cape are lead-zinc deposits with sphalerite.

Figure 778 Sphalerite crystals with dolomite. Pering mine, South Africa. Field of view 12.5 cm. BRUCE CAIRNCROSS SPECIMEN AND PHOTO.

Figure 779 Sphalerite crystals with yellow baryte, 5.1 cm. Randfontein Estates gold mine, South Africa. BRUCE CAIRNCROSS SPECIMEN AND PHOTO.

Figure 780 Polished brecciated grey dolomite with cross-cutting veins of white dolomite, filled with amber-coloured sphalerite, 17 cm. Bushy Park lead-zinc deposit, Griquatown district, South Africa. BRUCE CAIRNCROSS SPECIMEN AND PHOTO.

Figure 781 Sphalerite-siderite casts after calcite (see siderite), 38 cm. Aggeneys, South Africa. BRUCE CAIRNCROSS SPECIMEN AND PHOTO.

Figure 782 Three large sphalerite specimens from the Free State Geduld gold mine, South Africa, all collected in the early 1960s: **A** 7.6 cm; **B** 3.2 cm; **C** 6.8 cm.
BRUCE CAIRNCROSS SPECIMENS AND PHOTOS.

Possibly the largest southern African sphalerite crystals, up to 5 cm on edge, occur at some of the Witwatersrand gold mines, notably the Free State Geduld mine, Welkom district, and from Mponeng mine, Carletonville (Cairncross, 2021d). They are often associated with quartz, galena, pyrrhotite, pyrite, calcite and kerogen. This suite of well-crystallized minerals is not part of the primary gold-bearing reef conglomerate, but was formed much later in cavities and on fault planes, or is associated with younger igneous dyke intrusions. These Witwatersrand goldfield sphalerite crystals are the largest known from southern Africa.

Sphalerite is found in most of the lead-zinc deposits in **Namibia**, commonly hosted in dolomites, in the Otavi mountainland, the Kaokoveld, Grootfontein district and other widely dispersed areas (Wartha and Genis, 1992). The Namib Lead, Tsumeb, Abenab West, Kombat and Rosh Pinah mines all contained sphalerite. At Rosh Pinah, specimen-grade crystals of sphalerite and galena are occasionally collected (Cairncross and Fraser, 2012). The typically very dark brown to black sphalerite is aesthetic, occurring in euhedral crystals up to 3 cm on edge. Attractive, small transparent to translucent yellow to orange sphalerite crystals come from the Aris quarry south of Windhoek.

Sphalerite has been commercially mined at many copper-lead-zinc deposits in **Zimbabwe**, such as the Copper Queen mine (Gokwe district). The mineral has been described from the Cactus Prospect (Kwekwe district), Die Krikel (Nyanga district) and the Copper King, Cam and Motor mines, among others (Anderson, C.B., 1980).

Sphalerite is found in dolomite as crystal clusters the size of oranges at the She mine at Forbes Reef, **Eswatini** (Hunter, 1962; Davies, 1964). It is associated with the baryte deposit near Oshoek, and is found in the unusual epidote vein outcropping east of Nkambeni hill.

Figure 783 Sphalerite crystal with galena and pyrite, 5.8 cm. Rosh Pinah mine, Namibia. BRUCE CAIRNCROSS SPECIMEN AND PHOTO.

Figure 784 Sphalerite with yellow baryte. Rosh Pinah mine, Namibia. Field of view 3.7 cm. BRUCE CAIRNCROSS SPECIMEN AND PHOTO.

Figure 785 A rare faceted gem from a vanadium mine, this 12.76-carat (1.3-cm) sphalerite specimen was cut from transparent material mined at Berg Aukas, Namibia. WARREN TAYLOR RAINBOW OF AFRICA COLLECTION, MARK MAUTHNER PHOTO.

Spinel ◆ MgAl$_2$O$_4$

Spinel crystallizes in the cubic system, has a hardness of 7.5 to 8, specific gravity of 3.58, white streak, and vitreous to dull lustre. It is easily identified by its distinctive octahedral crystals. They vary in colour and may be black, dark blue, green, beige, pink or red. Spinel occurs in metamorphic serpentinite, gneiss, calc-silicate and marble, as well as some mafic igneous rocks.

Waterworn pink-red grains of spinel are found in diamondiferous gravels at Lichtenburg in the North West province, **South Africa**, and small crystals have been recorded from marbles at Marble Delta in southern KwaZulu-Natal. The zinc-bearing spinel, gahnite, is found in substantial crystals at Aggeneys in the Northern Cape. In the Soutpansberg region in Limpopo, spinel is found on the farm Cavan 508 MS. The mineral also occurs on the farms Groot Hoek 256 KT and Thorncliffe 374 in the Lydenburg district in Mpumalanga. Chrome-spinel is found in some kimberlites.

Sharp-edged crystals of dark grey to black spinel up to 2 cm on edge are found close to Rössing, **Namibia**. Large crystals of spinel up to 3 cm on edge are found on the farm Okahua, about 30 km south-east of Otjiwarongo. These occur in a skarn deposit.

Figure 786 A naturally etched 5.2-cm octahedral spinel crystal from the Ais dome, Namibia. BRUCE CAIRNCROSS SPECIMEN AND PHOTO.

Figure 787 Spinel crystals on calcite: **A** field of view 3 cm; **B** 4.2 cm. Rössing region, Namibia. BRUCE CAIRNCROSS SPECIMENS AND PHOTOS.

Staurolite ◆ $(Fe^{2+},Mg,Zn)_2Al_9(Si,Al)_4O_{22}(OH)_2$

Staurolite crystallizes in the monoclinic system, has a hardness of 7 to 7.5, specific gravity of 3.65 to 3.83, white to pale grey streak, and vitreous to resinous lustre. Staurolite is dark brown to black and forms prismatic crystals. Staurolite often forms cruciform twins. These may be two crystals at right angles ('fairy crosses'), or twinned in the shape of an 'X'. Staurolite is a metamorphic mineral most often found in schist and other associated metamorphic rocks. Staurolite is common in heavy-mineral sand deposits, especially where metamorphic schist acts as the source rock.

In **South Africa**, staurolite is found as well-formed, prismatic, dark brown crystals and cruciform twins in schist at Aggeneys and other Namaqualand localities, for example west of Upington. Staurolite crystals up to a few centimetres come from the Hoogenoeg andalusite mine, Limpopo.

In **Namibia**, staurolite comes from micaceous schist outcrops near the old Gorob mine in the Namib Desert Park. In the early twentieth century, these staurolite crystals were hand-cobbed from schist and piled in heaps, perhaps

Figure 788 Staurolite crystals in matrix, 13.5 cm. Hoogenoeg mine, Limpopo, South Africa. BRUCE CAIRNCROSS SPECIMEN AND PHOTO.

Figure 789 The abandoned outbuildings, office and core shed at the old Gorob Mine, Namibia. The schist outcrops contain staurolite. BRUCE CAIRNCROSS PHOTO, 2017.

because they were deemed to have some value. These crystals are up to 8 cm long and many display the typical twinned form. During the 1970s there were still thousands of crystals in the heaps, but they have all completely disappeared.

Staurolite is common at the Kondo mine, **Zimbabwe**, where staurolite schist hosts the pegmatite. In general, staurolite is fairly common in schist in the Karoi, Mutoko and Mount Darwin districts. Alluvial deposits of staurolite crystals are found in the extreme north-west of Zimbabwe, east of Chirundu.

In **Eswatini**, staurolite is reported from a locality 11 km east of Goedgegun, close to Ferreira's station.

◄ **Figure 790**
Staurolite crystals in schist, 4 cm. Gorob mine, Namibia. BRUCE CAIRNCROSS SPECIMEN AND PHOTO.

Figure 791 ➤
A 3.7-cm crystal of staurolite studded with small rounded almandine garnets. Gorob mine, Namibia. BRUCE CAIRNCROSS SPECIMEN AND PHOTO.

Figure 792 Twinned staurolite crystals, largest 6 cm. Gorob mine, Namibia. BRUCE CAIRNCROSS SPECIMENS AND PHOTO.

Stibnite ◆ Sb_2S_3

Stibnite crystallizes in the orthorhombic system, has a hardness of 2, specific gravity of 4.33 to 4.66, a grey streak, and metallic to brilliant lustre. Crystals are silver-grey, prismatic and are usually elongate. Stibnite is relatively soft and characteristically soils the fingers when handled. Stibnite is the main ore mineral exploited for antimony, which is used in batteries and metal alloys. Stibnite most commonly forms in hydrothermal deposits and veins.

The largest stibnite deposit in southern Africa – the oldest known antimony deposit, hosted in three-billion-year-old rocks – is found close to Gravelotte in the Murchison greenstone belt in Limpopo, **South Africa** (Pearton and Viljoen, 1986). Stibnite is the main ore mineral often associated with gold. Massive stibnite is very common in the Consolidated Murchison, Athens and Monarch mines, whereas crystals are scarce and are usually confined to fractures, faults and small vugs. The crystals can reach 4 cm in length. Near Steynsdorp, at the Morning Mist antimony mine in the Barberton district, lenticular pockets of stibnite are interbedded in schist. Stibnite crystals up to 1 mm in length have been found on rare occasions at the Argent mine near Delmas in Gauteng.

Stibnite is found in many deposits in **Zimbabwe**, mostly in the Kwekwe district (Nutt and Bartholomew, 1987; Cairncross 2020a). Here, the antimony mineralization occurs in hydrothermal quartz veins, often with gold and other sulphides. The Cactus and Indarama antimony mines, where stibnite occurs in quartz-carbonate veins, are an example. At the Gothic mine in the same district, stibnite is found in association with native antimony, pyrite and arsenopyrite. Stibnite was abundant at the Cam and Motor antimony mine. At the Globe and Phoenix mine, gold and native antimony was exploited from quartz veins (Cairncross, 2020a). This mine is world famous for spectacular, over 30-cm-long, sword-like crystals of kermesite that have pseudomorphed stibnite crystals. Coarsely crystallized stibnite also occurs in the Kadoma district, Mashonaland West.

There are a few stibnite deposits in **Botswana** in the Tati schist belt and at Last Hope. Stibnite is found with gold in quartz veins at Signal Hill and also occurs at the Map and Rainbow gold mines.

Acicular crystals of stibnite are found at the She mine in the Forbes Reef district, **Eswatini**. Small stibnite crystals in cavities came from the Primrose and Avalanche mines.

Figure 793
Stibnite crystals, 5.1 cm. Consolidated Murchison mine, South Africa. BRUCE CAIRNCROSS SPECIMEN AND PHOTO.

Figure 794 Stibnite crystals with calcite, 12 cm. Consolidated Murchison mine, South Africa. ALLAN FRASER SPECIMEN, BRUCE CAIRNCROSS PHOTO.

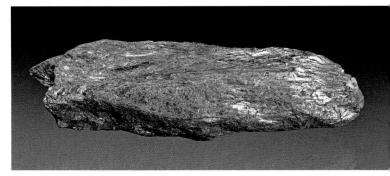

Figure 795 Massive stibnite, 11.4 cm. Kadoma district, Zimbabwe. BRUCE CAIRNCROSS SPECIMEN AND PHOTO.

Stichtite ◆ $Mg_6Cr_2(CO_3)(OH)_{16} \cdot 4H_2O$

Stichtite crystallizes in the trigonal system, has a hardness of 1.5 to 2, specific gravity of 2.16, pale lilac to white streak, and waxy to greasy lustre. Stichtite is one of the few purple minerals known. It is a member of the hydrotalcite group, and is very soft, with a soapy texture. Despite its softness, it has been used as a lapidary material, yielding very beautiful deep purple specimens. Stichtite is fairly rare, being restricted to certain ultramafic rocks. Southern Africa has famous deposits of this species. The mineral occurs in veins or lumps in greenstone belts associated with serpentinites.

Stichtite occurs in some clinochrysotile asbestos mines in the Barberton area of Mpumalanga, **South Africa**. It can be fashioned into ornaments and *objets d'art*. The most famous South African occurrence of stichtite is at Kaapsehoop overlooking the Kaap Valley west of Barberton, where it was found in veins and lumps up to 20 cm in diameter hosted in green serpentinites (Ashwal and Cairncross, 1997). Stichtite has also been recorded from the Consolidated Murchison mine, near Gravelotte.

Stichtite is known from the Shabani asbestos mine in the Zvishavane district, **Zimbabwe**.

Figure 796
Polished solid mass of stichtite with green serpentinite, 9.5 cm. Kaapsehoop, South Africa. BRUCE CAIRNCROSS SPECIMEN AND PHOTO.

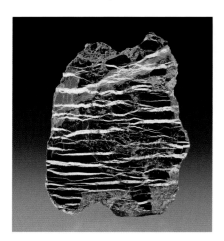

Figure 797
Purple stichtite in green serpentinite with veins of white chrysotile asbestos, 10.5 cm. Kaapsehoop, South Africa. BRUCE CAIRNCROSS SPECIMEN AND PHOTO.

Figure 798 Polished specimen of stichtite cross-cutting chrysotile asbestos in serpentinite, 13.2 cm. Kaapsehoop, South Africa. BRUCE CAIRNCROSS SPECIMEN AND PHOTO.

Stilbite ◆ $NaCa_2Al_5Si_{13}O_{36} \cdot 14H_2O$

Stilbite crystallizes in the monoclinic and orthorhombic systems, has a hardness of 3.5 to 4, specific gravity of 2.09 to 2.2, a white streak, and vitreous to pearly lustre. Stilbite-Ca and stilbite-Na now form the two end members with the Ca variety being the most common. Stilbite is one of the zeolite group of minerals. It has a distinctive pearly lustre and often forms bowtie-shaped crystals or fan-like sprays. It is characteristically white, but can be cream to orange. Stilbite can be found in cavities in volcanic rocks, notably basalt.

Stilbite occurs as salmon-pink to white, opaque, 'wheat-sheaf' bundles in Drakensberg basalt lavas in the Barkly East district, Eastern Cape, **South Africa** (Lock *et al.*, 1974). 'Wheat-sheaf' stilbite can be found with heulandite, another zeolite species, in road cuttings on Carlisle's Hoek road and near the top of Naude's Nek Pass. Stilbite is also found in basalts in the Witsieshoek area (Dunlevey *et al.*, 1993). Stilbite was relatively common at the Palabora mine and was found on fluorapophyllite-(K) in groups of small single crystals (as opposed to the more typical sheaf-like aggregates). Stilbite crystals, some over 1 cm in length, were also found on calcite. Stilbite occurs at the Mooinooi chrome mine associated with small chromite crystals and, rarely, with heulandite. At the Nababeep West mine, Northern Cape, stilbite occurred as 1-cm-long, straw-coloured crystals on calcite, orthoclase feldspar and chalcopyrite.

Figure 799 Stilbite crystals. Mooinooi mine, Brits district, South Africa. Field of view 1.6 cm. BRUCE CAIRNCROSS SPECIMEN AND PHOTO.

Figure 801 Small stilbite crystals with fluorapophyllite-(K). Field of view 3.2 cm. Palabora mine, South Africa. BRUCE CAIRNCROSS SPECIMEN AND PHOTO.

Figure 800 Stilbite with grains of metallic chromite, 4.5 cm. Mooinooi mine, Brits district, South Africa. BRUCE CAIRNCROSS SPECIMEN AND PHOTO.

Figure 802 Panoramic view of the road leading to the Grootberg Pass, Namibia. The hills are composed of Etendeka lavas that host stilbite and other minerals. BRUCE CAIRNCROSS PHOTO, 2017.

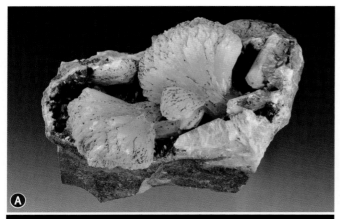

Figure 803 A large museum-sized specimen of stilbite with quartz, 16 cm. Butha-Buthe, Lesotho. BRUCE CAIRNCROSS SPECIMEN AND PHOTO.

Figure 804 Typical rural scene in Lesotho. The layered basalts that host stilbite and associated minerals are clearly visible. HERMAN DU PLESSIS PHOTO, 2017.

Figure 805 Stilbite specimens from Grootberg Pass, Namibia: **A** an 11.4-cm vug contains sprays of yellow-stained stilbite; **B** two quartz stalagmites/stalactites with stilbite crystals attached, left specimen 13.5 cm. BRUCE CAIRNCROSS SPECIMENS AND PHOTOS.

In **Namibia**, fine crystals of stilbite come from the basalts that outcrop in the Grootberg Pass, north of the Huab River, about 90 km east of Khorixas on the road to Sesfontein (Von Bezing *et al.*, 2014). Classic opaque and white bowtie crystals and wheat-sheaf forms up to 15 cm long are found here. They are associated with other zeolite species such as scolecite, heulandite, chabazite and analcime.

Stilbite is found south-west of Bulawayo, **Zimbabwe**, and at the Jessie copper mine in the vicinity of Victoria Falls. Crystals of stilbite fill joints and fractures in amphibolite that outcrops in the Mazowe River, 5 km upstream from the Southern Cross tungsten mine in the Rushinga district (Warner, 1972).

From 2012, and particularly late 2016 to early 2017, some of the finest stilbite specimens to be found in southern Africa were collected in the Butha-Buthe district of **Lesotho**. These were all collected from surface outcrops (Cairncross and Du Plessis, 2018). As in neighbouring South Africa, vugs and geodes in the Drakensberg basalts are host to the stilbite, together with associated apophyllite, calcite, laumontite and quartz. The stilbite crystals have a bright lustre, and colour varies from white or off-white to an attractive salmon-pink. Individual crystal size varies from less than 1 cm to 7 cm. The smaller crystals can be partially transparent, while larger crystals are opaque. Specimens occur as individual crystals that are doubly terminated floaters, interlocking crystals forming stand-alone groups with or without matrix, where the density of stilbite crystals present varies from a few scattered crystals, to surfaces virtually covered with stilbite and few associated species. Classic twinned bowtie forms are relatively common. Small stilbite crystals that are separately and randomly attached to the elongate drusy quartz 'fingers' form very attractive miniature and thumbnail-sized specimens. Cabinet-sized matrix specimens consisting of cascading stilbite crystals, many doubly terminated and openly stacked, also constitute aesthetic pieces.

Figure 806 Stilbite from Butha-Buthe, Lesotho: **A** highly lustrous robust crystals, 9.2 cm; **B** rare salmon-pink crystals on drusy quartz, 9.4 cm; **C** large single crystal on quartz, 4 cm; **D** mass of drusy quartz with stilbite scattered about, 9.1 cm. BRUCE CAIRNCROSS SPECIMENS AND PHOTOS.

Sturmanite ◆ $Ca_6(Fe^{3+},Al,Mn^{2+})_2(SO_4)_2(B[OH]_4)(OH)_{12} \cdot 25H_2O$

Sturmanite crystallizes in the trigonal system, has a hardness of 2.5, specific gravity of 1.85, a light yellow streak, and vitreous lustre. Sturmanite is a rare mineral of the ettringite group that was discovered in 1983 at the Kalahari manganese mines. Specimens from these mines are to be found in collections worldwide. Sturmanite is soft and forms either prismatic hexagonal or dipyramidal crystals. The chemistry of sturmanite is complex. Sturmanite is closely related to ettringite, charlesite, jouravskite and despujolsite. Visual identification of these species is virtually impossible and, although collectors tend to call the amber crystals sturmanite, the prismatic yellow crystals ettringite, the granular yellow crystals jouravskite and composite yellow forms charlesite, this is an oversimplification. Only sophisticated analyses are sufficient to identify these four species correctly.

Sturmanite is a type-locality species first discovered in **South Africa** (Peacor *et al.*, 1983). Commonly associated minerals include andradite, baryte, calcite, celestine, grossular, gypsum, hausmannite, hematite, kutnohorite, manganite and rhodochrosite. Thousands of specimens of sturmanite have come from the Kalahari manganese field (Cairncross *et al.*, 1997; Cairncross and Beukes, 2013). Numerous pockets in the N'Chwaning I and II mines have produced crystals up to 40 cm on edge and prismatic 4-cm-long crystals are relatively common. They vary in colour from white to pale apricot, yellow, lemon-yellow, dark orange and yellow-brown. Larger crystals display colour zoning, with a thin yellow outer layer coating a white to colourless core. The crystals are soft and brittle. Many discolour over time to a dull brown and partially alter to gypsum. Crystals that are exposed to moisture are particularly vulnerable to this alteration; those stored in a dry climate tend to show minimal change.

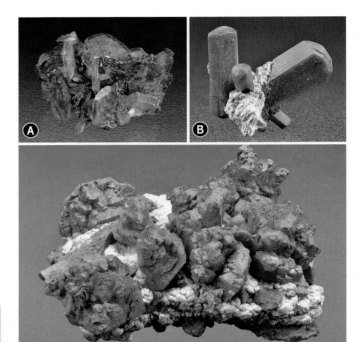

Figure 807 Various habits of sturmanite: **A** 2.7 cm; **B** 2.8 cm; **C** 3.4 cm, one of the first sturmanites collected in 1980. N'Chwaning II mine, South Africa. BRUCE CAIRNCROSS SPECIMENS AND PHOTOS.

Figure 808 Three large sturmanite crystals: **A** 7.7 cm, with manganite, the white section is a second attached crystal with gypsum(?) core; **B** 10.2 cm; **C** 14 cm. N'Chwaning II mine, South Africa. DESMOND SACCO SPECIMEN C. BRUCE CAIRNCROSS SPECIMENS (A) & (B) AND PHOTOS.

Figure 809 A very large 15.5 x 10.2-cm sturmanite crystal with several smaller crystals attached. N'Chwaning II mine, South Africa. BRUCE CAIRNCROSS SPECIMEN AND PHOTO.

Sugilite ◆ $K,Na_2(Fe^{2+},Mn^{2+},Al)_2Li_3Si_{12}O_{30}$

Sugilite crystallizes in the hexagonal system, has a hardness of 6 to 6.5, specific gravity of 2.74, white streak and vitreous lustre. Crystals of sugilite are very rare, and the mineral is most often found as solid lumps and coatings on other rocks. Although the mineral was originally discovered in Japan, South African sugilite created quite a stir when it was discovered in the Kalahari manganese field (Dunn *et al.*, 1980). Sugilite is light to very dark purple in colour and transparent pieces, which are rare, are very much in demand as gemstones. Massive sugilite is used for stone carving and lapidary purposes (Cairncross, 2017d). Sugilite has been called by several other names, such as 'wesselite' (after the Wessels mine) and 'royalazel' (referring to its royal purple colour and to Hotazel, a town near the Wessels mine).

Hundreds of tonnes of massive sugilite have been produced since the mid-1970s in **South Africa**, most coming from the Wessels mine and much less from N'Chwaning II and III mines in the Kalahari manganese field. Superb crystals of sugilite have been recovered, but in very limited quantities (Cairncross and Beukes, 2013). These can be associated with quartz, pectolite and hydroxyapophyllite-(K). Most sugilite from the Kalahari manganese field is interlayered with fine-grained aegirine, pectolite and manganese oxides. Gem-quality crystals up to 2 cm are known, but these are exceptional and extremely rare. Most measure less than 2 mm. Some quartz crystals contain 'phantoms' defined by fine laminae of sugilite. Other specimens have clear quartz on a sugilite matrix, imparting a beautiful purple colour to the quartz. In late 2013, unusual fibrous sugilite was found. Specimens consist of silky, matted fibres on brown matrix, some clustered together to form whorls.

Sugilite is associated with the alkali and alkali-calcium-manganese silicates, hence its occurrence with wollasonite, pectolite and johannsenite (Dixon, 1989). Interestingly, several of the rare Kalahari manganese field type-locality species are hosted by or found together with sugilite, namely cairncrossite (Giester *et al.*, 2016), colinowensite (Rieck

et al., 2015), effenbergerite (Giester and Rieck, 1994), hennomartinite and kornite (Armbruster *et al.*, 1993), hydroxymcglassonite (Miyawaki *et al.*, 2021a), lavinskyite (Yang *et al.*, 2014), meieranite (Yang *et al.*, 2019), lipuite (Gu *et al.*, 2019), scottyite (Yang *et al.*, 2013), strontioruizite and taniajacoite (Yang *et al.*, 2021), wesselsite (Giester and Rieck, 1996), and yuzuxiangite (Miyawaki *et al.*, 2021b).

Figure 810
Fibrous sugilite crystals on matrix, 11.1 cm with a 15.5-mm polished sugilite cabochon. N'Chwaning III mine, South Africa. JIM AND GAIL SPANN COLLECTION, TOM SPANN PHOTO.

Figure 811
A thick mass of fibrous sugilite crystals, 3.9 cm. N'Chwaning III mine, South Africa. BRUCE CAIRNCROSS SPECIMEN AND PHOTO.

Figure 812
Sugilite crystals are rare. These specimens consist of vuggy samples containing small but well-formed crystals: **A** 13 cm; **B** field of view 1.2 cm. Wessels mine, South Africa. BRUCE CAIRNCROSS SPECIMENS AND PHOTOS.

▲ **Figure 813** Faceted and polished sugilite, 15 mm. The mineral shows interesting semi-circular features. Wessels mine, South Africa. BRUCE CAIRNCROSS SPECIMEN AND PHOTO.

◄ **Figure 814** Gem-quality sugilite crystals with white pectolite. Wessels mine, South Africa. Field of view 2.8 cm. DESMOND SACCO SPECIMEN, BRUCE CAIRNCROSS PHOTO.

Figure 815 The most common type of sugilite is the massive variety that is fashioned into various lapidary, jewellery and *objets d'art*. These are examples of three different varieties: **A** a polished 8.8-cm slab displaying various shades of purple; **B** brecciated, angular grey clasts of banded iron formation are surrounded and cemented by sugilite, 8 cm; **C** highly folded and contorted sugilite and iron formation attesting to geological deformation, 10.6 cm. BRUCE CAIRNCROSS SPECIMENS AND PHOTOS.

Talc ◆ $Mg_3Si_4O_{10}(OH)_2$

Talc crystallizes in the monoclinic and triclinic systems, has a hardness of 1, specific gravity of 2.58 to 2.83, white streak, and pearly to dull lustre. Talc is the softest known mineral, with a hardness of 1 on the Mohs Scale. Rocks that are made up mainly of talc are referred to as 'soapstone' because they have a greasy, soapy feel. They are very easily carved and fashioned into *objets d'art* and lapidary items. Talc is widely used in the cosmetic and pharmaceutical industries, but also has industrial applications as a paint additive and in ferro castings and ceramics. Talc is usually found in hydrothermally altered ultramafic rocks and metamorphosed siliceous dolomites. Although talc occurs in many rocks in southern Africa, cosmetic-grade talc is not found in the region.

Talc is common in **South Africa** in rocks of the Archaean greenstone belts that outcrop in Limpopo, Mpumalanga, Gauteng and the North West (Astrup and Horn, 1998). Talc schists are commonly associated with Barberton greenstone belt ultramafic rocks in the Jamestown schist belt and in the vicinity of the Scotia talc mine to the east. Several talc mines operated in the Barberton district, for example Anne's Talc Workings, Scotia Talc mine and Southern Talc. Talc is also found in metamorphic marbles, as at Marble Delta in KwaZulu-Natal, and in the Tugela River valley in the Nqutu district.

Ultramafic talc serpentinites and talc schists occur in the Windhoek district, **Namibia**. Talc, together with tremolite, is found in dolomite east and south-east of Windhoek. Radiating white to red rosettes of talc up to 18 cm in diameter and several centimetres thick occur in dolomite on the farm Verloren 32. Talc is also found in the Gobabis district and in schists at many localities.

Talc is mined from siliceous dolomites, talc serpentinites and chlorite schists in **Zimbabwe**, where over 2,000 tonnes have been mined from the Athi mine (Bubi district) and the Gray mine (Nyanga district). Several other deposits of talc have been exploited in other districts, including Kwekwe, Makoni and Wedza.

Talc schists outcrop in the eastern part of **Botswana**.

Economic deposits of talc are found 3 km west of Sicunusa, **Eswatini**. The talc occurs in chlorite and chlorite-carbonate schist.

Figure 816
Chrome-rich talc specimens, backlit to show the translucency: **A** 6.4 cm; **B** 6.6 cm. Mtoroshanga, Mashonaland West, Zimbabwe. BRUCE CAIRNCROSS SPECIMENS AND PHOTOS.

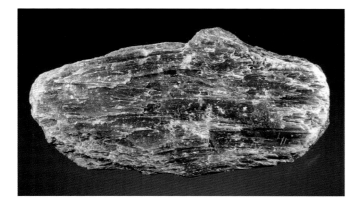

Figure 817 Micaceous talc, 18 cm. Barberton district, South Africa. DEPARTMENT OF GEOLOGY SPECIMEN, UNIVERSITY OF JOHANNESBURG, SOUTH AFRICA, BRUCE CAIRNCROSS PHOTO.

Figure 818 Vibrant green chrome-rich talc from Mtoroshanga, Mashonaland West, Zimbabwe, 6.5 cm. The black at the top is a chromite seam. The specimen comes from the Great Dyke, the main geological entity exploited for chromite in Zimbabwe. BRUCE CAIRNCROSS SPECIMEN AND PHOTO.

Tantalite-(Fe) ♦ $Fe^{2+}Ta_2O_6$

Tantalite-(Fe) crystallizes in the orthorhombic system, has a hardness of 6 to 6.5, specific gravity of 8.2, a brown-black streak and weak metallic lustre. 'Tantalite' is on obsolete mineral name for minerals with the general formula $(Mn,Fe)(Ta,Nb)_2O_6$. The two end members are tantalite-(Fe) and tantalite-(Mn), with the iron-bearing variety the most common. It typically forms black metallic crystals, but is more commonly disseminated as shapeless lumps. Crystals of 'tantalite' could belong to the columbite series of minerals; without chemical analysis they are difficult to distinguish visually. Because they are frequently found together, the colloquial, commercial term for these minerals is coltan, an amalgamation of columbite and tantalite. These minerals are exploited for tantalum. Tantalum has an extremely high melting point (almost 3,000°C) and is used in the manufacture of electronic and chemical products such as oscillators, amplifiers and alloys. Tantalite-(Fe) occurs in granitic pegmatites.

Tantalite-(Fe) occurs in large crystals in some pegmatites in Limpopo and Mpumalanga, **South Africa**. The closely allied species, columbite-(Fe), is more common in these pegmatites, particularly in the Northern Cape on the farms Steinkopf 22 and Vioolsdrif 226. Pegmatites at Palakop in Limpopo, 20 km west of Giyani, have been worked for tantalum and niobium (Boelema, 1998). They contain different suites of minerals: one, for instance, consists of an assemblage of beryl-columbite-(Fe)–tantalite-(Fe)-apatite; another contains spodumene, and a third is a kyanite-quartz pegmatite. Quartz, muscovite, plagioclase feldspar and beryl have been reported at the Kubannek quarry, which was worked for tantalite-(Fe).

Figure 819 Tantalite-(Fe) from an unnamed pegmatite in Namaqualand, South Africa, 8.2 cm. JOHANNESBURG GEOLOGICAL MUSEUM SPECIMEN, BRUCE CAIRNCROSS PHOTO.

Figure 820 Tantalite-(Fe) crystal, 2 cm, attached to white quartz. Collected in the early 1960s at Nyoka Kop, Mica district, South Africa. BRUCE CAIRNCROSS SPECIMEN AND PHOTO.

Figure 821 Tantalite-(Fe), 9.5 cm. Grietjie mine, South Africa. KARL MESSNER SPECIMEN, BRUCE CAIRNCROSS PHOTO.

In **Namibia**, tantalite-(Fe) is found in many tin-bearing pegmatites and may be associated with columbite-(Fe). Tantalite Valley in the south is named for its extensive coltan deposits. The farms Umeis 110 and Kindrzitt 132, 30 km south of Warmbad, have tantalite-columbite-bearing pegmatites in granite gneiss. Four economic pegmatites are: Witkop, White City, Homestead and Lepidolite. The pegmatite region between Karibib and Brandberg West is also an area of tantalite-(Fe) mineralization. Tantalite-(Fe) crystals up to 3 cm in diameter were found at the Rubikon mine, and crystals up to 16 cm in diameter were associated with cleavelandite feldspar at the Helikon mine.

Over 100 pegmatites in **Zimbabwe** have been exploited for coltan, commonly associated with beryl, cassiterite and lepidolite (Anderson, 1981). Most of these deposits are concentrated in the Karoi and Mutoko districts.

Alluvial gravels in the Forbes Reef area, **Eswatini**, contain tantalite-(Fe), as do tin-bearing gravels near Mbabane and at the Star mine near Sinceni.

Figure 822
A 3-cm subhedral tantalite-(Fe) crystal from Tantalite Valley, Namibia. BRUCE CAIRNCROSS SPECIMEN AND PHOTO.

Tarbuttite ◆ Zn₂(PO₄)(OH)

Tarbuttite crystallizes in the triclinic system, has a hardness of 3.5, specific gravity of 4.12 to 4.19, a white streak, and pearly lustre. Tarbuttite is a phosphate species and relatively rare worldwide but one locality in southern Africa has produced noteworthy specimens. It forms as a secondary mineral in certain zinc deposits, or can also form as a secondary mineral from bone breccias. Crystals are pale green to apple-green and commonly occur stacked together in sheaves and rosettes. Individual crystals can be translucent to transparent.

The Skorpion mine in southern **Namibia** has been a source of the finest tarbuttite crystals in southern Africa and, perhaps, globally (Cairncross, 2019b). Crystals are an attractive bright green, pale green, grey to white and form curved fan-like sprays of semi-parallel bladed crystals. The crystals are translucent and sometimes coated by granular hydrozincite. Specimens range from a few centimetres to large matrix pieces with multiple radiating clusters of tarbuttite. Some specimens consist of pseudomorphs of fluorapatite-(Ca) after tarbuttite.

Figure 823 Whorls of green tarbuttite adorn this large matrix tarbuttite specimen, 19 cm. Skorpion mine, Namibia. DESMOND SACCO SPECIMEN, BRUCE CAIRNCROSS PHOTO.

Figure 824 Fan-like sprays of tarbuttite, 4 cm. Skorpion mine, Namibia. BRUCE CAIRNCROSS SPECIMEN AND PHOTO.

Figure 825
Tarbuttite coated by white hydrozincite, 5.2 cm. Skorpion mine, Namibia.
BRUCE CAIRNCROSS SPECIMEN AND PHOTO.

Figure 826 Interlocking sprays of tarbuttite, 16 cm. DESMOND SACCO SPECIMEN, BRUCE CAIRNCROSS PHOTO.

Thaumasite ◆ Ca₃(SO₄)[Si(OH)₆](CO₃)·12H₂O

Thaumasite crystallizes in the hexagonal system, has a hardness of 3.5, specific gravity of 1.87, a white streak, and vitreous to silky lustre. Thaumasite is a member of the ettringite group of minerals and, although relatively common globally, the finest crystals come from South Africa and the mineral is therefore included here. Crystal habit can vary from fine, hair-like acicular crystals and crystal clusters to well-formed euhedral, hexagonal crystals. Colour ranges from white to pale yellow. Some of the larger hexagonal crystals are transparent.

Transparent, hexagonal thaumasite crystals have been found at Wessels and N'Chwaning mines, **South Africa**. In 1987, pale yellow, transparent crystals, up to 5 cm, came from N'Chwaning II (Cairncross and Beukes, 2013; Cairncross *et al.*, 2017). These are the best-known examples of this species in the world. Most are without matrix, and 1-cm-long crystals were common. Colourless, clear crystals were found at Wessels in 1992. Some analyses of pale yellow ettringite have revealed the crystals to be thaumasite, and micro-crystals have been found associated with a few olmiite-bultfonteinite specimens from N'Chwaning II mine. Small (under 5 mm) colourless ettringite that came out during the 2000s was found to be thaumasite.

Figure 827 Three doubly terminated thaumasite crystals, far right specimen 1 cm. N'Chwaning II mine, South Africa. BRUCE CAIRNCROSS SPECIMENS AND PHOTO.

Figure 828 A 2.4-cm cluster of pale yellow thaumasite. N'Chwaning II mine, South Africa. BRUCE CAIRNCROSS SPECIMEN AND PHOTO.

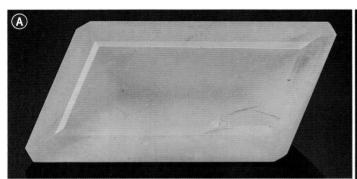

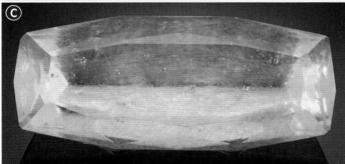

Figure 829 Rarely seen as faceted stones, these thaumasite gems are from N'Chwaning II mine, South Africa: **A** 8.01 carats (1.3 cm); **B** 5.86 carats (1.5 cm); **C** 2.15 carats (1.4 cm). WARREN TAYLOR RAINBOW OF AFRICA COLLECTION, MARK MAUTHNER PHOTOS.

Titanite ◆ CaTiSiO$_5$

Titanite crystallizes in the monoclinic system, has a hardness of 5 to 5.5, specific gravity of 3.4 to 3.5, a white streak, and adamantine lustre. Titanite forms wedge-shaped or prismatic crystals that are characteristically yellow, orange or brown, but may also be grey or black. Titanite does not normally form transparent gemstone crystals, although these are occasionally found in Zimbabwe and Namibia. Crystals often assume the shape of 'fishtail' twins. Titanite used to be called sphene. It is relatively scarce in southern Africa. Titanite occurs in granites, some metamorphosed calc-silicate rocks, and as an accessory mineral in other igneous rocks.

Titanite has been found as well-formed crystals up to 10 cm on edge in a few pegmatites in the Northern Cape, **South Africa**. Good specimens also occur in granite in the Pilgrim's Rest district near Mount Anderson, in flattened prisms up to 8 cm on edge. Twinned yellow crystals up to 3 cm on edge have been found in the Richtersveld. Titanite also occurs in the Pilanesberg rocks in the North West and in the Goudini carbonatite.

Very large and attractive green titanite crystals up to 9 cm on edge come from the farm Tantus in the Gamsberg region in **Namibia**. Some of these crystals can be faceted. The titanite is associated with quartz, rutile, albite ('pericline') and adularia – a typical alpine cleft assemblage of minerals. Titanite was found at the Krantzberg mine and occurs as an accessory mineral in carbonatites in the Kaokoveld Complex. The Ais dome skarn north of Uis contains titanite. Titanite crystals occurred in the altered portions of the copper orebody at the old Khan mine, and veins of massive titanite are found together with rutile in schist and amphibolite on the farm Eisgaubib 31, west of Windhoek.

Titanite is found as an accessory mineral in granite, gneiss, syenites and carbonatites such as the Shawa and Dorowa carbonatites, **Zimbabwe**. Gem-quality yellow titanite was extracted from the Ju Jube mine (Karoi district) and from epidote-rich veins 4 km east of the Ball mine (Mutoko district). Titanite occurred as yellow, green and brown crystals in calc-silicate rocks in the Makuti area (Warner, 1972).

Figure 830 Partly translucent titanite crystal, 1.3 cm. Richtersveld, South Africa. BRUCE CAIRNCROSS SPECIMEN AND PHOTO.

Figure 832 Titanite crystals partly embedded in granite. Mount Anderson, Lydenburg district, South Africa. Field of view 5.2 cm. COUNCIL FOR GEOSCIENCE, PRETORIA SPECIMEN, BRUCE CAIRNCROSS PHOTO.

Figure 831 Titanite crystal, 2.3 cm. Gamsberg region, Namibia. BRUCE CAIRNCROSS SPECIMEN AND PHOTO.

Figure 833 Titanite crystal, 2.3 cm, from the Khan mine, Namibia. BRUCE CAIRNCROSS SPECIMEN AND PHOTO.

Figure 834 Faceted titanite, 3.51 carats (1.2 cm). Gamsberg district, Namibia.

Figure 835 Yellow titanite crystal, 5.2 cm. Northern Zimbabwe.

Figure 836 Twinned titanite crystal, 2.1 cm. Tantus farm 30, Gamsberg area, Namibia.

Topaz ◆ Al$_2$SiO$_4$(F,OH)$_2$

Topaz crystallizes in the orthorhombic system, has a hardness of 8, specific gravity of 3.49 to 3.57, a white streak, and vitreous lustre. It has been known and used as a gemstone for millennia. Topaz varies from colourless (so-called silver topaz) to sherry-amber, yellow, pink, blue, violet or orange. One of the most popular colours is golden imperial topaz found in Brazil, Russia and Zambia. Topaz is found in granitic pegmatites, high-temperature quartz veins and silica-rich volcanic rocks, and is thus associated with other common pegmatite minerals such as quartz, microcline and schorl. Some individual crystals, such as those from Brazil, can weigh hundreds of kilograms.

Topaz is relatively rare in **South Africa**. It has been recorded from a few pegmatites in the Northern Cape. At the Baviaanskrans No. 1 and 2 pegmatites near Kakamas, well-formed white to pale pink crystals, as small as 5 cm and as enormous as 1 m in length, have been found. These were associated with beryl, bismuth, bismutite, cleavelandite, purple fluorite, pyrochlore, quartz and zircon. An unusual tin-zinc-tungsten deposit in the Northern Cape at Rhenosterkop, close to Augrabies Falls, also contains topaz. Small cassiterite-bearing greisens outcrop about 90 km north-east of Pretoria, on the farm Vlaklaagte 221 JR. They consist of quartz, sericite, cassiterite and topaz.

The oldest historic locality for topaz in **Namibia**, first recorded in 1889, is Klein Spitzkoppe, an impressive inselberg composed of an attractive light yellow to light brown alkali granite that is about 135 million years old. Pegmatite veins and miarolitic cavities in this granite host the famous Namibian topaz and aquamarine specimens. Gem-quality silver topaz, has been mined commercially and via artisanal diggers at Klein Spitzkoppe in Namibia (Cairncross et al., 1998). Topaz crystals were collected here and used to determine the mineral's crystallographic form and habits. In more recent times, fine topaz crystals up to several centimetres have been collected from the Erongo Mountains. Even larger, pale green topaz crystals, some almost the size of soccer balls, were discovered in the Brandberg Mountain north of Uis.

Figure 837 Some examples of the different complex forms of topaz from Klein Spitzkoppe, Namibia: **A** 1.5 cm; **B** 6.2 cm; **C** 3.2 cm. BRUCE CAIRNCROSS SPECIMENS AND PHOTOS A & B, WARREN TAYLOR RAINBOW OF AFRICA COLLECTION C, MARK MAUTHNER PHOTO C.

Figure 838 Two different modes of topaz mining at Klein Spitzkoppe, Namibia: **A** alluvial diggings in the sediment eroded from the adjacent granite that hosts topaz; **B** excavations in the original hard-rock granite, in a miarolitic cavity containing crystals. BRUCE CAIRNCROSS PHOTOS, A 2014, B 2017.

Figure 839 A large topaz crystal displaying interesting growth textures on the termination, 12.2 cm. Klein Spitzkoppe, Namibia. BRUCE CAIRNCROSS SPECIMEN AND PHOTO.

Figure 843 An 8-cm topaz crystal attached to a 13.5-cm smoky quartz. Klein Spitzkoppe, Namibia. DESMOND SACCO SPECIMEN, BRUCE CAIRNCROSS PHOTO.

Figure 840 A large, 20.5-cm, matrix specimen consisting of twinned orthoclase associated with two large topaz crystals and several smaller ones, and smoky quartz. Klein Spitzkoppe, Namibia. ROB SMITH COLLECTION, BRUCE CAIRNCROSS PHOTO.

Figure 841 A 27.93-carat faceted topaz with a 4.2-cm natural crystal alongside. CUT STONE MASSIMO LEONE SPECIMEN, CRYSTAL BRUCE CAIRNCROSS SPECIMEN AND PHOTO.

Figure 842 Topaz crystal associated with muscovite, 3.6 cm. Erongo Mountains, Namibia. BRUCE CAIRNCROSS SPECIMEN AND PHOTO.

Figure 844 A topaz crystal with numerous brown clay inclusions, 5.4 cm. Erongo Mountains, Namibia. BRUCE CAIRNCROSS SPECIMEN AND PHOTO.

Figure 845
A Cluster of topaz, 3.5 cm; **B** fluorescing yellow when viewed under 365 nm long-wave ultraviolet light. Erongo Mountains, Namibia. BRUCE CAIRNCROSS SPECIMEN AND PHOTOS.

Excellent dark blue topaz has been mined in north-west **Zimbabwe**. Blue topaz crystals, over 10 cm on edge, come from the St Ann's mine, Mwami area, in the Karoi district. The locality, discovered in the mid-1950s, is world famous for these beautiful crystals, . Gem-quality blue and colourless topaz was extracted from the ancient alluvial Somabula deposits in the Gweru district, and colourless topaz comes from several other pegmatites in the district. Topaz crystals are found at the Pope mine in the Goromonzi district and the Sarah beryl pegmatite in the Harare district (Warner, 1972).

Topaz is found in pegmatites in the foothills of the Sinceni Mountains, **Eswatini**.

Figure 846 A very large 141.5-carat (3.4-cm) fine blue topaz (possibly heat treated). St Ann's mine, Zimbabwe. WARREN TAYLOR RAINBOW OF AFRICA COLLECTION, MARK MAUTHNER PHOTO.

Figure 847 A large 10.9-cm (885-g) well-formed transparent topaz crystal. Crystals of this size are rare. St Ann's mine, Zimbabwe. BRUCE CAIRNCROSS SPECIMEN AND PHOTO.

Tourmaline ◆ see elbaite, schorl

Tremolite ◆ $Ca_2(Mg,Fe^{2+})_5Si_8O_{22}(OH)_2$

Tremolite is a monoclinic member of the amphibole group and has a hardness of 5 to 6, specific gravity of 2.9 to 3.2, and vitreous lustre. Tremolite forms either bladed, elongate or fibrous columnar crystals with a splintery fracture pattern. These are typically white, pale grey, pale green or dark green. The crystals may form stellate groups. Tremolite is found in metamorphosed and contact metamorphosed calcium- and/or magnesium-rich rocks such as dolomite, limestone, mafic and ultramafic rocks. It can also be found in skarns in metamorphosed metallic ore deposits. It is relatively common in southern Africa, occurring in marbles, calc-silicates, schists and amphibolite.

In **South Africa**, fibrous bundles of tan-coloured crystals up to 23 cm long have been found on the farm Assegaai 143 HT in the eMkhondo district of Mpumalanga. Equally large, white, fibrous crystals have come from Jenkinskop in the Richtersveld. Many tremolite localities are known in the dolomites surrounding the Bushveld Complex and from metamorphosed limestones at Marble Delta in KwaZulu-Natal.

Namibia has tremolite deposits in the Windhoek district on the farms Nauams 177 and Alberta 175. These formed from the alteration of dolomite, producing crystals up to 40 cm. Marble-hosted tremolite occurs in the Windhoek district. The farm Natas 220 also has tremolite in dolomite. An unusual deposit of giant tremolite crystals (up to 20 m long) occurs in the Hakos Mountains, 80 km west/north-west of Rehoboth.

Figure 848 Platy tremolite crystals, 10.5 cm. Ais, Erongo Region, Namibia. BRUCE CAIRNCROSS SPECIMEN AND PHOTO.

Figure 849 Sprays of white tremolite crystals in weathered dolomite. Field of view 4.2 cm. North West, South Africa. DEPARTMENT OF GEOLOGY, UNIVERSITY OF JOHANNESBURG SPECIMEN, BRUCE CAIRNCROSS PHOTO.

Figure 850 Platy tremolite crystals. Field of view 4.7 cm. Ais, Erongo Region, Namibia. BRUCE CAIRNCROSS SPECIMEN AND PHOTO.

Vanadinite ◆ Pb₅(VO₄)₃Cl

Vanadinite crystallizes in the hexagonal system, has a hardness of 3, specific gravity of 6.88, white to pale yellow streak, and a sub-adamantine to vitreous lustre. Vanadinite is easily identified by its perfect hexagonal crystals, which are barrel-shaped or flat and tabular. The crystals are invariably a vibrant red, but can be various shades, from amber to brown-red, as well as yellow-brown. Vanadinite is usually found in vanadium-rich deposits, often associated in southern Africa with some lead deposits.

In **South Africa**, beautiful vanadinite crystals, up to 1 cm in length, have been found at some of the defunct lead-zinc workings in the Ottoshoop district in the North West province (Wagner and Marchand, 1920). Their colour varies from yellow-brown, brown and orange-red to brown-red. Associated minerals include cerussite, galena, minium, pyromorphite and massicot. Vanadinite is also found in a number of other lead deposits as small, micromount crystals, notably at the Kindergoed deposit in the Marico district. At the old Argent silver mine near Delmas, Gauteng, vanadinite forms drusy crusts, tabular hexagonal plates or stubby barrel-shaped crystals (Atanasova *et al.*, 2016). The crystals can be clear to opaque and usually occur with pyromorphite.

The most important vanadium deposits in **Namibia** are in the Otavi mountainland (Wartha and Schreuder, 1992). The largest vanadinite crystals in the world, over 12 cm long, were collected at the Abenab mine in the Otavi mountainland, Namibia, which also produced specimens of descloizite and smithsonite (Cairncross, 1997). The geology and mineralogy of the Abenab mine and the neighbouring Berg Aukas mine are

similar, with rich vanadium deposits having been exploited at both. The Abenab mine was worked from 1922 to 1958. The orebody was a steeply dipping, pipe-like structure that was filled with brecciated country rock. This carrot-shaped pipe extended to a depth of 250 m below surface. Most of the cigar-sized vanadinite crystals that were found at the Abenab mine tended to have an outer coating of grey-green descloizite surrounding an inner red vanadinite core. Beautiful, small, bright red crystals of vanadinite were found on occasion at the Namib lead mine. The mineral also occurs on the farm Uitsab 654, 25 km west of Grootfontein. Unusual green vanadinite crystals have been found at Otjitheka in northern Namibia.

At the OBE mine near Bulawayo, **Zimbabwe**, brecciated quartz veins contained vugs and drusy quartz fragments coated with beautiful, red vanadinite crystals. These were associated with cerussite, galena and quartz. The vanadinite was collected from 1965 to 1966 and was sold as mineral specimens.

Figure 851 Platy red vanadinite crystals with yellow-green pyromorphite. Field of view 2 cm. BRUCE CAIRNCROSS SPECIMEN AND PHOTO.

Figure 853 A large vanadinite crystal, 4.1 cm. Abenab mine, Namibia. BRUCE CAIRNCROSS SPECIMEN AND PHOTO.

Figure 852 Brilliant red hexagonal vanadinite crystals on quartz. North West, South Africa. Field of view 2.4 cm. BRUCE CAIRNCROSS SPECIMEN AND PHOTO.

Figure 854 Two vanadinite specimens, Namib lead mine, Namibia: **A** associated with white calcite, field of view 3.3 cm; **B** field of view 1.6 cm. BRUCE CAIRNCROSS SPECIMENS AND PHOTOS.

Figure 855 A sawn and polished section cut perpendicular to red vanadinite crystals rimmed by dark green descloizite, 8.5 cm. The white matrix is dolomite. Berg Aukas mine, Namibia. BRUCE CAIRNCROSS SPECIMEN AND PHOTO.

Figure 856 ➤
Vanadinite, 10.5 cm, the largest crystal is 4.8 cm. Abenab mine, Namibia. COUNCIL FOR GEOSCIENCE PRETORIA COLLECTION, BRUCE CAIRNCROSS PHOTO.

▲ **Figure 857** Unusual green vanadinite, possibly coated by mottramite, 2.4 cm. Otjitheka, Kaokoveld, Namibia. BRUCE CAIRNCROSS SPECIMEN AND PHOTO.

Vesuvianite ◆ $Ca_{19}Fe(Mg,Al)_8Al_4(SiO_4)_{10}(Si_2O_7)_4(OH)_{10}$

Vesuvianite crystallizes in the tetragonal system, has a hardness of 6 to 7, specific gravity of 3.3 to 3.4, white streak, and a resinous to vitreous lustre. Vesuvianite forms attractive, prismatic crystals that may be black, green, purple, white or red. It may be confused with epidote. Vesuvianite occurs in altered and metamorphosed limestones and igneous rocks such as syenites.

In **South Africa**, vesuvianite crystals up to a few centimetres long and in various shades of blue and green have been found in the Kalahari manganese field. Dark red crystals (coloured by trace amounts of manganese) up to 2 cm long were found at the Wessels mine in the Kalahari manganese field in November 1994 and in late 2002. These were initially thought to be manganese-rich vesuvianite, but research showed sufficient quantities of manganese for the mineral to be classified as a new species, manganvesuvianite.

Vesuvianite is reported from the Kombat mine, **Namibia**. It is associated with scheelite at the Otjua skarn, 30 km north of Omaruru, and with scheelite as well as fluorapatite, titanite and zircon at the Ais dome skarn north of Uis. Large crystals, up to 15 cm long and 8 cm thick, came from the Ais skarn.

Vesuvianite occurs in metamorphosed limestones, with scheelite, garnet and epidote at the Beardmore mine in the Bikita district, **Zimbabwe**. It is also known from the Harare Portland Cement quarry and the Karoi district.

Figure 858 Blue and green vesuvianite with dark red andradite, 11 cm. Wessels mine, South Africa. BRUCE CAIRNCROSS SPECIMEN AND PHOTO.

Figure 859 A sawn specimen of prismatic vesuvianite, green diopside and white scapolite, 13.8 cm. Ais, Erongo Region, Namibia. BRUCE CAIRNCROSS SPECIMEN AND PHOTO.

Willemite ◆ Zn_2SiO_4

Willemite crystallizes in the trigonal system, has a hardness of 5.5, specific gravity of 3.89 to 4.19, white streak, and vitreous to resinous lustre. It forms as a secondary supergene mineral in some zinc deposits from the weathering of sphalerite. Its colour is variable and can range from pure white or colourless, to green, blue, yellow, red-brown, pink or black, depending on the amounts of trace mineral present in the zinc silicate.

The Tsumeb mine in **Namibia**, internationally known for its wide array of rare and beautiful mineral specimens, has produced willemite that occurs in several colours (Von Bezing *et al.*, 2016). Cadmium-rich willemite is bright yellow, while blue, brown and green varieties also occur. Barrel-shaped crystals tend to be small (less than 5 mm), well-formed and free standing, but usually clustered together. Botryoidal clusters are another form found at Tsumeb.

Berg Aukas has produced excellent willemite specimens (Cairncross, 2021c). Crystals are typically pseudohexagonal with rhombohedral or flat terminations. The crystals are highly striated parallel to the *c*-axis. Botryoidal clusters of willemite are formed by radiating crystals arranged in spherical shapes, a habit that is frequently seen in microscopic ore samples, but seldom as free-standing specimens. Crystals can range from transparent to opaque colourless, white, pale green, brown or tan, with individual crystals measuring less than 1 mm to up to 10 mm, although the larger crystals are rare. The lustre is typically vitreous. Commonly associated minerals are cerussite, descloizite and smithsonite.

◀ **Figure 860**
A Glassy blue willemite with white dolomite, 5.2 cm; **B** the willemite fluoresces bright yellow under 365 nm long-wave ultraviolet light. Tsumeb mine, Namibia. BRUCE CAIRNCROSS SPECIMEN AND PHOTO.

Figure 861 ➤
Bright yellow cadmium-rich willemite. Tsumeb mine, Namibia. Field of view 1.5 cm. BRUCE CAIRNCROSS SPECIMEN AND PHOTO.

Figure 862 Two translucent gel-like blue willemite specimens: **A** associated with white dolomite, 6.7 cm; **B** with translucent calcite, 2.8 cm. Tsumeb mine, Namibia. BRUCE CAIRNCROSS SPECIMENS AND PHOTOS.

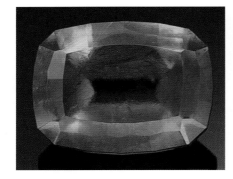

Figure 863
Faceted 1.45-carat
(7-mm) blue
willemite. Tsumeb
mine, Namibia.
WARREN TAYLOR RAINBOW
OF AFRICA COLLECTION,
MARK MAUTHNER PHOTO.

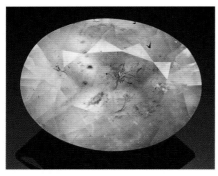

Figure 864 White
willemite, 5.2 carats
(1.2 cm). Tsumeb
mine, Namibia.
WARREN TAYLOR RAINBOW
OF AFRICA COLLECTION,
MARK MAUTHNER PHOTO.

Figure 865 Somewhat unusual spheres of willemite. Field of
view 3 cm. Berg Aukas mine, Namibia. BRUCE CAIRNCROSS SPECIMEN
AND PHOTO.

Figure 866 Highly
lustrous bundles
of willemite.
Berg Aukas mine,
Namibia. Field of
view 7.2 cm. BRUCE
CAIRNCROSS SPECIMEN
AND PHOTO.

Wulfenite ♦ PbMoO$_4$

Wulfenite crystallizes in the tetragonal system, has a hardness of 2.75 to 3, specific gravity of 6.5 to 7, white streak, and adamantine to resinous lustre. Wulfenite, a lead molybdate, is a relatively rare mineral that typically forms very distinctive, extremely flat, tabular crystals with bevelled edges. It is usually bright yellow or orange. Wulfenite forms in the secondary oxidation zones of certain lead deposits.

Wulfenite is rare from **South Africa**. Minute crystals of wulfenite have been reported from the Leeuwenkloof and Argent mines. Beautiful, albeit tiny yellow crystals occur in some of the Bushveld Complex-related deposits, such as at the Houtenbeck, Slipfontein and Argent mines (Atanasova *et al.*, 2016). Wulfenite is also found in sandstones in the Mpendle district in KwaZulu-Natal, associated with molybdenite and ilmenite.

Wulfenite crystals from Tsumeb mine, **Namibia**, are some of the finest known in the world (Wilson, 1977). The colour of the crystals found at the Tsumeb mine ranges from grey to cream, canary-yellow, red and (rarely) blue. Some crystals were over 2.5 cm thick and 5 cm on edge. A mass of intergrown, paper-thin wulfenite crystals was discovered in the late 1960s, some of which measured an astonishing 50 cm on edge. Wulfenite crystals of a light caramel colour were found at the Khusib Springs deposit to the west of Tsumeb. In the Kaokoveld, beautiful butter-yellow wulfenite crystals also occur, some associated with shattuckite (Bowell *et al.*, 2013; Von Bezing *et al.*, 2014).

Wulfenite is found at the Osborne's wulfenite deposit and in the Selukwe region of **Zimbabwe**.

Figure 867
Dipyramidal wulfenite crystal with internal reflections. Houtenbeck mine, South Africa. Field of view 1.9 mm.
WOLF WINDISCH SPECIMEN AND PHOTO.

Figure 868
Tiny but well-formed wulfenite crystals perched on the edge of a purple fluorite. Houtenbeck mine, South Africa. Field of view 1.9 mm.
WOLF WINDISCH SPECIMEN AND PHOTO.

Figure 870
Bright yellow wulfenite crystals. Tsumeb mine, Namibia. Field of view 1.6 cm.
BRUCE CAIRNCROSS SPECIMEN AND PHOTO.

Figure 869   A tabular wulfenite crystal, 2.3 cm. Tsumeb mine, Namibia. BRUCE CAIRNCROSS SPECIMEN AND PHOTO.

Figure 871   Several red wulfenite crystals with quartz and calcite, 4.7 cm. Tsumeb mine, Namibia. BRUCE CAIRNCROSS SPECIMEN AND PHOTO.

Figure 872 Wulfenite crystals associated with azurite and malachite, 4.4 cm. Tsumeb mine, Namibia. BRUCE CAIRNCROSS SPECIMEN AND PHOTO.

Figure 873 Wulfenite crystals on sulphide ore, 8.7 cm. Khusib Springs mine, Namibia. BRUCE CAIRNCROSS SPECIMEN AND PHOTO.

Figure 874 Wulfenite crystals and calcite, 6.2 cm. Tsumeb mine, Namibia. DESMOND SACCO SPECIMEN, BRUCE CAIRNCROSS PHOTO.

Figure 875 Bright red wulfenite crystals with quartz. Kaokoveld, Namibia. Field of view 16 mm. BRUCE CAIRNCROSS SPECIMEN AND PHOTO.

▲ **Figure 876** Wulfenite on weathered matrix. Neuhoff 100, Maltahohe district, Namibia. Field of view 1.2 cm. BRUCE CAIRNCROSS SPECIMEN AND PHOTO.

◄ **Figure 877** Wulfenite crystals on calcite, 12.5 cm. Tsumeb mine, Namibia. DESMOND SACCO SPECIMEN, BRUCE CAIRNCROSS PHOTO.

Zircon ♦ ZrSiO₄

Zircon crystallizes in the tetragonal system, has a hardness of 7.5, specific gravity of 4.6 to 4.7, a white streak, and vitreous to pearly lustre. Zircon crystals are invariably prismatic and caramel, tan, brown, red-brown to black in colour. They are usually opaque, but may be translucent to transparent. Zircon sand is used as a refractory product and in mouldings, and some superconductors contain zirconium. Zirconium is used in control rods in nuclear power stations. The chemical industry uses zirconium for a variety of purposes, including glass, fuel cells, deodorants and ceramics, and it is steadily replacing toxic lead as a drying agent in paint. A synthetic gemstone, cubic zirconia, is used as a substitute for diamonds in jewellery. Zircon is most commonly found in pegmatites and granites and is an important mineral species in heavy-mineral beach sands. Zircon is relatively common in some kimberlite pipes.

In KwaZulu-Natal, **South Africa**, zircon is found as brown, opaque crystals in decomposed granite along the Umhlatuzi River and its tributaries on Bull's Run Estate 12987. Some transparent, cinnamon-brown crystals have been found here. Zircon is also a major component of the heavy-mineral beach sands at Richards Bay and all along the north-east coast. These sands also contain ilmenite, rutile, monazite and garnet (Hira, 1998). In the Vanrhynsdorp district, zircon is found in the monazite deposits at Steenkampskraal, Uitklip and Roodewal (Hira, 1998). A similar association occurs in the Free State in the Bothaville district. Crudely formed to well-formed yellow to grey-brown crystals up to 7 cm on edge have been found in pegmatites in the Kenhardt and Gordonia districts. Zircon is associated with gadolinite and fergusonite in the Japie pegmatite at Bokvasmaak and with bismuth, gadolinite and pyrochlore in the Baviaanskrans pegmatite. Some large fragments and crystals of zircon in kimberlite pipes have been faceted as gemstones weighing several carats. The crystals tend to be a pale yellow-brown. Rare tetragonal zircon crystals have been found at the Palabora mine. These tend to be relatively small.

In **Namibia**, zircon is found in granite, granite gneiss and syenites, and well-formed crystals occur in pegmatites, for example at the Neu Schwaben deposit in the Karibib district. Zircon is an accessory mineral in kimberlite in the Gibeon district. Crystals have been found in the Uis tin pegmatites, and the De Rust pegmatite north of the Brandberg. Heavy-mineral sands along the coast are a source of detrital zircon, as are Namib Desert dunes.

In **Zimbabwe**, zircon is associated with tantalite at the Benson mine (Mudzi district), lepidolite and tantalite at the Fungwe Gem mine, and cassiterite, beryl, tantalite and spodumene at the Good Days beryl pegmatite.

Zircon is found in granites and some pegmatites in **Eswatini**, but not in major concentrations. Zircon is sometimes found in alluvial gravels east of the Lebombo Mountains.

Zircon is found with rutile, ilmenite, garnet and other heavy minerals in the beach sands and dune deposits along the southern **Mozambique** coastline.

Figure 878 A cluster of zircon crystals associated with phlogopite(?), and magnetite (not in view). Palabora mine, South Africa. Field of view 1.7 cm. BRUCE CAIRNCROSS SPECIMEN AND PHOTO.

ACKNOWLEDGEMENTS

A book of this nature is rarely the result of a sole effort, and the following people and institutions are thanked for their assistance in attaining the final product.

For the use of their photographs, I thank Maria Atanasova (SEM images), David Carter, Herman du Plessis, Simon Harrison, George Henry, Gerhard Louw, Mark Mauthner, Demetrius Pohl, John Rakovan, Jeff Scovil, Tom Spann, Anthony Tumo Sebolai, Martin Slama, Simon Taylor, the late Wolf Windisch and Debbie Woolf. The following collectors allowed their specimens to be photographed: Uli Bahmann, Paul Balayer, Schalk Barnard, Paul Botha, Council for Geoscience, Department of Geology at University of Johannesburg, Eric Farquharson, Allan Fraser, Philip Hitge, Johannesburg Geological Museum, Massimo Leone, McGregor Museum Kimberley, Ronnie McKenzie, Karl Messner, Paul Meulenbeld, Desmond Sacco, Rob Smith African Gems and Minerals, and Warren Taylor's Rainbow of Africa Collection. All other mineral specimens and photographs are the author's. Mineral specimens and gemstones are credited to their respective owners at the time the photographs were taken. Some may have subsequently changed hands. For security reasons, institutions who loaned their diamond and gold specimens for photography requested anonymity.

Paul Balayer has sourced some of the finest minerals from South Africa, particularly from the Kalahari manganese field and Riemvasmaak. I am most grateful to Paul for the numerous times he hosted me in the field and for many fruitful discussions. Michael Cooper is thanked for providing information and samples of quartz from relatively unknown localities in KwaZulu-Natal, South Africa. Graham and Nicky Harrison of Cape Minerals were the source of Botswana agates. Lizzie Tau sourced the scenic Botswana photo. Mindat.org, the most important mineral website, is gratefully acknowledged as a source of invaluable information.

Thanks to Pippa Parker, Publisher at Struik Nature, and the production team of Colette Alves and Gillian Black, who as always have been a pleasure to work with and have once again produced a well-crafted and well-designed book.

Notwithstanding the above, the final content of the book remains the responsibility of the author.

REFERENCES AND FURTHER READING

Many of these publications are referred to in the text. Others are listed here as a source of additional information.

Ackermann, K.J., Branscombe, K.C., Hawkes, J.R. and Tidy, A.J.L. 1966. The geology of some beryl pegmatites in Southern Rhodesia. *Transactions of the Geological Society of South Africa* 69(1), pp. 1–38.

Agangi, A., Hofmann, A., Hegner, E., Xie, H., Teschner, C., Slabunov, A. and Svetov, S.A. 2020. The Mesoarchaean Dominion Group and the onset of intracontinental volcanism on the Kaapvaal craton – geological, geochemical and temporal constraints. *Gondwana Research* 84, pp. 131–150.

Agangi, A., Hofmann, A., Ossa Ossa, F., Paprika, D. and Bekker, A. 2021. Mesoarchaean acidic volcanic lakes: a critical ecological niche in early land colonization. *Earth and Planetary Science Letters* 556:116725, pp. 1–11.

Amm, F.L. 1946. The geology of the lower Gwelo gold belt. *Southern Rhodesia Geological Survey, Bulletin* No. 37. Salisbury. 76 pp.

Anderson, C.B. 1979. Tungsten in Rhodesia. *Annals of the Zimbabwe Geological Survey* 4, pp. 76–82.

Anderson, C.B. 1980. Lead and zinc in Zimbabwe. *Annals of the Zimbabwe Geological Survey* 5, pp. 43–53.

Anderson, C.B. 1981. The production of tantalum minerals in Zimbabwe. *Annals of the Zimbabwe Geological Survey* 6, pp. 35–36.

Anderson, S.M. 1975a. A note on the occurrence of emerald at Mayfield Farm. *Annals of the Zimbabwe Geological Survey* 1, pp. 60–61.

Anderson, S.M. 1975b. Additions to the 'Check list of minerals of Rhodesia'. *Annals of the Zimbabwe Geological Survey* 1, pp. 62–64.

Anderson, S.M. 1976. Notes on the mineralogy and occurrence of emerald in Rhodesia. *Annals of the Zimbabwe Geological Survey* 2, pp. 50–55.

Anderson, S.M. 1979a. Additions to the 'Check list of minerals of Rhodesia'. *Annals of the Zimbabwe Geological Survey* 4, p. 75.

Anderson, S.M. 1979b. Euclase. *Annals of the Zimbabwe Geological Survey* 4, pp. 64–74.

Anderson, S.M. 1980. Gemstones and ornamental rocks in Zimbabwe. *Chamber of Mines Journal* 23(11), pp. 29–59.

Anhaeusser, C.R. 1976a. Archean metallogeny in Southern Africa. *Economic Geology* 71(1), pp. 16–43.

Anhaeusser, C.R. 1976b. The nature of chrysotile asbestos occurrences in southern Africa: a review. *Economic Geology* 71(1), pp. 96–116.

Anhaeusser, C.R. 1986a. Archaean gold mineralization in the Barberton Mountain Land. In: Anhaeusser, C.R. and Maske, S. (eds), *Mineral Deposits of Southern Africa*, vols I & II. Johannesburg: Geological Society of South Africa. pp. 113–154.

Anhaeusser, C.R. 1986b. The geological setting of chrysotile asbestos occurrences in southern Africa. In: Anhaeusser, C.R. and Maske, S. (eds), *Mineral Deposits of Southern Africa*, vols I & II. Johannesburg: Geological Society of South Africa. pp. 359–376.

Anhaeusser, C.R. and Maske, S. 1986. *Mineral Deposits of Southern Africa*, vols I & II, Johannesburg: Geological Society of South Africa. 2,335 pp.

Armbruster, T., Oberhänsli, R., Bermanec, V. and Dixon, R. 1993. Hennomartinite and kornite, two new Mn 3+ rich silicates from the Wessels mine, Kalahari, South Africa. *Schweizerische mineralogische und petrographische Mitteilungen* 73, pp. 349–355.

Ashwal, L.D. and Cairncross, B. 1997. Mineralogy and origin of stichtite in chromite-serpentinites. *Contributions to Mineralogy and Petrology* 127, pp. 75–86.

Ashworth, L. 2014. *Mineralised pegmatites of the Damara Belt, Namibia: Fluid inclusion and geochemical characteristics with implications for postcollisional mineralization.* PhD thesis, School of Geoscience, University of the Witwatersrand, Johannesburg. 318 pp.

Astrup, J. and Hammerbeck, E.C.I. 1998. Iron. In: Wilson, M.G.C. and Anhaeusser, C.R. *Mineral Resources of South Africa*, edn 6. *Handbook* 16, pp. 402–416. Pretoria: Council for Geoscience.

Astrup, J. and Horn, G.F.J. 1998. Talc and pyrophyllite. In: Wilson, M.G.C. and Anhaeusser, C.R. *Mineral Resources of South Africa*, edn 6. *Handbook* 16, pp. 599–603. Pretoria: Council for Geoscience.

Atanasova, M.T., Cairncross, B. and Windisch, W. 2016. Microminerals of the Bushveld Complex, South Africa. *Council for Geoscience Popular Series* No. 6. 441 pp.

Bahnemann, K.P. 1986. A review of the geology of the Messina copper deposits, northern Transvaal. In: Anhaeusser, C.R. and Maske, S. (eds), *Mineral Deposits of Southern Africa*, vols I & II. Johannesburg: Geological Society of South Africa. pp. 1671–1688.

Baldock, J.W. 1977. Resources inventory of Botswana: metallic minerals, mineral fuels and diamonds. *Geological Survey Department Mineral Resources Report* No. 4. Lobatse: Geological Survey. 69 pp.

Baldock, J.W., Hepworth, J.V. and Marengwa, B.S. 1976. Gold, base metals, and diamonds in Botswana. *Economic Geology* 71(1), pp. 139–156.

Barry, J.J. 2021. The mineral industry of Eswatini. *2017– 2018 Minerals Yearbook Eswatini*, p. 17.1. US Department of the Interior / US Geological Survey.

Bartholomew, D.S. 1990a. Base metal and industrial mineral deposits of Zimbabwe. *Mineral Resources Series* No. 22. Harare: Zimbabwe Geological Survey. 153 pp.

Bartholomew, D.S. 1990b. Gold deposits of Zimbabwe. *Mineral Resources Series* No. 23. Harare: Zimbabwe Geological Survey. 75 pp.

Beukes, G.J., Slabbert, M.J., De Bruiyn, H., Botha, B.J.V., Schoch, A.E. and Van der Westhuizen, W.A. 1987. Ti-dumortierite from the Keimoes area, Namaqua mobile belt, South Africa. *Neues Jahrbuch für Mineralogie, Abhandlungen* 157, pp. 303–318.

Beukes, N.J. and Dreyer, C.J.B. 1986. Amosite asbestos deposits of the Penge area. In: Anhaeusser, C.R. and Maske, S. (eds), *Mineral Deposits of Southern Africa*, vols I & II. Johannesburg: Geological Society of South Africa. pp. 901–910.

Blenkinsop, T.G., Martin, A., Jelsma, H.A. and Vinyu, M.L. 1997. The Zimbabwe Craton. In: De Wit, M.J. and Ashwal, L.D. (eds), *Greenstone Belts*. New York: Oxford University Press. pp. 567–580.

Boelema, R. 1998. Tantalum and niobium. In: Wilson, M.G.C. and Anhaeusser, C.R. (eds), *Mineral Resources of South Africa*, edn 6. *Handbook* 16, pp. 604–606. Pretoria: Council for Geoscience.

Böllinghaus, T., Cairncross, B. and Van Nieuwenhuizen, J.S.J. 2007. Rauchquarz auf Milchquarz: Ein schooner neufund aus dem Namaqualand, Südafrika. *Lapis* 32(9), pp. 37–41.

Bonazzi, P., Bindi, L., Medenbach, O., Pagano, R., Lampronti, G.I. and Menchetti, S. 2007. Olmiite, CaMn[SiO$_3$(OH)](OH), the Mn-dominant analogue of poldervaartite, a new mineral species from Kalahari manganese fields (Republic of South Africa). *Mineralogical Magazine* 71(2), pp. 229–238.

Bowell, R. and Cook, R.B. 2009. Connoisseur's Choice: Shattuckite Kunene District Kaokoveld, Namibia. *Rocks & Minerals* 84(6), pp. 544–550.

Bowell, R.J., Ermolina, O., Van der Plas, W., Van Us, J. and Steiner, M. 2013. Minerals of the Kaokoveld District, Kunene Region, Namibia. *Mineralogical Record* 44, pp. 485–504.

Bowles, M. 1988. Tungsten mineralization in the Namaqualand Bushmanland region, northwest Cape, South Africa. *Geological Survey of South Africa Memoir* 74. 75 pp.

Brandl, G. and De Wit, M.J. 1997. The Kaapvaal Craton, South Africa. In: De Wit, M.J. and Ashwal, L.D. (eds), *Greenstone Belts*. New York: Oxford University Press. pp. 581–607.

Burchell, W.J. 1822. *Travels in the Interior of Southern Africa*. London: Longman, Hurst, Rees, Orme & Brown. 582 pp.

Cabral, A.R., Moore, J.M., Mapani, B.S., Koubová, M. and Sattler, C.D. 2011. Geochemical and mineralogical constraints on the genesis of the Otjosondu ferromanganese deposit, Namibia: hydrothermal exhalative versus hydrogenetic (including snowball-Earth) origins. *South African Journal of Geology* 114(1), pp. 57–76.

Cairncross, B. 1991. The Messina mining district, South Africa. *Mineralogical Record* 22, pp. 187–199.

Cairncross, B. 1997. The Otavi Mountain Land Cu-Pb-Zn-V deposits, Namibia. *Mineralogical Record* 28, pp. 109–130.

Cairncross, B. 2002. Merenskyite, type-mineral from the Bushveld Complex. *Rocks & Minerals* – Special Issue on 'Minerals of Africa', 77, pp. 48–50.

Cairncross, B. 2004a. *Field Guide to Rocks & Minerals of Southern Africa*. Cape Town: Struik. 297 pp.

Cairncross, B. 2004b. History of the Okiep copper district, Namaqualand, Northern Cape Province, South Africa. *Mineralogical Record* 35, pp. 289–317.

Cairncross, B. 2005a. Famous mineral localities: Klein Spitzkoppe, Namibia. *Mineralogical Record* 36, pp. 317–335.

Cairncross, B. 2005b. Beryl from Southern Africa. *ExtraLapis* No. 23, pp. 78–83. Connecticut, USA.

Cairncross, B. 2006. Fluorite from Southern Africa. Fluorite – The Collector's Choice. *ExtraLapis* No. 9, pp. 104–107. Connecticut, USA.

Cairncross, B. 2009. Fluorite from Riemvasmaak, Northern Cape Province, South Africa. *Mineralogical Record* 40, pp. 307–324.

Cairncross, B. 2010a. Smithsonite from Tsumeb and Berg Aukas, Namibia. *ExtraLapis* English No. 13, pp. 48–59. Special Smithsonite Issue.

Cairncross, B. 2010b. *A Pocket Guide to Rocks & Minerals of Southern Africa.* Cape Town: Struik Nature. 159 pp.

Cairncross, B. 2011. The National Heritage Resource Act 1999: can legislation protect South Africa's rare geoheritage resources? *Resources Policy* 36, pp. 204–213.

Cairncross, B. 2012a. Namibia's famous amethyst. *ExtraLapis*, English No. 16, Amethyst: uncommon vintage. pp. 84–93.

Cairncross, B. 2012b. Southern African minerals. *Rocks & Minerals* 87, pp. 424–429.

Cairncross, B. 2012c. Vonbezingite: Who's who in mineral names. *Rocks & Minerals* 87, pp. 439–441.

Cairncross, B. 2014. South African diamonds: a photographic personal perspective. *Rocks & Minerals* 89, pp. 76–86.

Cairncross, B. 2016a. Ajoite: Connoisseur's Choice. *Rocks & Minerals* 91, pp. 426–432.

Cairncross, B. 2016b. Shigaite: Connoisseur's Choice. *Rocks & Minerals* 91, pp. 150–153.

Cairncross, B. 2016c. Who's who in mineral names: Alpheus Fuller Williams (1874–1953). *Rocks & Minerals* 91, pp. 366–367.

Cairncross, B. 2016d. African treasures – Gemstones and minerals of Africa. In: Anhaeusser, C.R., Viljoen, M.J. and Viljoen, R.P. (eds), *Africa's Top Geological Sites.* Cape Town: Struik Nature. pp. 281–290.

Cairncross, B. 2017a. Bultfonteinite: The where of mineral names. *Rocks & Minerals* 92(6), pp. 578–581.

Cairncross, B. 2017b. Leiteite: Connoisseur's Choice. *Rocks & Minerals* 92(3), pp. 264–269.

Cairncross, B. 2017c. Nchwaningite: The where of mineral names. *Rocks & Minerals* 92(3), pp. 290–292.

Cairncross, B. 2017d. Sugilite: Connoisseur's Choice. *Rocks & Minerals* 92(6), pp. 550–555.

Cairncross, B. 2017e. Tsumebite: The where of mineral names. *Rocks & Minerals* 92(5), pp. 466–470.

Cairncross, B. 2017f. Tsumcorite: Connoisseur's Choice. *Rocks & Minerals* 92(5), pp. 454–461.

Cairncross, B. 2018a. Hausmannite: Connoisseur's Choice. *Rocks & Minerals* 93(3), pp. 244–249.

Cairncross, B. 2018b. Iowaite: The where of mineral names. *Rocks & Minerals* 93(3), pp. 271–273.

Cairncross, B. 2018c. Mountainite: Who's who in mineral names. *Rocks & Minerals* 93(3), pp. 276–278.

Cairncross, B. 2018d. Namibite: The where of mineral names. *Rocks & Minerals* 93(2), pp. 184–187.

Cairncross, B. 2018e. Okorusu fluorite mine, Namibia. *Mineralogical Record* 49(3), pp. 375–398.

Cairncross, B. 2018f. Hydrocerussite: Connoisseur's Choice. *Rocks & Minerals* 93(2), pp. 150–156.

Cairncross, B. 2018g. Skorpionite: The where of mineral names. *Rocks & Minerals* 93(6), pp. 562–564.

Cairncross, B. 2019a. Connoisseur's Choice. Duftite after wulfenite. *Rocks & Minerals* 94(1), pp. 54–59.

Cairncross, B. 2019b. Tarbuttite: Connoisseur's Choice. *Rocks & Minerals* 94(2), pp. 150–155.

Cairncross, B. 2019c. Nimite: The where of minerals names. *Rocks & Minerals* 94(2), pp. 191–193.

Cairncross, B. 2019d. Connoisseur's Choice: Vonbezingite. *Rocks & Minerals* 94(3), pp. 250–253.

Cairncross, B. 2019e. Gamagarite: The where of mineral names. *Rocks & Minerals* 94(3), pp. 280–285.

Cairncross, B. 2019f. Senegalite: The where of mineral names. *Rocks & Minerals* 94(5), pp. 460–462.

Cairncross, B. 2019g. Roymillerite: Who's who in mineral names. *Rocks & Minerals* 94(5), pp. 475–476.

Cairncross, B. 2020a. Connoisseur's Choice: Kermesite, Globe and Phoenix mine, Kwekwe District, Zimbabwe. *Rocks & Minerals* 95(5), pp. 440–446.

Cairncross, B. 2020b. Connoisseur's Choice: Nambulite, Kombat Mine, Grootfontein, Otjozondjupa Region, Namibia. *Rocks & Minerals* 95(6), pp. 530–534.

Cairncross, B. 2020c. Karibibite: The where of mineral names. *Rocks & Minerals* 95(2), pp. 174–179.

Cairncross, B. 2020d. Nigerite: The where of mineral names. *Rocks & Minerals* 95(4), pp. 377–379.

Cairncross, B. 2020e. Taniajacoite: Who's who in mineral names. *Rocks & Minerals* 95(2), pp. 180–182.

Cairncross, B. 2020f. The where of mineral names: Hotsonite, Hotson 6 Mine, Koenabib Farm (Hotson 42 Farm), Khâi-Ma, Namakwa, Northern Cape, South Africa. *Rocks & Minerals* 95(6), pp. 567–570.

Cairncross, B. 2020g. Wesselsite: The where of mineral names. *Rocks & Minerals* 95(3), pp. 282–285.

Cairncross, B. 2021a. Southern lights: Fluorescent minerals from southern Africa. *Rocks & Minerals* 96(1), pp. 44–53.

Cairncross, B. 2021b. Connoisseur's Choice: Boltwoodite, Goanikontes Claim, Arandis, Erongo Region, Namibia. *Rocks & Minerals* 96(3), pp. 238–246.

Cairncross, B. 2021c. The minerals of Berg Aukas, Otavi Mountainland, Namibia. *Rocks & Minerals* 96(2), pp. 110–147.

Cairncross, B. 2021d. The Witwatersrand goldfield, South Africa. *Rocks & Minerals* 96(4), pp. 296–351.

Cairncross, B. 2022. Connoisseur's Choice: Papagoite, Messina mine, Limpopo Province, South Africa. *Rocks & Minerals* 97(2), pp. 152–158.

Cairncross, B. and Anhaeusser, C.R. 1992. Gold in South Africa. *Mineralogical Record* 23, pp. 209–226.

Cairncross, B. and Bahmann, U. 2006a. Famous mineral localities, Erongo, Namibia. *Mineralogical Record* 37, pp. 361–470.

Cairncross, B. and Bahmann, U. 2006b. Minerals from the Goboboseb Mountains, Brandberg region, Namibia. *Rocks & Minerals* 81, pp. 442–457.

Cairncross, B. and Bahmann, U. 2007. Zepterquartz und prehnite vom Tafelkop, Namibia. *Lapis* 32(6), pp. 31–36.

Cairncross, B., Bahmann, U. and Knoper, M. 2004. Spektakuläre 'Kaktusquarze' aus Südafrika. *Lapis* 29, pp. 16–25.

Cairncross, B. and Beukes, N.J. 2013. *The Kalahari Manganese Field: the Adventure Continues*. Cape Town: Struik Nature. 384 pp.

Cairncross, B., Beukes, N.J. and Gutzmer, J. 1997. *The Manganese Adventure*. Johannesburg: Associated Ore & Metal Corporation. 250 pp.

Cairncross, B., Beukes, N.J., Moore, T. and Wilson, W.E. 2017. The N'Chwaning mines, Kalahari Manganese Field, Northern Cape Province, South Africa. *Mineralogical Record* 48(1), pp. 13–114.

Cairncross, B. and Dixon, R. 1995. *Minerals of South Africa*. Johannesburg: Geological Society of South Africa. 289 pp.

Cairncross, B. and Du Plessis, H. 2018. Stilbite and associated minerals from Lesotho. *Rocks & Minerals* 93(4), pp. 306–319.

Cairncross, B. and Fraser, A. 2012. The Rosh Pinah lead-zinc mine, Namibia. *Rocks & Minerals* 87, pp. 398–407.

Cairncross, B., Fraser, A. and McGregor, S. 2016. The Thabazimbi mine cave, Limpopo Province, South Africa. *Rocks & Minerals* 91, pp. 322–331.

Cairncross, B., Huizenga, J.M. and Campbell, I.C. 1998. Topaz, aquamarine, and other beryls from Klein Spitzkoppe, Namibia. *Gems & Gemology* 34, pp. 114–125.

Cairncross, B., Kramers, J. and Villa, I.M. 2018. Unusual speleothem formation in the Thabazimbi mine cave, Limpopo Province, South Africa, and its chronology. *South African Journal of Geology* 121(3), pp. 261–270.

Cairncross, B. and McCarthy, T.S. 2015. *Understanding Minerals & Crystals*. Cape Town: Struik Nature. 312 pp.

Cairncross, B. and Moir, S. 1996. The Onganja mining district, Namibia. *Mineralogical Record* 27, pp. 85–97.

Cairncross, B. and Rademeyer, B. 2001. Large barite crystals from the Elandsrand gold mine, South Africa. *Mineralogical Record* 32, pp. 177–180.

Cairncross, B., Tsikos, H. and Harris, C. 2000. Prehnite from the Kalahari manganese field, South Africa, and its possible implications. *South African Journal of Geology* 103, pp. 231–236.

Cairncross, B., Windisch, W., Smit, H., Fraser, A. and Gutzmer, J. 2008. The Vergenoeg Fluorite mine, Gauteng Province, South Africa. *Rocks & Minerals* 83, pp. 410–421.

Carney, J.N., Aldiss, D.T. and Lock, N.P. 1994. The geology of Botswana. *Geological Survey Department Mineral Department, Bulletin* 0037.

Cawthorn, R.G., Lee, C.A., Schouwstra, R.P. and Mellowship, P. 2002. Relationship between PGE and PGM in the Bushveld Complex. *Canadian Mineralogist* 40, pp. 311–328.

Chamberlain, S.C. 2014. Cairncrossite: Who's who in mineral names. *Rocks & Minerals* 89(6), pp. 545–547.

Chatupa, J.C. 1999. Gold prospects and occurrences in the greenstone belts of Botswana. *Mineral Resource Series* No. 14, Gold monograph. Botswana Geological Survey.

Cole, D.I., Ngcofe, L. and Halenyane, K. 2014. Mineral commodities in the Western Cape Province, South Africa. *Council for Geoscience Report* No. 2014-0012. Bellville: Western Cape Regional Office. 85 pp.

Cook, R.B. 1999. Connoisseur's Choice: Prehnite, Brandberg, Namibia. *Rocks & Minerals* 74(3), pp. 178–180.

Crocker, I.T. 1979. Metallogenic aspects of the Bushveld granites: fluorite, tin and associated rare metal carbonate mineralization. *Geological Society of South Africa, Special Publication* 5, pp. 275–295.

Crocker, I.T. 1985. The volcanogenic fluorite hematite deposits and associated rock suite at Vergenoeg, Bushveld Complex, South Africa. *Economic Geology* 80, pp. 1181–1200.

Davies, D.N. 1964. The nickel-tungsten mineralization at Forbes Reef. *Swaziland Geological Survey and Mines Department, Bulletin* No. 4, pp. 45–50. Mbabane.

Davies, D.N., Urie, J.G., Jones, D.H. and Winter, P.E. 1964. The alumina, pyrophyllite and silica deposits in the Insuzi Series, Mahlangatsha and Mkopeleli areas, Shiselweni and Manzini districts. *Swaziland Geological Survey and Mines Department, Bulletin* No. 4, pp. 5–22. Mbabane.

De Villiers, S.B. 1976. Corundum. In: Coetzee, C.B. (ed.), Mineral resources of the Republic of South Africa. *Geological Survey of South Africa, Handbook* 7, pp. 315–320.

De Wit, M. and Main, M. 2016. The Tsodilo Hills of Botswana. In: Anhaeusser, C.R., Viljoen, M.J. and Viljoen, R.P. (eds), *Africa's Top Geological Sites*. Cape Town: Struik Nature. pp. 176–180.

Denny, G.A. 1897. *The Klerksdorp Gold Fields being a Description of the Geologic and of the Economic Conditions Obtaining in (sic) the Klerksdorp District, South African Republic*. London: Macmillan. pp. 93–98.

Diehl, B.J.M. 1992a. Tin. In: *The Mineral Resources of Namibia*, edn 1. Windhoek: Ministry of Mines and Energy, Geological Survey. 2.8-1–2.8-24.

Diehl, B.J.M. 1992b. Tungsten. In: *The Mineral Resources of Namibia*, edn 1. Windhoek: Ministry of Mines and Energy, Geological Survey. 2.9-1–2.9-10.

Diehl, B.J.M. 1992c. Thorium, yttrium and rare earth elements. In: *The Mineral Resources of Namibia*, edn 1. Windhoek: Ministry of Mines and Energy, Geological Survey. 6.20-1–6.20-5.

Dixon, R.D. 1989. Sugilite and associated metamorphic silicate minerals from Wessels mine, Kalahari manganese field. *Geological Survey of South Africa, Bulletin* 93. 47 pp.

Duff, A. 2020. *The Story of the Millwood/Knysna Goldfield.* Private publication. 102 pp.

Dunlevey, J.N., Ramluckan, V.R. and Mitchell, A.A. 1993. Secondary mineral zonation in the Drakensberg Basalt Formation, South Africa. *South African Journal of Geology* 96, pp. 215–220.

Dunn, P.J., Brummer, J.J. and Belsky, H. 1980. Sugilite, a second occurrence: Wessels mine, Kalahari Manganese Field, Republic of South Africa. *The Canadian Mineralogist* 18, pp. 37–39.

Ehlers, D.L. 2003. A preliminary report on the exploitation of amethyst at Boekenhouthoek village in the Mkobola District of Mpumalanga. *Council for Geoscience Report* No. 2003-0169. 10 pp.

Falster, A.U., Simmons, W.B. and Weber, K. 2018. Mineralogy and geochemistry of the Erongo sub-volcanic granite-miarolitic-pegmatite complex, Erongo, Namibia. *The Canadian Mineralogist* 56, pp. 425–449.

Ferguson, J.C. and Wilson, T.H. 1937. The geology of the country around the Jumbo mine, Mazoe district. *Southern Rhodesia Geological Survey, Bulletin* No. 33. Salisbury. 137 pp.

Foster, R.P., Mann, A.G., Stowe, C.W. and Wilson, J.F. 1986. Archaean gold mineralization in Zimbabwe. In: Anhaeusser, C.R. and Maske, S. (eds), *Mineral Deposits of Southern Africa*, vols I & II. Johannesburg: Geological Society of South Africa. pp. 43–112.

Fraser, A. 2013. Collector profile: Bruce Cairncross and his collection. *Mineralogical Record* 44, pp. 201–214.

Frommurze, H.F., Gevers, T.W. and Rossouw, P.J. 1942. *The Geology and Mineral Deposits of the Karibib Area, South West Africa. Explanation of Sheet No. 79 (Karibib, S.W.A.).* Pretoria: Geological Survey, Department of Mines. 180 pp.

Gebhard, G. (1999.) *Tsumeb II.* Grossenseifen, Germany: GG Publishing. 328 pp.

Gevers, T.W. 1929. The tin-bearing pegmatites of the Erongo area, South-West Africa. *South African Journal of Geology* 32(1), pp. 111–149.

Gevers, T.W. 1969. The tin-bearing pegmatites of the Erongo area, South-West Africa. In: Newhouse, W.H. (ed.), *Ore Deposits as Related to Structural Features*. New York, London: Hafner. pp. 138–140.

Gevers, T.W. and Frommurze, H.F. 1930. The tin-bearing pegmatites of the Erongo area, South-West Africa. *Transactions of the Geological Society of South Africa* 32, pp. 111–150.

Giester, G., Lengauer, C.L., Pristacz, H., Rieck, B., Topa, D. and Von Bezing, K-L. 2016. Cairncrossite, a new Ca-Sr (-Na) phyllosilicate from the Wessels mine, Kalahari Manganese Field, South Africa. *European Journal of Mineralogy* 28, pp. 495–505.

Giester, G. and Rieck, B. 1994. Effenbergerite, $BaCu[Si_4O_{10}]$, a new mineral from the Kalahari manganese field, South Africa: description and crystal structure. *Mineralogical Magazine* 58, pp. 663–670.

Giester, G. and Rieck, B. 1996. Wesselsite, $SrCu[Si_4O_{10}]$, a further new gillespite-group mineral from the Kalahari Manganese Field, South Africa. *Mineralogical Magazine* 60, pp. 795–798.

Giuliani, G. and Groat, L.A. 2020. Geology of corundum and emerald gem deposits: a review. *Gems & Gemology* 55(4), pp. 464–489.

Gliddon, J.P. and Braithwaite, R.S. 1991. Zeolites and associated minerals from the Palabora mine, Transvaal. *Mineralogical Record* 22, pp. 255–262.

Gliozzo, E., Cairncross, B. and Venneman, T. 2019. A geochemical and micro-textural comparison of basalt-hosted chalcedony from the Jurassic Drakensberg and Neoarchaean Ventersdorp Supergroup (Vaal River Alluvial Gravels), South Africa. *International Journal of Earth Sciences* 108, pp. 1857–1877.

Goldberg, I. 1976. A preliminary account of the Otjihase copper deposit, South West Africa. *Economic Geology* 71(1), pp. 384–390.

Grice, J.D., Lussier, A.J., Friis, H., Rowe, R., Poirer, G.G. and Fihl, Z. 2019. Discreditation of the pyroxenoid mineral name 'marshallsussmanite' with a reinstatement of the name schizolite, $NaCaMnSi_3O_8(OH)$. *American Mineralogist* 83, pp. 473–478.

Gu, X., Yang, H., Xie, X., Van Nieuwenhuizen, J.J., Downs, R.T. and Evans, S.H. 2019. Lipuite, a new manganese phyllosilicate mineral from the N'Chwaning III mine, Kalahari Manganese Fields, South Africa. *Mineralogical Magazine* 83(5), pp. 645–654.

Gübelin, E.J. 1958. Emeralds from Sandawana. *The Journal of Gemmology* 6(8), pp. 340–354.

Gurney, J.J. 1990. The diamondiferous roots of our wandering continent. *South African Journal of Geology* 93, pp. 425–437.

Gurney, J.J., Levinson, A.A. and Smith, H.S. 1991. Marine mining of diamonds off the west coast of southern Africa. *Gems & Gemology* 27(4), pp. 206–219.

Gutzmer, J., Beukes, N.J. and Cairncross, B. 2003. New interpretation of the origin of tiger's-eye: comment and reply. *Geology*, pp. e44-e45.

Gutzmer, J. and Cairncross, B. 2002. Spectacular minerals from the Kalahari manganese field, South Africa. *Rocks & Minerals* – Special Issue on 'Minerals of Africa', 77, pp. 94–107.

Hall, A.L. 1913. Notes on the tin deposits of Embabaan and Forbes Reef in Swaziland. *Transactions of the Geological Society of South Africa* 16, pp. 142–146.

Hall. A.L. 1924. On 'jade' (massive garnet) from the Bushveld in the western Transvaal. *Transactions of the Geological Society of South Africa* 27, pp. 39–55.

Hammerbeck, E.C.I. 1970. On the genesis of lead-zinc and fluorspar deposits in the southwestern Marico district, Transvaal. *Annals of the Geological Survey of South Africa* 8, pp. 102–110.

Hammerbeck, E.C.I. 1986. Andalusite in the metamorphic aureole of the Bushveld Complex. In: Anhaeusser, C.R. and Maske, S. (eds), *Mineral Deposits of Southern Africa*, vols I & II. Johannesburg: Geological Society of South Africa. pp. 993–1004.

Harding, R.R. and Jobbins, E.A. 1984. Verdite and rubyverdite from Zimbabwe. *The Journal of Gemmology* 19(2), pp. 150–159.

Hatch, F.H. 1910. *Report on the Mines and Mineral Resources of Natal*. London: Richard Clay. 155 pp.

Haughton, S.H. 1936. *The Mineral Resources of the Union of South Africa*. Geological Survey of South Africa. Pretoria: Department of Mines, Government Printer. 454 pp.

Haughton, S.H., Frommurze, H.F., Gevers, T.W., Schwellnus, C.M. and Rossouw, P.J. 1939. *The Geology and Mineral Deposits of the Omaruru Area, South West Africa. Explanation of Sheet 71 (Omaruru, S.W.A.)*. Pretoria: Geological Survey, Department of Mines. 160 pp.

Heron, H.D.C. 1989. An explanation of the quartz formations from near Seven Oaks in the Umvoti Valley, Natal. *South African Lapidary Magazine* 21(1), pp. 18–21.

Hira, G. 1998. Zirconium and hafnium. In: Wilson, M.G.C. and Anhaeusser, C.R. (eds), *Mineral Resources of South Africa*, edn 6. *Handbook* 16, pp. 682–685. Pretoria: Council for Geoscience.

Hirsch, M.F.H. and Genis, G. 1992a. Gold. In: *The Mineral Resources of Namibia*, edn 1. Windhoek: Ministry of Mines and Energy, Geological Survey. 4.1-1–4.1-18.

Hirsch, M.F.H. and Genis, G. 1992b. Silver. In: *The Mineral Resources of Namibia*, edn 1. Windhoek: Ministry of Mines and Energy, Geological Survey. 4.2-1–4.2-22.

Hu, K. and Heaney, P.J. 2010. A microstructural study of pietersite from Namibia and China. *Gems & Gemology* 46(4), pp. 280–286.

Hugo, P.J. 1962. Fluorspar deposits on Pyp Klip West and Wit Vlei, Kenhardt district, Cape Province. *Annals of the Geological Survey of South Africa* 1, pp. 119–126.

Hugo, P.J. 1970. The pegmatites of the Kenhardt and Gordonia districts, Cape Province. *Geological Survey of South Africa Memoir* 58. 94 pp.

Hugo, P.J. 1986. Some deposits of feldspar, mica, and beryl in the north Western Cape Province. In: Anhaeusser, C.R. and Maske, S. (eds), *Mineral Deposits of Southern Africa*, vols I & II. Johannesburg: Geological Society of South Africa. pp. 1651–1662.

Hugo, V.E. and Cornell, D.H. 1991. Altered ilmenites in Holocene dunes from Zululand, South Africa: petrographic evidence for multistage alteration. *South African Journal of Geology* 94(5/6), pp. 365–378.

Hunter, D.R. 1961. *The Geology of Swaziland*. Mbabane: Swaziland Geological Survey and Mines Department. Reprinted 1991. 104 pp.

Hunter, D.R. 1962. *The Mineral Resources of Swaziland*. Mbabane: Swaziland Geological Survey and Mines Department. Bulletin No. 2. 111 pp.

Jacobsen, J.B.E., McCarthy, T.S. and Laing, G.J.S. 1976. The copper-bearing breccia pipes of the Messina district, South Africa. *Mineralium Deposita* 11, pp. 33–45.

Jahn, S. 2000. Das blaue Wundervom Erongo – Auf Aquamarin-Jagd im Innern Nambias. In: Jahn, S., Medenbach, O., Niedermayr, G. and Schneider, G. (eds), *Namibia Zauberwelt edler Steine und Kristalle*. Haltern, Germany: Rainer Bode. pp. 72–79.

Jahn, S. and Bahmann, U. 2000. Die Miarolen in Erongo Granit – ein Eldorado für Aquamarin, Schorl & Co. In: Jahn, S., Medenbach, O., Niedermayr, G. and Schneider, G. (eds), *Namibia Zauberwelt edler Steine und Kristalle*. Haltern, Germany: Rainer Bode. pp. 80–96.

Jahn, S., Medenbach, O., Niedermayr, G. and Schneider, G. (eds) 2006. *Namibia. Zauberwelt edler Steine und Kristalle*, edn 2. Haltern, Germany: Bode Verlag. 288 pp.

Janse, A.J.A. 1995. A history of diamond sources in Africa: Part 1. *Gems & Gemology* 31(4), pp. 228–255.

Johnson, M.R., Anhaeusser, C.R. and Thomas, R.J. (eds) 2006. *The Geology of South Africa*. Johannesburg: Geological Society of South Africa / Pretoria: Council for Geoscience. 691 pp.

Jones, D.H. 1962. *Report on the Devil's Reef gold mine area, lapsed Mineral Concession No. 32B, Pigg's Peak district*. Unpublished report, Geological Survey of Swaziland, Stencil No. 442, pp. 281–285.

Keep, F.E. 1929. The geology of the Shabani mineral belt, Belingwe district. *Southern Rhodesia Geological Survey, Bulletin* No. 12. Salisbury. 193 pp.

Kershaw, D., Cairncross, B., Freese, B. and De Vries, M. 2003. Secondary minerals from the Carletonville gold mines, Witwatersrand goldfield, South Africa. *Rocks & Minerals* 78, pp. 390–399.

Kleyenstüber, A.S.E. 1984. The mineralogy of the manganese-bearing Hotazel Formation, of the Proterozoic Transvaal Sequence in Griqualand West, South Africa. *Transactions of the Geological Society of South Africa* 87, pp. 257–272.

Knoper, M.W. 2010. The Mesoproterozoic Steenkampskraal rare-earth element deposit in Namaqualand, South Africa. In: *2010 GSA, Denver Annual Meeting (31 October–3 November 2010)*, Paper No. 132–134.

Krige, A.V. 1921. The nature of the tin deposits near Kuils River, Stellenbosch district and their relation to other occurrences in the neighbourhood. *Transactions of the Geological Society of South Africa* 24, pp. 53–70.

Lächelt, S. 2004. *Geology and Mineral Resources of Mozambique*. Mineral Resources Management Capacity Building Project, World Bank Credit No. 3486. Maputo: Direcção Nacional de Geologia Moçambique. 515 pp.

Laubscher, D.H. 1986. Chrysotile asbestos in the Zvishavane (Shabani) and Mashava (Mashaba) areas, Zimbabwe. In:

Anhaeusser, C.R. and Maske, S. (eds), *Mineral Deposits of Southern Africa*, vols I & II. Johannesburg: Geological Society of South Africa. pp. 377–393.

Lensing-Burgdorf, M., Watenphul, A., Schlüter, J. and Mihailova, B. 2017. Crystal chemistry of tourmalines from the Erongo Mountains, Namibia, studied by Raman spectroscopy. *European Journal of Mineralogy* 29, pp. 257–267.

Lenthall, D.H. 1974. Tin production from the Bushveld Complex. *Economic Geology Research Unit Information Circular* No. 93. Johannesburg: University of the Witwatersrand. 15 pp.

Lock, B.E., Reid, D.R. and Broderick, T.J. 1974. Stratigraphy of the Karroo volcanic rocks in the Barkly East District. *Transactions of the Geological Society of South Africa* 77, pp. 117–129.

Lockett, N.H. 1979. The geology of the country around Dett. *Rhodesia Geological Survey, Bulletin* 85. 198 pp.

Lombaard, A.F. 1986. The copper deposits of the Okiep district, Namaqualand. In: Anhaeusser, C.R. and Maske, S. (eds), *Mineral Deposits of Southern Africa*, vols I & II. Johannesburg: Geological Society of South Africa. pp. 1421–1445.

Lowe, D.R, Drabon, N. and Byerly, G.R. 2019. Crustal fracturing, unconformities, and barite deposition, 3.26–3.23 Ga, Barberton greenstone belt, South Africa. *Precambrian Research* 327, pp. 34–46.

Lum, J.E., Viljoen, K.S. and Cairncross, B. 2016a. Mineralogical and geochemical characteristics of emeralds from the Leydsdorp area, South Africa. *South African Journal of Geology* 119, pp. 359–378.

Lum, J.E., Viljoen, F., Cairncross, B. and Frei, D. 2016b. Mineralogical and geochemical characteristics of beryl (aquamarine) from the Erongo Volcanic Complex, Namibia. *Journal of African Earth Sciences* 124, pp. 104–125.

Lynn, M.D., Wipplinger, P.E. and Wilson, M.G.C. 1998. Diamonds. In: Wilson, M.G.C. and Anhaeusser, C.R. (eds), *Mineral Resources of South Africa*, edn 6. *Handbook* 16, pp. 232–258. Pretoria: Council for Geoscience.

MacGregor, A.M. 1921. The geology of the diamond-bearing gravels of the Somabula Forest. *Geological Survey of Southern Rhodesia, Bulletin* No. 8, pp. 1–38. Salisbury.

MacGregor, A.M. 1928. The geology of the country around the Lonely mine, Bubi district. *Southern Rhodesia Geological Survey, Bulletin* No. 1. Salisbury. 96 pp.

Macintosh, E.K. 1990. *Rocks, Minerals and Gemstones of Southern Africa: a Collector's Guide*. Cape Town: Struik. 120 pp.

Maritz, J.H. and Uludag, S. 2019. Developing a mining plan for restarting the operation at Uis mine. *Journal of the South African Institute of Mining and Metallurgy* 119(7), pp. 621–630.

Massey, N.W.D. 1973. Resources inventory of Botswana: industrial rocks and minerals. *Geological Survey Department Mineral Resources Report* No. 3. Lobatse: Geological Survey. 39 pp.

Matsimbe, K.K. 2019. *Secondary copper minerals from the Messina mine, Limpopo Province, South Africa*. Geology Honours Research Project (unpublished). Department of Geology, University of Johannesburg. 48 pp.

Maufe, H.B. 1920. Geology of the Enterprise mineral belt. *Southern Rhodesia Geological Survey, Bulletin* No. 7. Salisbury. 52 pp.

Maufe, H.B., Lightfoot, B. and Zealley, A.E.V. 1919. The geology of the Selukwe mineral belt. *Southern Rhodesia Geological Survey, Bulletin* No. 3. Salisbury. 96 pp.

Mayer, P. and Moore, T.P. 2016. Recent finds of ajoite-included quartz in the Artonvilla mine, Musina district, South Africa. *Mineralogical Record* 47, pp. 461–473.

McCarthy, T. and Cairncross, B. 2021. *Minerals & Crystals: Morphology Properties Identification*. Cape Town: Struik Nature. 312 pp.

McIver, J.R. 1966. *Gems, Minerals and Rocks in Southern Africa*. Johannesburg: Purnell. 267 pp.

McIver, J.R. and Mihálik, P. 1975. Stannian andradite from Davib Ost, South West Africa. *The Canadian Mineralogist* 13(3), pp. 217–221.

McLachlan, I., Tsikos, H. and Cairncross, B. 2001. Glendonites (pseudomorphs after ikaite) in late Carboniferous marine Dwyka sedimentary strata in southern Africa. *South African Journal of Geology* 104, pp. 265–272.

Mellor, E.T. 1907. Note on the field relations of the Transvaal cobalt lodes. *Transactions of the Geological Society of South Africa* 10, pp. 36–43.

Menge, G.F.W. 1986. Sodalite carbonatite deposits of Swartbooisdrif, South West Africa / Namibia. In: Anhaeusser, C.R. and Maske, S. (eds), *Mineral Deposits of Southern Africa*, vols I & II. Johannesburg: Geological Society of South Africa. pp. 2261–2268.

Metson, N.A. and Taylor, A.M. 1977. Observations on some Rhodesian emerald occurrences. *The Journal of Gemmology* 15(8), pp. 422–434.

Middleton, R.C. 1976. The geology of the Prieska Copper Mines Limited. *Economic Geology* 71, pp. 328–350.

Miller, R.McG. 1969. The geology of the Etiro pegmatite, Karibib district, S.W.A. *Annals of the Geological Survey of South Africa* 7, pp. 125–130.

Mineral Resources of Namibia – see *The Mineral Resources of Namibia*.

Minnaar, H. and Theart, H.F.J. 2006. The exploitability of pegmatite deposits in the lower Orange River area (Vioolsdrif – Henkries – Steinkopf). *South African Journal of Geology* 109, pp. 341–352.

Miyawaki, R., Hatert, F., Pasero, M. and Mills, S.J. 2021a. IMA Commission on New Minerals, Nomenclature and Classification (CNMNC) – Newsletter 59. *European Journal of Mineralogy* 33, pp. 139–143.

Miyawaki, R., Hatert, F., Pasero, M. and Mills, S.J. 2021b. IMA Commission on New Minerals, Nomenclature and Classification (CNMNC) – Newsletter 60. *European Journal of Mineralogy* 33, pp. 203–208.

Moore, J.M. 2010. Comparative study of the Onganja copper mine, Namibia: a link between Neoproterozoic mesothermal Cu-(Au) mineralization in Namibia and Zambia. *South African Journal of Geology* 113(4), pp. 445–460.

Moore, T.P. 2013. What's new? *Mineralogical Record* 44(3), pp. 329–350.

Morgan, S.C. 1929. The geology of the Gaika gold mine, Que Que, S. Rhodesia. *Southern Rhodesia Geological Survey, Bulletin* No. 14. Salisbury. 42 pp.

Morrison, E.R. 1970. Barium minerals in Rhodesia. *Mineral Resources Series* No. 15. Rhodesia Geological Survey. 16 pp.

Morrison, E.R. 1972. Corundum in Rhodesia. *Mineral Resources Series* No. 16. Rhodesia Geological Survey. 24 pp.

Morrison, E.R. 1979. Prospecting for uranium in Rhodesia. *Annals of the Zimbabwe Geological Survey* 4, pp. 83–87.

Mountain, E.D. 1942. Bubbles in polyhedral geodes from Swaziland. *Transactions of the Royal Society of South Africa* 29(1), pp. 1–7.

Muchemwa, E. 1987. Graphite in Zimbabwe. *Mineral Resources Series* No. 20. Harare: Zimbabwe Geological Survey. 15 pp.

Mugumbate, F. 1990. Rare earth elements in Zimbabwe. *Annals of the Zimbabwe Geological Survey* 14, pp. 35–36.

Mugumbate, F. 1997. Emeralds in Zimbabwe: occurrences, geology and mineralization. *Annals of the Zimbabwe Geological Survey* 11, pp. 52–59.

Mugumbate, F., Oesterlen, P.M., Masiyambiri, S. and Dube, W. 2001. Industrial minerals and rock deposits of Zimbabwe. *Mineral Resources Series* No. 27. Harare: Zimbabwe Geological Survey. 159 pp.

Nel, C.J., Beukes, N.J. and De Villiers, J.P.R. 1986. The Mamatwan manganese mine of the Kalahari manganese field. In: Anhaeusser, C.R. and Maske, S. (eds), *Mineral Deposits of Southern Africa*, vols I & II. Johannesburg: Geological Society of South Africa. pp. 963–978.

Nel, H.J. 1949. Papers on the mineralogy of South Africa. *Geological Survey of South Africa Memoir* 43. 74 pp.

Nel, T.N., Jacobs, H., Allan, J.T. and Bozzoli, G.R. 1937. 'Wonderstone'. *Geological Survey of South Africa, Bulletin* 8. 44 pp.

Nutt, T.H.C. and Bartholomew, D.S. 1987. Antimony in Zimbabwe. *Annals of the Zimbabwe Geological Survey* 11, pp. 56–62.

Page, D.C. 1970. *The mineralogy of South African jade and the associated rocks in the district of Rustenburg, western Transvaal.* MSc thesis (unpublished), University of Pretoria. 74 pp.

Peacor, D.R., Dunn, P.J. and Duggan, M. 1983. Sturmanite, a ferric iron, boron analogue of ettringite. *The Canadian Mineralogist* 21, pp. 705–709.

Pearton, T.N. 1986. The Monarch cinnabar mine, Murchison greenstone belt. In: Anhaeusser, C.R. and Maske, S. (eds), *Mineral Deposits of Southern Africa*, vols I & II. Johannesburg: Geological Society of South Africa. pp. 339–348.

Pearton, T.N. and Viljoen, M.J. 1986. Antimony mineralization in the Murchison greenstone belt – an overview. In: Anhaeusser, C.R. and Maske, S. (eds), *Mineral Deposits of Southern Africa*, vols I & II. Johannesburg: Geological Society of South Africa. pp. 293–320.

Phaup, A.E. 1937. The geology of the Umtali gold belt. *Southern Rhodesia Geological Survey, Bulletin* No. 32. Salisbury. 186 pp.

Phaup, A.E. and Dobell, F.O.S. 1938. The geology of the lower Umfuli gold belt, Hartley and Lomagundi districts. *Southern Rhodesia Geological Survey, Bulletin* No. 34. Salisbury. 150 pp.

Philpott, G.D. and Ainslie, L.C. 1986. Lead mineralization on Leeuwbosch 129 KQ, Thabazimbi district. In: Anhaeusser, C.R. and Maske, S. (eds), *Mineral Deposits of Southern Africa*, vols I & II. Johannesburg: Geological Society of South Africa. pp. 861–866.

Pirajno, F. 1994. Mineral resources of anorogenic alkaline complexes in Namibia: a review. *Australian Journal of Earth Sciences* 41, pp. 157–168.

Pirajno, F. and Jacob, R.E. 1987. Sn-W metallogeny in the Damara orogen, South West Africa / Namibia. *South African Journal of Geology* 90, pp. 239–255.

Pirajno, F. and Schlögl, H.U. 1987. The alteration-mineralization of the Krantzberg tungsten deposit, South West Africa / Namibia. *South African Journal of Geology* 90, pp. 499–508.

Prendergast, M.D. and Wilson, A.H. 1989. The Great Dyke of Zimbabwe II: Mineralization and mineral deposits. In: Prendergast, M.D. and Jones, M.J. (eds), *Magmatic Sulphides – the Zimbabwe Volume.* London: Institution of Mining and Metallurgy. pp. 21–42.

Pretorius, D.A. 1976. The stratigraphic, geochronologic, ore-type, and geologic-environment sources of mineral wealth in the Republic of South Africa. *Economic Geology* 71(1), pp. 5–15.

Raal, F.A. 1969. A study of some gold mine diamonds. *American Mineralogist* 54, pp. 292–296.

Rakovan, J., Gaillou, E., Post, J.E., Jaszczak, J.A. and Betts, J.H. 2014. Optically sector-zoned (star) diamonds from Zimbabwe. *Rocks & Minerals* 89(2), pp. 173–178.

Reeks, G. 1996. The Kruisrivier cobalt mine – Transvaal (South Africa). *Mineralogical Record* 27(6), pp. 417–428.

Reimer, T.O. 1980. Archaean sedimentary baryte deposits of the Swaziland Supergroup (Barberton Mountain Land, South Africa). *Precambrian Research* 12(4), pp. 393–410.

Reynolds, I.M. 1986. The mineralogy and ore petrography of the Bushveld titaniferous magnetite-rich layers. In: Anhaeusser, C.R. and Maske, S. (eds), *Mineral Deposits of Southern Africa*, vols I & II. Johannesburg: Geological Society of South Africa. pp. 1267–1286.

Richardson, S.H., Chinn, I.L. and Harris, J.W. 1999. Age and origin of eclogitic diamonds from the Jwaneng kimberlite, Botswana. In: Gurney, J.J., Dawson, J.B. and Nixon, P.H. (eds), *Proceedings of the VIIth International Kimberlite Conference*, vol. 2 (authors L–Z). Cape Town. pp. 709–713.

Rieck, B., Pristacz, H. and Giester, G. 2015. Colinowensite, $BaCuSi_2O_6$, a new mineral from the Kalahari Manganese Field, South Africa, and new data on wesselsite, $SrCuSi_4O_{10}$. *Mineralogical Magazine* 79(7), pp. 1769–1778.

Rijks, H.R.P. and Van der Veen, A.H. 1972. The geology of the tin-bearing pegmatites in the eastern part of the Kamativi district, Rhodesia. *Mineralium Deposita* 7, pp. 383–395.

Robb, L.J. and Robb, V.M. 1986. Archaean pegmatite deposits in the north eastern Transvaal. In: Anhaeusser, C.R. and Maske, S. (eds), *Mineral Deposits of Southern Africa*, vols I & II. Johannesburg: Geological Society of South Africa. pp. 437–449.

Robertson, I.D.M. 1972. Mercury in Rhodesia. *Mineral Resources Series* No. 17. Rhodesia Geological Survey. 17 pp.

Roesener, H. and Schreuder, C.P. 1992. Molybdenite. In: *The Mineral Resources of Namibia*, edn 1. Windhoek: Ministry of Mines and Energy, Geological Survey. 2.7-1–2.7-5.

Rogers, A.W. 1924. An occurrence of diaspore in the Transvaal on Mooikopje 58, Middelburg. *Transactions of the Geological Society of South Africa* 27, pp. 71–75.

Rozendaal, A., Toros, M.S.C. and Anderson, J.R. 1986. The Rooiberg tin deposits, west central Transvaal. In: Anhaeusser, C.R. and Maske, S. (eds), *Mineral Deposits of Southern Africa*, vols I & II. Johannesburg: Geological Society of South Africa. pp. 1307–1327.

Rubidge, R.N. 1857. On the copper mines of Namaqualand. *Quarterly Journal of the Geological Society of London* (13)1, pp. 233–239.

Ryan, P.J., Lawrence, A.L., Lipson, R.D., Moore, J.M., Paterson, A., Stedman, D.P. and Van Zyl, D. 1986. The Aggeneys base metal sulphide deposits, Namaqualand district. In: Anhaeusser, C.R. and Maske, S. (eds), *Mineral Deposits of Southern Africa*, vols I & II. Johannesburg: Geological Society of South Africa. pp. 1447–1473.

Schmetzer, K., Stocklmayer, S., Stocklmayer, V. and Malsy, A-K. 2011. Alexandrites from the Novello alexandrite-emerald deposit, Masvingo District, Zimbabwe. *The Australian Gemmologist* 24(6), pp. 133–147.

Schnaitmann, E.A. and Jahn, S. 2010. Mineralien aus Schürfen im Kaokoveld, Namibia. *Mineralien-Welt* 21(2), pp. 78–90.

Schneider, G.I.C. 1992a. Bismuth. In: *The Mineral Resources of Namibia*, edn 1. Windhoek: Ministry of Mines and Energy, Geological Survey. 3.2-1–3.2-3.

Schneider, G.I.C. 1992b. Manganese. In: *The Mineral Resources of Namibia*, edn 1. Windhoek: Ministry of Mines and Energy, Geological Survey. 2.6-1–2.6-9.

Schneider, G.I.C. and Genis, G. 1992a. Graphite. In: *The Mineral Resources of Namibia*, edn 1. Windhoek: Ministry of Mines and Energy, Geological Survey. 6.12-1–6.12-3.

Schneider, G.I.C. and Genis, G. 1992b. Gypsum. In: *The Mineral Resources of Namibia*, edn 1. Windhoek: Ministry of Mines and Energy, Geological Survey. 6.13-1–6.13-3.

Schneider, G.I.C. and Miller, R.McG. 1992. Diamonds. In: *The Mineral Resources of Namibia*, edn 1. Windhoek: Ministry of Mines and Energy, Geological Survey. 5.1-1–5.1-32.

Schneider, G.I.C. and Seeger, K.G. 1992a. Barite. In: *The Mineral Resources of Namibia*, edn 1. Windhoek: Ministry of Mines and Energy, Geological Survey. 6.5-1–6.5-4.

Schneider, G.I.C. and Seeger, K.G. 1992b. Copper. In: *The Mineral Resources of Namibia*, edn 1. Windhoek: Ministry of Mines and Energy, Geological Survey. 2.3-1–2.3-118.

Schneider, G.I.C. and Seeger, K.G. 1992c. Fluorite. In: *The Mineral Resources of Namibia*, edn 1. Windhoek: Ministry of Mines and Energy, Geological Survey. 6.10-1–6.10-9.

Schneider, G.I.C. and Seeger, K.G. 1992d. Semi-precious stones. In: *The Mineral Resources of Namibia*, edn 1. Windhoek: Ministry of Mines and Energy, Geological Survey. 5.2.-1–5.2-16.

Schneider, G.I.C. and Watson, N.I. 1992. Andalusite-kyanite-sillimanite. In: *The Mineral Resources of Namibia*, edn 1. Windhoek: Ministry of Mines and Energy, Geological Survey. 6.2-1–6.2-3.

Scott, P. 1950. Mineral Development in Swaziland. *Economic Geography* 26(3), pp. 196–213.

Shor, R., Weldon, R., Janse, A.J.A., Breeding, C.M. and Shirey, S.B. 2015. Letseng's unique diamond proposition. *Gems & Gemology* 51(3), pp. 280–299.

Sims, D.H.R. (undated). A historical review of the Swaziland gold mining industry. *Swaziland Geological Survey and Mines Department, Bulletin* No. 9. Mbabane. 86 pp.

Smart, K.A., Tappe, S., Stern, R.A., Webb, S.J. and Ashwal, L.D. 2016. Early Archean tectonics and mantle redox recorded in Witwatersrand diamonds. *Nature Geoscience* 9, pp. 255–259.

Smit, K.V., Myagkaya, E., Persaud, S. and Wang, W. 2018. Black diamond from Marange (Zimbabwe): a result of natural irradiation and graphite inclusions. *Gems & Gemology* 54(2), pp. 132–148.

Snyman, J.E.W. 1998. Gemstones. In: Wilson, M.G.C. and Anhaeusser, C.R. (eds), *Mineral Resources of South Africa*, edn 6. *Handbook* 16, pp. 282–293. Pretoria: Council for Geoscience.

Söhnge, A.P.G. 1950. The tungsten mine near Nababeep, South Africa. *American Mineralogist* 35, pp. 931–940.

Söhnge, A.P.G. 1986. Mineral provinces of southern Africa. In: Anhaeusser, C.R. and Maske, S. (eds), *Mineral Deposits of Southern Africa*, vols I & II. Johannesburg: Geological Society of South Africa. pp. 1–23.

Söhnge, P.G. 1945. The geology of the Messina copper mines and surrounding country. *Geological Survey of South Africa Memoir* 40. 272 pp.

Southwood, M. and Cairncross, B. 2017. The minerals of Palabora mine. *Rocks & Minerals* 92, pp. 426–452.

Southwood, M. and Robison, J. 2016. Dolomite 'casts' and epimorphs from Tsumeb, Namibia. *Rocks & Minerals* 91(4), pp. 334–345.

Southwood, M.J. 1984. A preliminary study of the mineralogy of the nickel occurrence at Mabilikwe Hill, northern Transvaal. *MINTEK Report* No. M145. Randburg: Council for Mineral Technology. 6 pp.

Southwood, M.J. 1986. The mineralogy of the Pering zinc-lead deposit, Cape Province, with special reference to supergene alteration. In: Anhaeusser, C.R. and Maske, S. (eds), *Mineral Deposits of Southern Africa*, vols I & II. Johannesburg: Geological Society of South Africa. pp. 875–890.

Southwood, M.J. and Viljoen, E.A. 1986. Lead chromate minerals from the Argent lead silver mine, Transvaal, South Africa: crocoite, vauquelinite, and a possible second occurrence of embreyite. *Mineralogical Magazine* 50, pp. 728–730.

Stagman, J.G. 1978. An outline of the geology of Rhodesia. *Geological Survey of Rhodesia, Bulletin* No. 80. Salisbury. 126 pp.

Stocklmayer, V.R. 1981. Diamonds in Zimbabwe. *Annals of the Zimbabwe Geological Survey* 6, pp. 27–34.

Strydom, J.H. 1998. Magnesite. In: Wilson, M.G.C. and Anhaeusser, C.R. (eds), *Mineral Resources of South Africa*, edn 6. *Handbook* 16, pp. 444–449. Pretoria: Council for Geoscience.

The Mineral Resources of Namibia. 1992, edn 1. Windhoek: Ministry of Mines and Energy, Geological Survey.

Thomas, R.J., Bullen, W.D., De Klerk, I. and Scogings, A.J. 1990. The distribution and genesis of precious and base metal mineralization in the Natal Metamorphic Province, South Africa. *South African Journal of Geology* 93, pp. 683–695.

Trumbull, R.B. 1995. Tin mineralization in the Archean Sinceni rare element pegmatite field, Kaapvaal Craton, Swaziland. *Economic Geology* 90, pp. 648–657.

Tyndale-Biscoe, R. 1933. The geology of the central part of the Mazoe valley gold belt. *Southern Rhodesia Geological Survey, Bulletin* No. 22. Salisbury. 120 pp.

Tyndale-Biscoe, R. 1951. The geology of the Bikita tin-field, Southern Rhodesia. *Transactions of the Geological Society of South Africa* 54, pp. 11–25.

Urie, J.G. 1964. Pyrophyllite occurring in the Mozaan Series. *Swaziland Geological Survey and Mines Department, Bulletin* No. 4, pp. 25–28. Mbabane.

Urie, J.G. 1964. The Mhlosheni fluorspar prospect, mineral concession No. 31, Shiselweni district. *Swaziland Geological Survey and Mines Department, Bulletin* No. 4, pp. 39–44. Mbabane.

Vafeas, N.A., Blignaut, L.C., Viljoen, K.S. and Meffre, S. 2018. New evidence for the early onset of supergene alteration along the Kalahari unconformity. *South African Journal of Geology* 121(2), pp. 157–170.

Van Vuuren, C.J.J. 1986. Regional setting and structure of the Rosh Pinah zinc-lead deposit, South West Africa / Namibia. In: Anhaeusser, C.R. and Maske, S. (eds), *Mineral Deposits of Southern Africa*, vols I & II. Johannesburg: Geological Society of South Africa. pp. 1593–1607.

Vermaak, C.F. and Von Gruenewaldt, G. 1986. Introduction to the Bushveld Complex. In: Anhaeusser, C.R. and Maske, S. (eds), *Mineral Deposits of Southern Africa*, vols I & II. Johannesburg: Geological Society of South Africa. pp. 1021–1030.

Vertriest, W. and Pardieu, V. 2016. Update on gemstone mining in northern Mozambique. *Gems & Gemology* 52(4), pp. 404–409.

Vertriest, W. and Saeseaw, S. 2019. A decade of ruby from Mozambique: a review. *Gems & Gemology* 55(2), pp. 162–183.

Verwoerd, W.J. 1986. Mineral deposits associated with carbonatites and alkaline rocks. In: Anhaeusser, C.R. and Maske, S. (eds), *Mineral Deposits of Southern Africa*, vols I & II. Johannesburg: Geological Society of South Africa. pp. 2173–2191.

Verwoerd, W.J. 1990. The Salpeterkop ring structure, Cape Province, South Africa. *Tectonophysics* 171, pp. 275–285.

Viljoen, M.J., Bernasconi, A., Van Coller, N., Kinloch, E. and Viljoen, R.P. 1976. The geology of the Shangani nickel deposit, Rhodesia. *Economic Geology* 71(1), pp. 76–95.

Von Bezing, K.-L., Dixon, R.D., Pohl, D. and Cavallo, G. 1991. The Kalahari manganese field – an update. *Mineralogical Record* 22, pp. 279–297.

Von Bezing, K.-L. and Kotze, J. 1993. The Jan Coetzee copper mine, Namaqualand, South Africa. *Mineralogical Record* 24, pp. 39–40.

Von Bezing, L., Bode, R. and Jahn, S. 2008. *Namibia Minerals and Localities*, edn Schloss Freudenstein. Haltern, Germany: Bode Verlag GmbH. 856 pp.

Von Bezing, L., Bode, R. and Jahn, S. 2014. *Namibia Minerals and Localities I*, edn Schloss Freudenstein. Haltern, Germany: Bode Verlag GmbH. 608 pp.

Von Bezing, L., Bode, R. and Jahn, S. 2016. *Namibia Minerals and Localities II*, edn Schloss Freudenstein. Haltern, Germany: Bode Verlag GmbH. 664 pp.

Von Gruenewaldt, G. 1977. The mineral resources of the Bushveld Complex. *Minerals, Science and Engineering* 9, pp. 83–95.

Wagener, J.H.F. and Wiegand, J. 1986. The Sheba gold mine, Barberton greenstone belt. In: Anhaeusser, C.R. and Maske, S. (eds), *Mineral Deposits of Southern Africa*, vols I & II. Johannesburg: Geological Society of South Africa. pp. 155–162.

Wagner, P.A. 1909. Notes on the tin deposits in the vicinity of Cape Town. *Transactions of the Geological Society of South Africa* 12, pp. 102–111.

Wagner, P.A. 1914. *The Diamond Fields of Southern Africa*. Cape Town: C. Struik. 355 pp.

Wagner, P.A. 1918. Corundum in the Zoutpansberg fields and its matrix. *Transactions of the Geological Society of South Africa* 21, pp. 37–42.

Wagner, P.A. 1921. The Mutue Fides Stavoren tinfields. *Geological Survey of South Africa Memoir* 16. 192 pp.

Wagner, P.A. 1927. Crystals of sperrylite from the Potgietersrust platinum fields. *Proceedings of the Geological Society of South Africa (January to December, 1926)*, pp. xxxix–xl.

Wagner, P.A. 1929. *The Platinum Deposits and Mines of South Africa*. Edinburgh: Oliver & Boyd. 338 pp.

Wagner, P.A. and Marchand, B. de C. 1920. A new occurrence of vanadinite in the Marico District, Transvaal. *Transactions of the Geological Society of South Africa* 23, pp. 59–63.

Warner, S.M. 1972. Check list of the minerals of Rhodesia. *Rhodesia Geological Survey Bulletin* No. 69. Salisbury. 101 pp.

Wartha, R.R. and Genis, G. 1992. Lead and zinc. In: *The Mineral Resources of Namibia*, edn 1. Windhoek: Ministry of Mines and Energy, Geological Survey. 2.5-1–2.5-43.

Wartha, R.R. and Schreuder, C.P. 1992. Vanadium. In: *The Mineral Resources of Namibia*, edn 1. Windhoek: Ministry of Mines and Energy, Geological Survey. 2.10-1–2.10-14.

Weldon, R. and Shor, R. 2014. Botswana's scintillating moment. *Gems & Gemology* 50(2), pp. 96–113.

Wheatley, C.J.V., Friggens, P.J. and Dooge, F. 1986a. The Bushy Park carbonate-hosted zinc-lead deposit, Griqualand West. In: Anhaeusser, C.R. and Maske, S. (eds), *Mineral Deposits of Southern Africa*, vols I & II. Johannesburg: Geological Society of South Africa. pp. 891–900.

Wheatley, C.J.V., Whitfield, G.G., Kenny, K.J. and Birch, A. 1986b. The Pering carbonate-hosted zinc-lead deposit, Griqualand West. In: Anhaeusser, C.R. and Maske, S. (eds), *Mineral Deposits of Southern Africa*, vols I & II. Johannesburg: Geological Society of South Africa. pp. 867–874.

Wiles, J.W. 1961. The geology of the Miami mica field. *Geological Survey of Southern Rhodesia, Bulletin* No. 51. 235 pp.

Wilke, D.P. 1965. Magnesite deposits north of the Soutpansberg, Transvaal. *Geological Survey of South Africa, Bulletin* 44. 52 pp.

Willemse, J., Schwellnus, C.M., Brandt, J.W., Russell, H.D. and Van Rooyen, D.P. 1944. Lead deposits in the Union of South Africa and South West Africa. *Geological Survey of South Africa Memoir* 39. 186 pp.

Williams, A.F. 1932. *The Genesis of the Diamond*, vols I & II. London: Ernest Benn. 635 pp.

Wilson, M.G.C. and Anhaeusser, C.R. (eds) 1998. *Mineral Resources of South Africa*, edn 6. *Handbook* 16. Pretoria: Council for Geoscience. 740 pp.

Wilson, W.E. 1977. Tsumeb. *Mineralogical Record* 8(3), pp. 4–129.

Wilson, W.E. 2010. Sperrylite from the Tweefontein farm, Limpopo Province, South Africa. *Mineralogical Record* 41(2), pp. 145–155.

Wilson, W.E. and Dunn, P.J. 1978. Famous mineral localities: the Kalahari manganese field. *Mineralogical Record* 9, pp. 137–153.

Wilson, W.E., Johnston, C.W. and Swoboda, E. 2002. Jeremejevite from Namibia. *Mineralogical Record* 33(4), pp. 289–301.

Wilson-Moore, C. and Wilmer, W.H.C. 1893. *The Minerals of Southern Africa*. Johannesburg: Witwatersrand Chamber of Mines. 119 pp.

Worst, B.G. 1960. The Great Dyke of Southern Rhodesia. *Geological Survey of Southern Rhodesia, Bulletin* No. 47. 234 pp.

Yager, T.R. 2019. The mineral industry of Mozambique. *USGS 2015 Minerals Yearbook*. Reston, Virginia, USA: US Geological Survey. pp. 30.1–30.8.

Yang, H., Downs, R.T., Evans, S.H. and Pinch, W.W. 2013. Scottyite, the natural analogue of synthetic $BaCu_2Si_2O_7$, a new mineral from the Wessels mine, Kalahari manganese fields, South Africa. *American Mineralogist* 98, pp. 478–484.

Yang, H., Downs, R.T., Evans, S.H. and Pinch, W.W. 2014. Lavinskyite, $K(LiCu)Cu_6(Si_4O11)_2(OH)_4$, isotypic with planchéite, a new mineral from the Wessels mine, Kalahari manganese fields, South Africa. *American Mineralogist* 99, pp. 525–530.

Yang, H., Gu, X., Cairncross, B., Downs, R.T. and Evans, S.H. 2021. Taniajacoite and strontioruizite, two new minerals isostructural with ruizite from the N'Chwaning III mine, Kalahari manganese field, South Africa. *The Canadian Mineralogist* 59(2), pp. 431–444, doi:10.3749/canmin.2000037.

Yang, H., Gu, X., Downs, R.T., Evans, S.H., Van Nieuwenhuizen, J.J., Lavinsky, R.M. and Xie, X. 2019. Meieranite, $Na_2Sr_3MgSi_6O_{17}$, a new mineral from the Wessels mine, Kalahari manganese fields, South Africa. *The Canadian Mineralogist* 57(4), pp. 457–466.

Zenz, J. 2005. *Agates*. Haltern, Germany: Bode Verlag. 656 pp.

Zwaan, J.C. 2006. Gemmology, geology and origin of the Sandawana emerald deposits, Zimbabwe. *Scripta Geologica* 131. 211 pp.

Zwaan, J.C., Kanis, J. and Petsch, E.J. 1997. Update on emeralds from the Sandawana mines, Zimbabwe. *Gems & Gemology* 33(2), pp. 80–100.

Zwaan, J.C., Mertz-Kraus, R., Renfro, N.D., McClure, S.F. and Laurs, B.M. 2018. Rhodochrosite gems: properties and provenance. *The Journal of Gemmology* 36(4), pp. 332–345.

INDEX

A page number in **bold** refers to a photograph.